Progress in Colloid & Polymer Science · Vol. 71

Progress in
Colloid & Polymer Science

Editors: H.-G. Kilian (Ulm) and A. Weiss (Munich)

Frontiers in Polymer Science

Guest Editor: W. Wilke (Ulm)

Springer-Verlag Berlin Heidelberg GmbH

ISBN 978-3-662-15716-9   ISBN 978-3-7985-1698-4 (eBook)
DOI 10.1007/978-3-7985-1698-4
ISSN 0340-255 X

© 1985 by Springer-Verlag Berlin Heidelberg
Originally published by Dr. Dietrich Steinkopff Verlag GmbH & Co. KG, Darmstadt in 1985
Softcover reprint of the hardcover 1st edition 1985

Copy editing: Cynthia Feast; Production: H. Frey

Contents

LAUDATIO

Unsere Einladung, Herrn Professor Kilian zum 60. Geburtstag Originalarbeiten zu widmen, fand große Resonanz, die die restlichen Hefte von Colloid & Polymer Science 1985 gesprengt oder die regulären Manuskripte verdrängt hätte. Da beides nicht zu verantworten war, entstand diese Festschrift als Sonderband „Frontiers in Polymer Science", die dem Wissenschaftler und Herausgeber Hanns-Georg Kilian gewidmet ist und daher auch eine kurze Würdigung seiner Person mit auf den Weg nehmen soll.

Geboren am 8. November 1925 in Dessau, mit Reifevermerk 1943 an die Front geschickt und 1948 aus der Gefangenschaft entlassen, machte Herr Kilian 1949 sein Abitur in Jena und studierte Physik in Gießen und Aachen, arbeitete bei Professor Jenckel und promovierte dort 1958. Dann ging er an die Universität Marburg zu Professor F. H. Müller und habilitierte sich 1965 für das Fach Physik. 1969 folgte er dem Ruf auf den ersten Lehrstuhl für Experimentelle Physik an die neugegründete Universität Ulm.

Schon während der zeitraubenden Aufbauarbeit begann Herr Professor Kilian seinen Forschungsschwerpunkt, die Physik makromolekularer Stoffe, in Ulm zu etablieren. Sein wohl wichtigstes Ziel ist, den Zusammenhang zwischen Struktur und Eigenschaften polymerer Stoffe aufzuklären und für ihre mannigfachen Anwendungen quantitativ zu beschreiben. Aufbauend auf vielseitigen Experimenten bemüht sich Herr Kilian um ein grundlegendes Verständnis dieses Zusammenhangs, wobei er methodisch die phänomenologische Thermodynamik reversibler und irreversibler Prozesse bevorzugt. Er folgt dabei der Einsicht, daß sich Experiment und Theorie gegenseitig befruchten — Experimente ohne theoretische Deutung und eine Theorie ohne experimentelle Prüfung gleichermaßen steril bleiben.

20 von 45 Arbeiten aus den letzten 10 Jahren sind der Entwicklung einer Zustandsgleichung realer Netzwerke aus Polymerketten gewidmet, die für kleine und große Deformationen anwendbar ist und von der Idee eines „Van der Waals Netzwerkes" mit endlicher Dehnbarkeit der Netzketten und ihrer Wechselwirkung über fluktuierende Netzpunkte ausgeht. Diese Zustandsgleichung erlaubt — zusammen mit allgemeinen thermodynamischen Betrachtungen — eine konsistente Beschreibung der thermomechanischen Phänomene im gesamten Deformationsbereich und bietet — in Verbindung mit Methoden der irreversiblen Thermodynamik — neuartige Möglichkeiten, auch Relaxationsphänomene zu diskutieren, dehnungsinduzierte Kristallisation zu behandeln oder gefüllte bzw. gequollene Netzwerke zu untersuchen.

Für diese Arbeiten erhielt Professor Kilian 1984 den Merckle-Forschungspreis der Universität Ulm. Die hohe Einschätzung seiner Forschungsleistungen geht auch aus häufigen Vortragseinladungen aus dem In-und Ausland, sowie aus seiner Tätigkeit als Sondergutachter der DFG und im Forschungsbeirat des ILL-Grenoble hervor. Seit 1982 ist Professor Kilian Herausgeber des Polymerteils der Zeitschrift „Colloid & Polymer Science" und seit 1983 Vorsitzender des DFG-Fachausschusses „Physik der Polymere".

Zur Forscherpersönlichkeit von Herrn Professor Kilian gehört aber auch die intensive Diskussion mit Diplomanden, Doktoranden und Mitarbeitern die er für seine Ideen zu begeistern versteht und denen er — wie auch seinen Kollegen — Vorbild ist in sachlichem Engagement, kritischem Hinterfragen, aber auch humanem Respekt vor anderen Ideen.

Wir wünschen ihm als Wissenschaftler und Herausgeber noch viele fruchtbare Jahre erfolgreicher Arbeit für unsere Polymerwissenschaft.

F. H. Müller   W. Pechhold   A. Weiss

On polymer solution thermodynamics*)

E. Nies[1]), R. Koningsveld, and L. A. Kleintjens

Chemistry Department, University of Antwerp, Wilrijk, Belgium
DSM Research & Patents, Geleen, The Netherlands

Abstract: Various solution properties have been selected for testing the usefulness of the classic lattice model. The analysis leads to the inclusion of improvements such as different contact numbers for molecules and repeat units, distinction between concentration regimes, dilute and concentrated in polymer. The extensions of the simple model supply an adequate description of the data provided the free enthalpy of mixing in the concentrated region is made temperature- and molar mass-dependent. Part of the latter contribution can be attributed to the chains bending back on themselves. The other part is not easily accommodated in the model and might be related to free volume effects that have been left out of the present consideration. The resulting free enthalpy expression quantitatively covers independent data like osmotic pressure measurements by Krigbaum and binodals in near ternary systems by Hashizume et al.

Key words: Free energy, interaction function, partial miscibility, prediction.

Introduction

Since its origination the rigid lattice model of polymer solutions has been the subject of much criticism [1]. It is the objective of the present paper to investigate the limits to which the applicability of the model can be stretched, in particular from the point of view of its predictive power. It is obvious that the main point of criticism – the inability to deal with excess volumes – cannot be removed in the present context, but we shall restrict ourselves to systems at constant, ambient pressure.

A further restriction is that we avoid the serious complications presented by polymolecularity [2–4] and limit the analysis to near binary systems in which the polymeric constituent has a very narrow molar mass distribution. The system studied is cyclohexane/polystyrene and we use light scattering and ultracentrifuge data by Scholte et al. [5–9], near binary binodals by Hashizume et al. [10] spinodal points [7, 11–13] and

critical miscibility data [14–18]. 'Predictive' calculations will then be tested i. a. with data on systems with moderate polymolecularity, viz. Krigbaum's osmotic pressure measurements on polystyrene fractions [19] and Hashizume et al.'s [10] near ternary coexistence curves. Prediction means calculation within experimental error of properties on the same system, not used in the data fitting.

The free enthalpy (Gibbs free energy ΔG) of mixing n_1 moles of solvent with n_2 moles of polymer with a single molar mass M_2 was derived independently by Flory [20, 21], Huggins [22, 23] and Staverman and Van Santen [24, 25]. We write

$$\Delta G/RT = n_1 \ln \phi_1 + n_2 \ln \phi_2 + \Gamma \qquad (1)$$

where RT has its usual meaning, $\phi_1 = n_1/N$, $\phi_2 = n_2 m_2/N$, $N = n_1 + n_2 m_2$, m_2 = number of lattice sites occupied by a polymer chain. The term Γ is used to describe all conceivable corrections needed to amend the first two combinatorial terms so that ΔG conforms with actual thermodynamic properties. In the following we develop Γ, starting from the simple Van Laar expression, led by the selected experimental data.

*) Affectionately dedicated to Prof. Dr. H.-G. Kilian on the occasion of his 60th birthday.
[1]) Present address: Polymer Technology, University of Technology, Eindhoven, Netherlands.

Within the framework of the model ϕ_i is the volume fraction of component i. It is in the spirit of the present study to calculate volume fractions from mass fractions on the assumption of additivity of volume on mixing, we ignore the obvious problems with the volume fraction [26]. Then, the 'relative chain length' m_2 is given by

$$m_2 = v_2 M_2 / V_1 \tag{2}$$

where v_2 and V_1 are the specific volume of the polymer in the liquid state, and the molar volume of the solvent, respectively.

Shultz-Flory method

The Van Laar expression for Γ reads

$$\Gamma / N = g \phi_1 \phi_2 = (g_S + g_H / T) \phi_1 \phi_2. \tag{3}$$

The simplest interpretation of the interaction function g defines it in terms of the interchange energy (see ref. [27]). Experimental data suggest interpretation as a free enthalpy to be preferable [28, 29]. Thus we have an entropic contribution g_S to g, in addition to the enthalpic term g_H (in a first approximation [30]).

Shultz and Flory used equations (1) and (3) to analyse miscibility gaps in several solvent/polymer systems for a series of polymer chain lengths [31]. At a critical point both spinodal and critical conditions are obeyed. In a binary system one has

$$\{(\partial^2 \Delta G / NRT) / \partial \phi_2^2\}_{p,T} = 0 \text{ (spinodal)} \tag{4}$$

$$\{(\partial^3 \Delta G / NRT) / \partial \phi_2^3\}_{p,T} = 0 \text{ (critical point)}. \tag{5}$$

Application of conditions (4) and (5) to equations (1) and (3) leads to

$$g_H / T_m = \frac{1}{2} - g_S + (m_2^{-1/2} + 1/(2m_2)). \tag{6}$$

Shultz and Flory identified the maximum separation temperature T_m with the critical point and plotted $1/T_m$ vs. the expression between brackets in equation (6) to find the parameters g_S and g_H.

Shultz-Flory plots are often found to be linear, in spite of the fact that the underlying assumptions (strictly binary systems and validity of eqs. (1) and (3)) are probably never realistic. It can be explained that, nevertheless, linearity is observed frequently, but slope

and intercept cannot be interpreted in terms of equation (6) [30, 32].

It is therefore not surprising that Shultz and Flory noted the 'predictive' power of their treatment to be extremely poor. This aspect is illustrated in figure 1 which shows the miscibility gaps calculated with equations (1) and (3), with the parameters from a Shultz-Flory plot. The maximum separation temperatures check with the experimental ones for obvious reasons, but the calculated gaps are much narrower and more

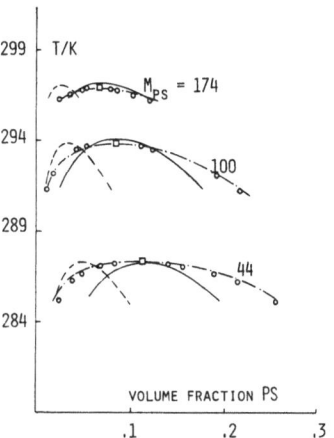

Fig. 1. Experimental binodals (coexisting phase compositions: O) as a function of temperature for indicated molar masses (kg/mole), and their representation by the Shultz-Flory method (eqs. (3) and (6): ---), by equation (8) (——) and by equations (18) and (19) (—·—·—). Data from Hashizume et al. [10]. Critical points: □. System cyclohexane/polystyrene

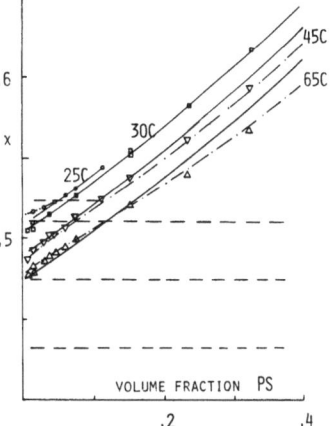

Fig. 2. Experimental values of χ (= $g - \phi_1 \partial g/\partial \phi_2$) at indicated temperature and and their representation by the Shultz-Flory method (eqs. (3) and (6): ---), by equation (8) (——) and by equations (18) and (19) (—·—·—). Data from Scholte et al. [5–9]. Molar mass: 166 kg/mole. System cyclohexane/polystyrene

asymmetric than the experimental curves. The discrepancy is illustrated even more clearly if measured values of the quantity $(g - \phi_1 \partial g/\partial \phi_2)$ are compared with those calculated with equations (1) and (3). The inadequacy of the procedure is demonstrated in figure 2.

In the present examples polymolecularity can hardly be expected to cause the discrepancy. Figure 2, however, indicates a major reason for the failure. Evidently, the parameter g depends markedly on concentration.

The quantity $(g - \phi_1 \partial g/\partial \phi_2)$ is usually known by the symbol χ, representing the interaction parameter in the expression for the chemical potential of the solvent.

Disparity of contact numbers

The major reason for g to depend on concentration is to be sought in the disparity in size and shape between solvent molecules and repeat units in the polymer. Staverman [33], working out earlier suggestions by Langmuir [34] and Butler [35], assumed the number of nearest neighbour contacts a molecule or segment can make to be proportional to its accessible surface area. Thus he obtained an expression for the enthalpy of mixing which, in terms of g, reads

$$g = B(T)/(1 - \gamma\phi_2) \tag{7}$$

where $B(T)$ summarizes the temperature dependence and $\sigma_2/\sigma_1 \, (= 1 - \gamma) =$ the ratio of the surface areas of polymer segments and solvent molecules.

The idea has been applied by other authors, with or without reference to the unit cell volume of the lattice [36, 37]. The viability of Staverman's straightforward approach is illustrated in figure 3. It is interesting to note that the curve in figure 3 b was calculated with Bondi's [38] σ_2/σ_1 value of 1.35, and passes right through the experimental points. Whether one uses the surface area ratio as such, or corrected for the cell volume, depends on how close one wishes to adhere to the notion of a rigid lattice. Remembering Rowlinson's remark that the lattice model might prove more useful than is often thought, provided it is seen as an abstraction, useful for the purpose of calculation, the straightforward treatment would not seem objectionable [39]. We shall adhere to it in this paper.

Fitting critical points for various molar masses in the system cyclohexane/polystyrene to equations (1), (3),

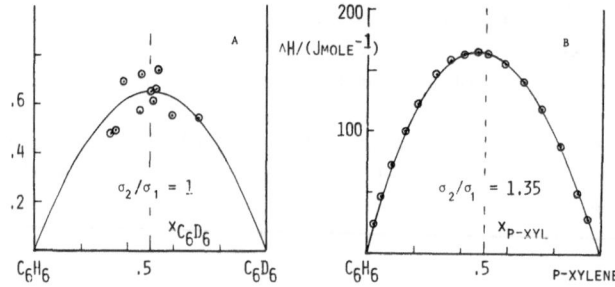

Fig. 3. Enthalpies of mixing (ΔH) for benzene with perdeutereous benzene (a, ref. [64]) and with *p*-xylene (b, ref. [65]). The curves were calculated according to Staverman [33] with molecular surface ratios of 1 (a) and 1.35 (b)

(4), (5) and (7) reveals the need of an extra, empirical parameter [40], viz.

$$g = B(T)/(1 - \gamma\phi_2) + C \tag{8}$$

where $B(T)$ is expressed as $B(T) = B_o + B_1/T$.

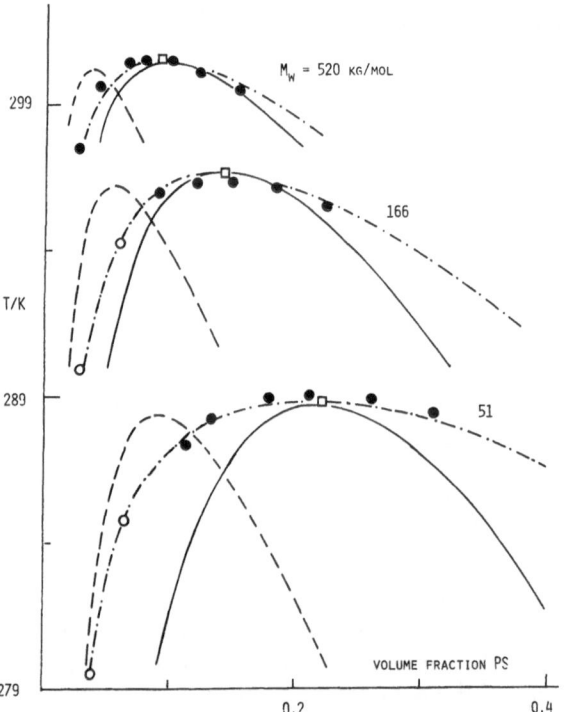

Fig. 4. Experimental spinodals and their representation by the Shultz-Flory method (eqs. (3) and (6): ———), by equation (8) (——) and by equations (18) and (19) (—·—·—). Light scattering data from Scholte et al. [5–9] (O) and from Gordon et al. [11–13] (●). System cyclohexane/polystyrene. Critical points: □

Table 1. Parameters values for equations (3) and (8)

g_S	=	− 0.26
g_H/K	=	234
B_0	=	0.57
B_1/K	=	108
γ	=	0.22
C	=	− 0.22

The binary binodals, calculated with the parameters so obtained, show a much better agreement with the measured coexistence curves (fig. 1). Both maximum separation temperature and concentration are now reproduced. Yet, the shape of the curves still deviates, the experimental curves being broader than the calculated ones. The improvement over equation (3) is most spectacular with the data in figure 2. The introduction of the surface area ratio shifts the calculated curves into the range of the measurements. It should be noted that the σ_2/σ_1 value following from the fitting of the critical points is 0.78 and compares favourably with the value calculated with Bondi's method, viz. 0.87.

The calculated spinodals show a quite comparable situation, as shows figure 4. The values of the parameters leading to this description and prediction are listed in table 1, together with those for the Shultz-Flory treatment. Figures 1, 2 and 4 present examples for one or a few chain lengths only, the other molar masses, ranging 50 to 520 kg/mole, show analogous behaviour.

The deviations between calculated and measured curves come out clearly in binodals and spinodals (figs. 1 and 4). Figure 2 (χ values) also clearly shows the deviations at high concentration but less convincingly at the dilute end. A different way of treating the data, and the issuing plots, does not only bring the differences out more clearly but also reveals that C must be a function of temperature. To this end we use the parameters B_0, B_1 and γ of table 1 and let the χ data (fig. 2) determine a value of C for each molar mass separately. The contribution of C to χ is given by

$$C' = C - \phi_1 \partial C/\partial \phi_2 . \tag{9}$$

Figure 5, showing a plot of C' vs. T for $M = 51$ kg/mole, reveals that, at concentrations above $\phi_2 \simeq 0.15$, C' hardly depends on concentration and is, to a good approximation, linear in T. At $\phi_2 < 0.15$ and $T > 300$ K, C' is also independent of concentration and linear in T. This linearity extends to $T < 300$ K for $\phi_2 > 0.15$. At

$\phi_2 < 0.15$ and $T < 300$ K we see a dependence of C' on concentration which increases upon a decrease of ϕ_2. Summarizing, one might write

$$C'(\phi_2 > 0.15) = A(T) = A_1 + A_2 T . \tag{10}$$

With those other molar masses for which a sufficient T range has been measured, one finds analogous results but different values for A_1 and A_2, both of which seem to be expressible in a linear relation to reciprocal molar mass (see fig. 6).

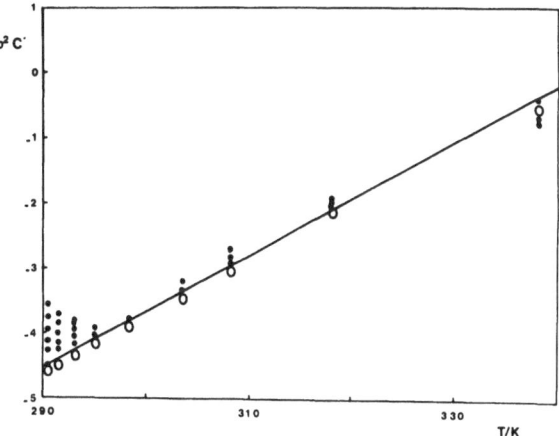

Fig. 5. The quantity C' (eqs. (8) and (9)) calculated from experimental values of the chemical potential of the solvent cyclohexane as a function of temperature, molar mass polystyrene 51 kg/mole. Polymer mass fraction $w_p \lesssim 0.15$: ●; $0.15 \lesssim w_p \lesssim 0.4$: ○. The drawn curve was calculated by regression analysis of C' data for $w_p \gtrsim 0.15$ (see text)

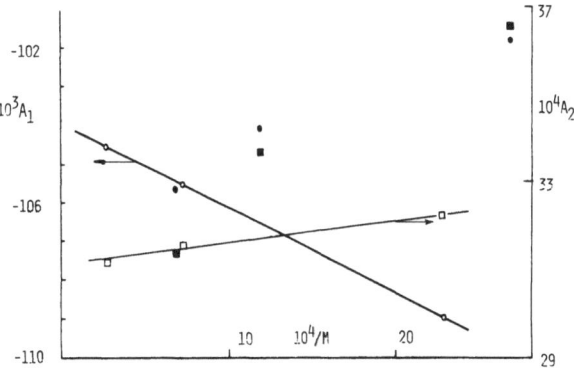

Fig. 6. Chain length dependence of the coefficients A_1 and A_2 in equation (10). The curves were fitted to the three data sets for molar masses 51, 166 and 520 kg/mole (○, □) for which A_1 and A_2 could be estimated from a wide temperature range. For the other three molar masses (●, ■) (45, 103 and 180 kg/mole) only a limited T-interval was available, which might explain the deviation of two of the points

From these findings we see that the free enthalpy expression still needs improvement on two points if a quantitative covering of the present sets of data is to be obtained. Corrections are necessary for the dilute solution regime as well as the range of higher concentrations. The next two sections deal with these aspects.

Before continuing we might reflect on a possible molecular origin of the obviously necessary contributions B_o and C. The term B_1 is well anchored in the molecular concept of the interchange energy associated with the breaking of contacts between like molecules and replacing them by unlike nearest neighbours (see e.g. ref. [27] or [28]).

Many years ago it was already pointed out by Staverman [41] that it does not suffice to consider only the energy of mixing on the basis of numbers of contacts between molecules. The entropy of mixing should then also be calculated for the numbers of contacts rather than the numbers of molecules, as is usually done. Today this is still very much the normal practice, it involves the assumption of a constant coordination number z on the lattice.

We relax that condition and allow the various species to differ in coordination number z_{ij}. If the two component molecules or repeat units differ in size but not much in shape, we have three coordination numbers, setting $z_{11} \simeq z_{22}$. If all z_{ij} are equal the number of arrangements on the lattice is well known [20–25, 27, 28, 36]. We denote this number by Ω_o. If $z_{11} = z_{22} \neq z_{12} \neq z_{21}$ we need to correct Ω_o for the over- and underestimations referring to z_{12} and z_{21}. Following a procedure suggested by Huggins [36] and Silberberg [42] we can write [43]

$$\Omega = \Omega_o \, (z_{12}/\bar{z})^{P_{12}} \, (z_{21}/\bar{z})^{P_{21}} \tag{11}$$

where $P_{12} \equiv P_{21}$ is the number of unequal contact pairs, and $\bar{z}$ is an average number depending on concentration.

Arbitrarily writing

$$\bar{z} \propto \sigma_1 \phi_1 + \sigma_2 \phi_2 \tag{12}$$

and using the regular solution approximation for P_{12}, one obtains

$$C = 2z_{22} \, (\ln Q)/Q \, ; \quad B_o = - z_{22} \ln (z_{12} \, z_{21}/z_{11}^2) \tag{13}$$

where $Q = 1 - \gamma \phi_2$.

It is seen that equation (13) thus supplies an after the fact explanation for the empirical parameters B_o and C, though in a qualitative sense. We ignore a possible

ϕ_2 dependence of C (by the term Q). It should be noted that Staverman has recently developed a rigorous treatment of contact statistics [44].

Non-uniform segment density

The concentration range covered by the experimental data employed here contains two essentially different regimes. As was pointed out by Flory as early as 1949, polymer solutions cannot, at high dilution, be looked upon as systems with uniformly distributed segments [45]. They consist of isolated coils separated by regions of pure solvent. The other regime ranges upward from concentrations where the coils overlap effectively, and the segment density can be considered approximately uniform.

Theoretical treatments of dilute solutions [45, 46] on the one hand, and concentrated systems on the other [1, 47, 48], are abundant, but a ΔG expression continuously covering both concentration regimes has had very little attention so far. However, it is needed for the description of liquid-liquid phase equilibria in polymer solutions where one of the phases is normally very dilute, and the other so concentrated that the coils overlap extensively.

To remedy the situation, Stockmayer et al. [49] wrote the interaction function as a sum of two terms, one for each concentration range:

$$g = g^*(T, m_2) \, P + g^c \tag{14}$$

where g^c, governing the concentrated regime, can be expressed by equation (8). The term $g^*(T, m_2)$ quantifies the differences relevant for the dilute regime compared with the uniform density state. The first term in equation (14) is attenuated by the probability factor P

$$P = \exp \left(- \lambda_o \, m_2^{1/2} \, \phi_2 \right) \tag{15}$$

where λ_o can be expressed in molecular parameters, independently determineable [49]. P stands for the probability that a given volume element in the solutions does not fall within any of the coil domains.

The function g^* was originally 'calibrated' versus the osmotic second virial coefficient, expressions for which are in ample supply [46]. However, it has become clear that a fully theoretical treatment of equation (14) leads to qualitatitve agreement with experiment only [2, 50]. It is therefore not surprising that we had to drop the theoretical definition of g^* in order to obtain the quantitative agreement aimed at.

A detailed analysis of the upward curvature in C' at small ϕ_2 confirms, within experimental accuracy, the theoretical probability factor P in its general exponential form, whereas we found g^* to be well represent by

$$g^* = \{g_1 + g_2(T - \theta)\}(T - \theta)/m_2 \qquad (16)$$

where g_1 and g_2 are adaptable parameters and θ is the Flory temperature. It is to be noted that equation (16) does not obey the theoretical boundary condition that the dilute solution effect should disappear for a mixture of small molecules:

$$\lim_{m_2 \to 1} g^*(T, m_2) = 0. \qquad (17)$$

This boundary condition can be included if, rather arbitrarily, we introduce a further chain length dependence:

$$g^* = \{g_1 + g_2(T - \theta)\}(T - \theta)(1 - 1/m_2)/m_2. \qquad (18)$$

All these considerations finally result in a semi-empirical expression for the interaction function g [4]:

$$g = (B_o + B_1/T)/(1 - \gamma\phi_2) + A(T, m_2)$$
$$+ g^*(T, m_2)P \qquad (19)$$

where

$$A(T, m_2) = A_{10} + A_{11}/m_2 + (A_{20} + A_{21}/m_2)(T - \theta).$$
$$(19\,a)$$

Expressing the temperature variable as $(T - \theta)$ proved advantageous in the fitting. We also noted that the additonal factor $(1 - 1/m_2)$ improved the overall agreement between semi-empirical theory and experiment.

The adjustable parameters in equation (19) were determined in a simultaneous optimisation of the complete data-set consisting of χ values, spinodal points, coexisting phase compositions and critical points, each with the relevant equation. A representative selection of the resulting descriptions is shown in figures 1, 2, 4 and 7–9. Full details and a comprehensive report on the procedure, including the step-wise approximations eventually leading to equation (19), can be found in [4]. Table 2 lists the values of the various parameters found to give the best fit to the complete data set.

Chain length dependence of the entropy of mixing

In most theoretical treatments of concentrated solutions the chain length dependence of the entropy of

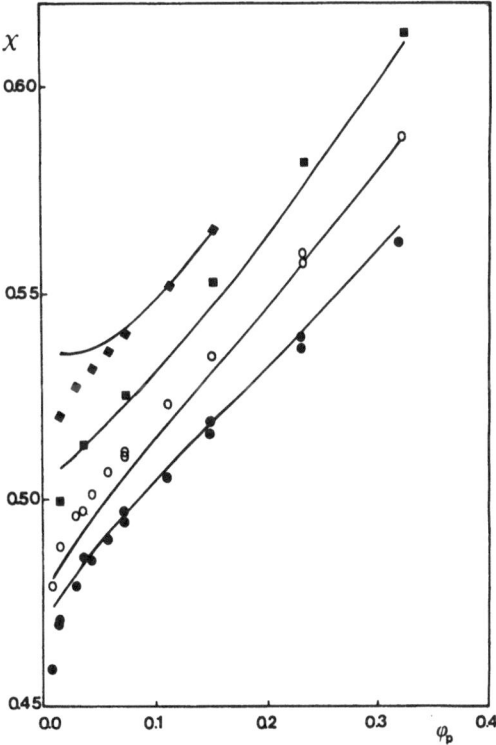

Fig. 7. Experimental χ data for polystyrene ($M = 51$ kg/mole) in cyclohexane as a function of concentration and temperature, and their description by equations (18) and (19). Data from Scholte et al. [5–9]

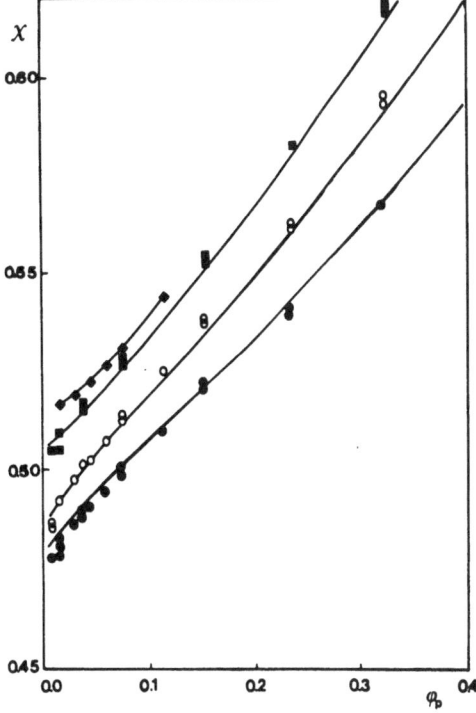

Fig. 8. As figure 7, $M = 166$ kg/mole

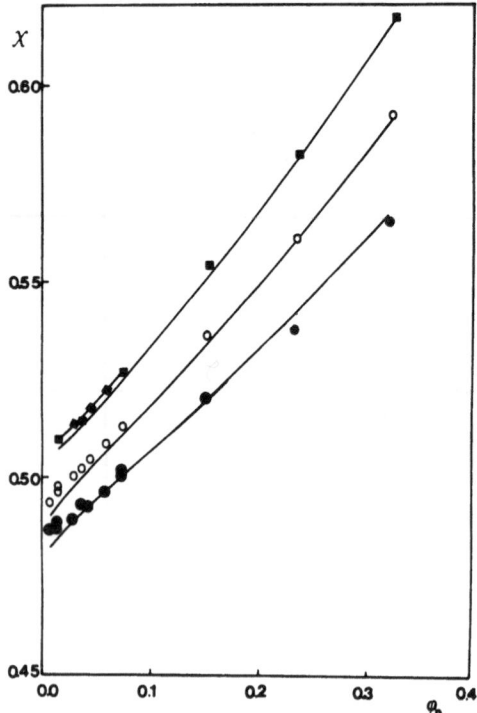

Fig. 9. As figure 7, $M = 520$ kg/mole

Table 2. Parameter values for equations (18) and (19)

B_0	=	-0.89
B_1/K	=	490
γ	=	0.29
A_{10}	=	-0.003
A_{11}	=	-2.10
A_{20}	=	0.00282
A_{21}	=	0.032
g_1	=	-0.0281
g_2	=	0.00069
λ_0	=	0.885

mixing is considered to be adequately accounted for by the second, combinatorial, term in equation (1). Writing ΔG per mole of lattice sites in equation (20) we see the reciprocal dependence on m_2, and the second term diminishes with increasing m_2, vanishing at $m_2 \to \infty$. For this reason it is sometimes neglected altogether

$$Z = \Delta G/NRT = \phi_1 \ln \phi_1 + (\phi_2/m_2) \ln \phi_2 + \Gamma/N \,.$$
$$(20)$$

Such simplifications are not commendable because, even if the second term is very small compared to the

first, it may still be too large to be negligible. The sum total of the many contributions to Z is itself, most of the time, relatively small. Minor contributions may thus influence the value of the sum. Since details in $Z(\phi_2)$ determine phase relationships sensitively, it is evident that none of the contributions should be ignored without checking. Also, both small and large terms should be assessed with the maximum possible precision.

The original ΔG expression (1) with g function (3) has the osmotic second virial coefficient α_2 independent of molar mass:

$$\alpha_2 \propto \left(\frac{1}{2} - \chi_1 \right) \qquad (21)$$

where

$$\chi_1 \equiv g \,. \qquad (21a)$$

Equation (21) is appropriate also if g depends on ϕ_2, then we have

$$\chi_1 = (g - \phi_1 \partial g/\partial \phi_2)_{\phi_2 = 0} \,. \qquad (21b)$$

The molar mass dependence of α_2 has been observed in early measurements already and has been successfully attributed to the non uniform segment density in dilute polymer solutions [45, 46]. The dilute solution theory can be summarized as

$$\alpha_2 \propto \left(\frac{1}{2} - \chi_1 \right) h(z) \qquad (21c)$$

where $h(z)$ is the excluded volume function [46] which introduces the chain length dependence in a qualitatively correct manner.

Recently, Tong et al. [51] have questioned the validity of equation (21c) and, thereby, dilute solution theory in general. The basis of their criticism was formed by extensive measurements of α_2 by light scattering at $T < \theta$ for a wide range of molar masses in the system cyclohexane/polystyrene.

We analysed Tong et al.'s data with our empirical g^* function and found it to provide a quite satisfactory description, although the molar mass and temperature dependencies had to be extended to include quadratic and higher terms:

$$g^*(T, m_2) = g_0(m_2) + g_1(m_2)(T - \theta)$$
$$+ g_2(m_2)(T - \theta)^2 \qquad (22)$$

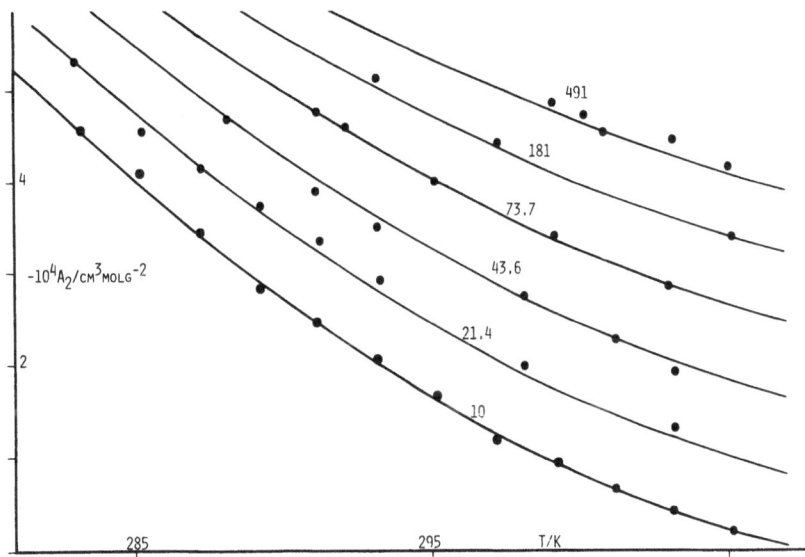

Fig. 10. Experimental second virial coefficients as a function of temperature. Molar masses in kg/mole indicated. Data from Tong et al. [51], curves fitted to the data with equation (22). All curves (except for $M = 10$ kg/mole) successively shifted vertically by 0.8×10^{-4} cm^3 molg^{-2}

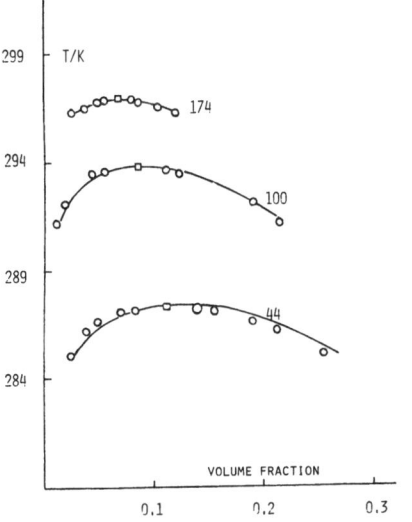

Fig. 11. As figure 1. Curves calculated with equations (19) and (22)

with

$$g_0(m_2) = (g_{01} + g_{02}/m_2)/m_2$$
$$g_1(m_2) = (g_{11} + g_{12}/m_2)/m_2$$
$$g_2(m_2) = (g_{21} + g_{22}/m_2 + g_{23}/m_2^2)/m_2 .$$

Figure 10 shows the representation of the experimental second virial coefficients so obtained. Using the values of the A and B coefficients in table 2 in combination with g^* function (22) we calculated binodals for molar masses of 44, 100 and 174 kg/mole, and obtained a de-

scription comparable in quality to that in figure 1 (see fig. 11). We conclude therefore that Tong et al.'s virial coefficient data do not contradict the ΔG function developed in this paper.

Nevertheless, Tong et al.'s criticism of the dilute solution theory is justified, albeit for a slightly different reason. The chain length dependence we established for the contribution $A(T, m_2)$ to ΔG necessitates a further amendment to the dilute solution expression (21 c). The term χ_1 contains a contribution from $A(T, m_2)$ and consequently becomes chain length dependent, an effect not accounted for in the theory. Hence, we should write

$$\alpha_2 \propto \left\{ \frac{1}{2} - \chi_1(m_2) \right\} h(z) \qquad (21\,d)$$

and it is seen that the excluded volume function $h(z)$ should be tested with equation (21 d) rather than with equation (21 c). Whether this situation prevails at temperatures further removed from θ, where the solvent quality increases, is an open question at this moment. In view of remarks in the following paragraph one might expect any near θ situation to be representative of equation (21 d) rather than equation (21 c).

For a molecular explanation of the molar mass dependence of the term A in g^c we turn to a calculation by Staverman [52]. He considered the effect of the chains bending back on themselves on the entropy of mixing for the athermal case. Summarizing his equations one could write

$$g^c = g^c(y, \delta, T); \quad \delta = \delta(m_2, T) \qquad (23)$$

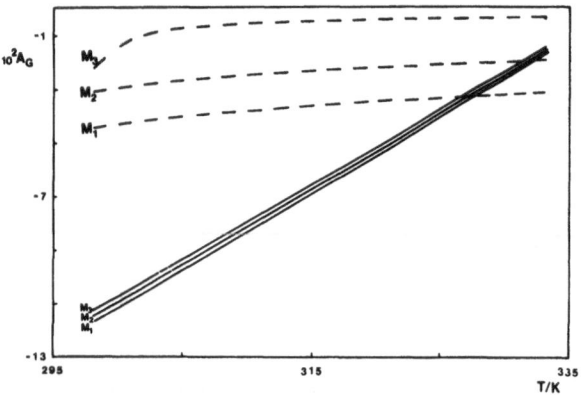

Fig. 12. Temperature dependence of the empirical free enthalpy contribution $A(T, m_2)$ (eqs. (10) and (19 a)) (——) and the same contribution as predicted by Staverman's model [4, 52] (– – –), calculated for molar masses M_1, M_2 and M_3 of 50, 100 and 1000 kg/mole, respectively

where δ is the fraction of external contacts on the chain lost to intramolecular contacts by back-bending. According to Staverman, δ must be expected to depend on the size of the polymer coil and this, in turn, depends on molar mass and temperature.

Extending his treatment to the non-athermal case we find the resulting equation for $A(T, m_2)$ to show the correct trend; magnitude and slope, however, are too small. Figure 12 illustrates the situation. Since the theory is in the process of further development at the moment [44] we let a further discussion of the $A(T, m_2)$ term rest. We only note that back bending is more likely to occur in near θ states than in good solvents.

Temperature dependence of the interaction function g

The ΔG function developed so far includes a semi-empirical expression for g with a rather complex temperature dependence. The original model, on the other hand, only allows for a linear dependence on $1/T$. This temperature function has often been and still is the ground for objections against the Flory-Huggins model, represented by equations (1) and (3), because of its alleged incapability to describe lower critical miscibility, in particular if it occurs together with upper critical miscibility in the same system. Such criticism is not justified [30].

In the first place there is no reason why g_H in equation (3) could not assume negative values, the interchange energy determining its sign. Lower critical miscibility goes with a negative g_H. Further, if the criticism

has no objections against the empirical transformation of g into an free enthalpy function by the additon of g_S, one can hardly refuse to accept the outcome of an application of some fundamental thermodynamic relations to g, viz:

$$\Delta H = \int \Delta c_p \, dT; \quad \Delta S = \int (\Delta c_p / T) \, dT \quad (24)$$

where ΔH, ΔS and Δc_p are the enthalpy, entropy and specific heat changes upon mixing.

The specific heat at constant pressure, c_p, of a liquid is known to depend on temperature. In additon, Δc_p must be expected to vary with concentration, e. g.

$$\Delta c_p = (c_0 + c_1 T) \phi_1 \phi_2. \quad (25)$$

Equations (24) and (25), together with equation (1), can now be understood to define the interaction function $g(T)$. One finds

$$g = g_a + g_b / T + g_c T + g_d \ln T \quad (26)$$

where g_a and g_b arise from integration constants (eq. (24)) and $g_c = - c_1 / 2NR$; $g_d = - c_0 / NR$. Use of the Flory-Huggins equation in the form of equations (1) and (3) unrealistically ignores the temperature dependence of Δc_p. The latter was formulated theoretically by Delmas et al. [53] who found Prigogine's model [1] to supply expressions for g_b and g_c, sufficient to deal with the occurrence of both upper and lower critical miscibility.

We thus see that the Flory-Huggins equation, properly used, is not to be criticised in this respect. We see also that a complex $g(T)$ function, like our equations (18) and (19), is to be expected. Whether the various coefficients in equation (26) are consistent with those in equation (19) could not be checked and may probably not be expected. The main point, however, is that the form found experimentally for $g(T)$ is conceivable and qualitatively consistent.

Predictive calculations

During the development of an appropriate free enthalpy function for the present set of data, the treatment has changed character from a model to a curve-fitting procedure. The number of parameters is large and, though the general form of the equations can be supported by molecular considerations, the procedure needs justification. This can be found in the predictive power of the equations used and parameter values obtained.

Krigbaum's extensive osmotic pressure data, covering molar masses ranging from 51 to 566 kg/mole [19], offer a suitable example. We calculated the reduced osmotic pressure by standard methods, using the values of the parameters in table 2. It is seen in figure 13 that the 'predicted' curves pass through the experimental points within the accuracy specified by Krigbaum.

Other examples can be drawn from the work of Kuwahara et al. [16–18] who reported near binary coexistence curves (binodals). It is seen in figure 14 that the calculated binodals agree in shape with the experimental ones, but the temperatures are about 0.5 °C off. A small shift of the predicted curves makes them coincide fairly well.

As a final test we introduce polydispersity and try to calculate ternary binodals, assuming that the ternary system can be treated as a simple superposition of three binary ones without any cross terms. Figure 15 demonstrates that Hashizume et al.'s data [10] are well reproduced by the free enthalpy equation presented in this paper.

When it comes to wide molar mass distributions and to the description of distribution coefficients as a function of chain length, the present model and correlation function have not yet been analysed in depth [3, 4]. It is conceivable that the simple superposition method would be inadequate for such cases. This problem is a matter of current study.

Limiting critical concentration

We now drop the precise representation of data striven after in the preceding sections, and focus on a peculiar phenomenon indicated several years ago by Flory and Daoust [54]. These authors pointed out that the critical concentration at infinite chain length does not necessarily have to be zero. Later, Dušek [55] analysed the situation and found that in such a case, if existing at all, the θ temperature, taken to be the temperature at which the osmotic second virial coefficient vanishes, is not identical to the critical temperature at infinite chain length. Kennedy [56] suggested making a

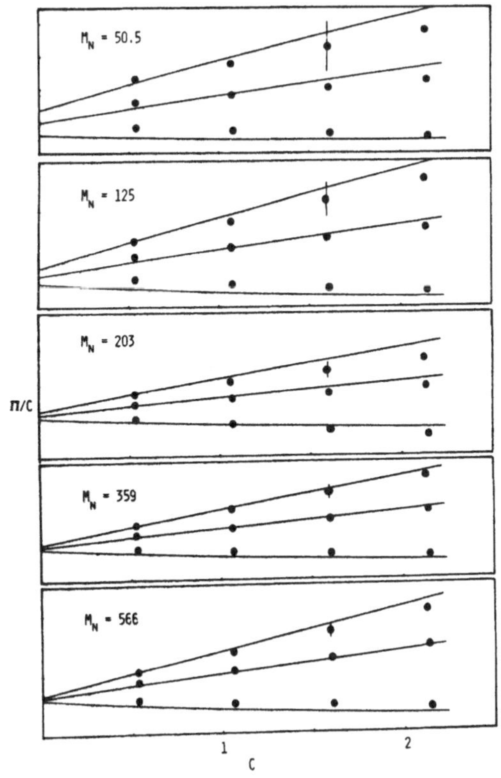

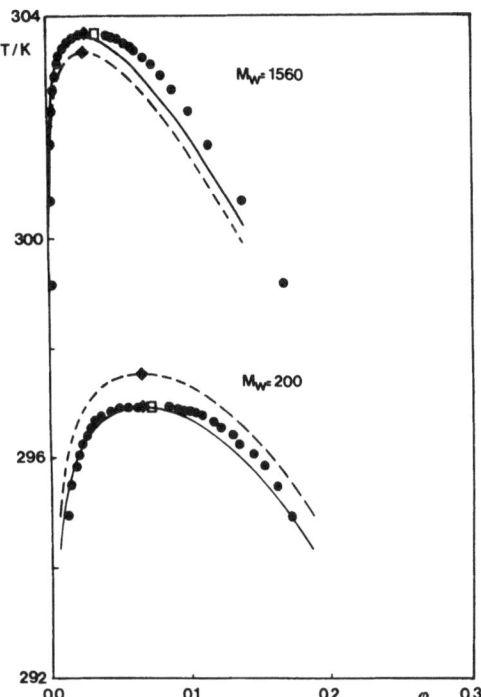

Fig. 13. Reduced osmotic pressures for various indicated molar masses (kg/mole) as a function of concentration and temperature (bottom, middle and top curves: 30, 40 and 50 °C, respectively) cyclohexane/polystyrene, data from Krigbaum [19]. Curves calculated with equations (18) and (19)

Fig. 14. Experimental binodals and their representation by equations (18) and (19) (−−−). System cyclohexane/polystyrene, data from Kuwahara et al. [16–18] (●). Critical points: □: experimental; ◆: calculated. Shifted calculated curves: ——

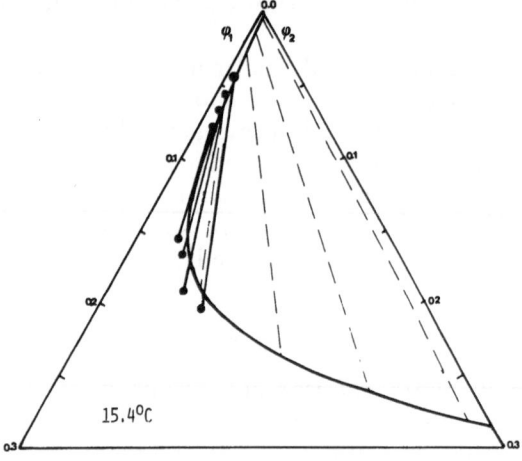

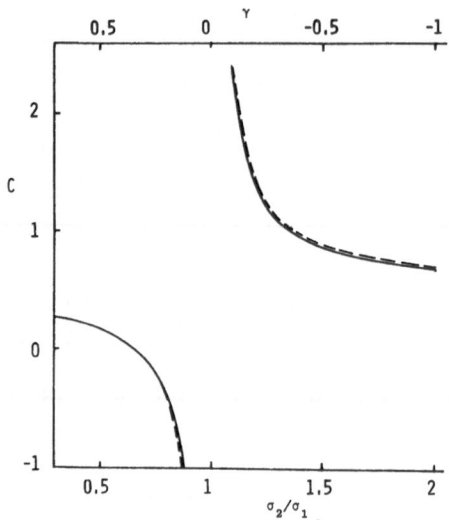

Fig. 16. Relation between C and y for a non-zero limiting critical concentration at infinite chain length. $\phi_{2c} = 0.02$: ——; $\phi_{2c} = 0.04$: ———

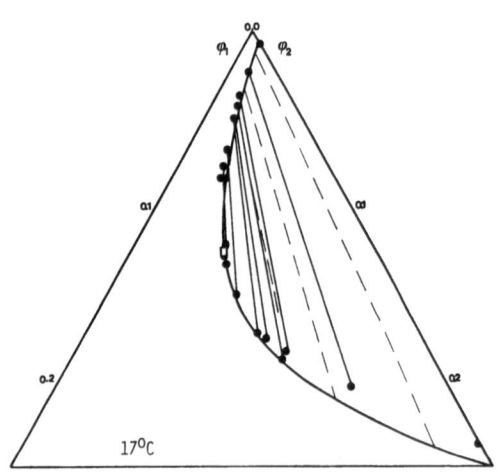

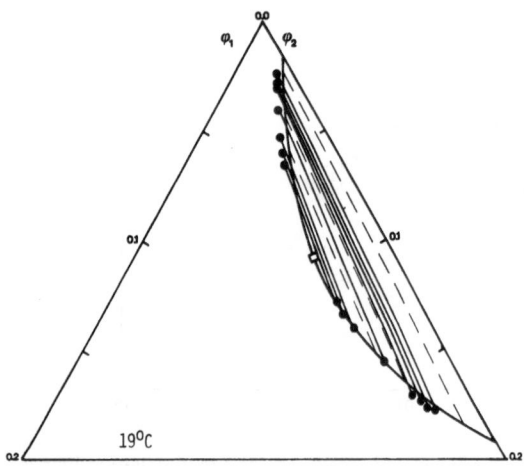

Fig. 15. Ternary phase relations for the system cyclohexane/polystyrene. Molar masses polymer constituents: 45 and 103 kg/mole. Data from Hashizume et al. [10]. Binodals (——) and tie lines (———) calculated with equations (18) and (19)

clear distinction between the θ state and the critical state for infinite m_2, denoting the latter by the term 'limiting critical state'. Experimental evidence of the existence of a limiting critical concentration, ϕ_{2L}, has been reported [57].

Since we are now interested in properties at $m_2 \to \infty$, much of the intricacies discussed above can be dropped. Equation (19) reduces to equation (8) in which we shall ignore a possible temperature dependence of C. Combining equations (1), (4), (5) and (8) we can write the critical condition as

$$1/m_2 = (\phi_2/\phi_1)^2 (Q_1 - 3y + 6C y \phi_1^2)/Q_1 \qquad (27)$$

where $Q_1 = 1 + 2y \phi_2$.

Equation (2) shows that, at $m_2 = \infty$, $\phi_{2c} = 0$ always represents a possible root. Another root (ϕ_{2L}) might arise from the second expression between brackets, if the values of the parameters allowed it to be between 0 and 1. Figure 16 demonstrates that the parameters C and y would have to be related in a special way. Since C values are often found to be close to zero the ratio σ_2/σ_1 of surface areas should be smaller than 1. The model thus predicts that experimental examples had best be sought in systems where the solvent molecule is bigger than the repeat units in the polymer.

Experiments on diphenylether/polyethylene [58], cyclohexane/polystyrene [15], and benzene/polyisobutylene [57] show that such a trend can indeed be observed. Recent measurements on diphenylether/polyisobutylene also indicate $\phi_{2L} > 0$ (see fig. 17). The

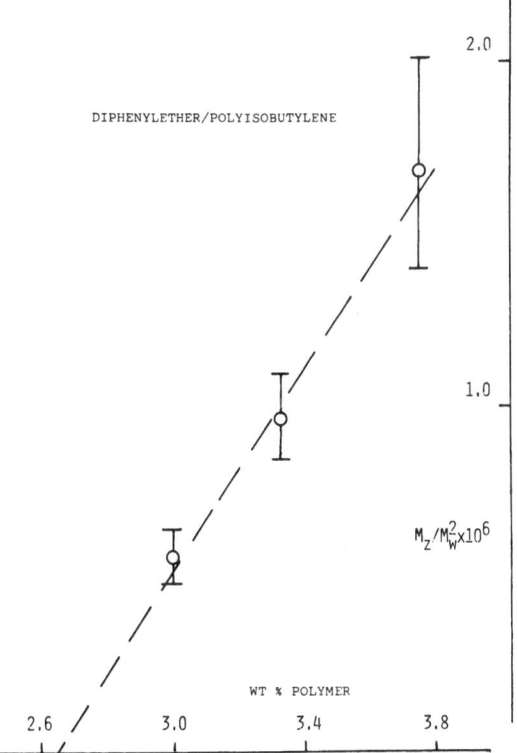

Fig. 17. Experimental indication for non-zero critical concentration ϕ_{2L} at infinite chain length. The abcissa represents value of M_z/M_w^2 to correct for differences in polydispersity. The experimental error in the critical concentration (determined with the phase-volume-ratio method [30]) is indicated by the horizontal lines on the error bars

Bondi rations σ_2/σ_1 are 0.28, 0.87, 0.93 and 0.57, respectively. Most samples had wide molar mass distributions and we should therefore have used the multicomponent versions of equations (4) and (5) to derive equation (27). This can be shown to lead to replacement of m_2 by m_w^2/m_z, where m_w and m_z are the mass- and z-average chain lengths [59]. Hence, this quantity, or M_w^2/M_z, has been plotted in figure 17 and the ratio M_z/M_w is assumed to remain finite when $M_w \to \infty$. We do not want to stretch the present argument too far, but feel justified in concluding that these examples furhter exemplify the importance of accounting for differences in size and shape between the constituent molecules and repeat units.

Discussion

In this paper we have tried to carry the description of thermodynamic properties of polymer solutions as

far as the rigid lattice model would permit. In the course of the development we relaxed the constancy of the coordination number, giving up the model in its strict, literal sense, but obviously greatly adding to its applicability. A further improvement includes the chain length dependence at a high degree of dilution. Finally, we saw that the back bending of chains onto themselves affects the free enthalpy of mixing, particularly in the concentrated regime where the coils overlap extensively. However, we are still left with a sizeable correction term to which the present model cannot assign a molecular basis.

It should be noted that the data sets used come from different sources and yet allow a comprehensive treatment. It must be admitted that the number of adaptable parameters seems excessive (table 2), but dropping one or more immediately worsens both description and predictions [4]. Therefore we feel that, at the moment, the presented procedure is the optimum treatment within the lattice model.

In spite of the multitude of parameters the molecular background behind most of the necessary correction terms is clear. We know that molecules and repeat units differ in size and shape, and will differ considerably in the number of nearest neighbour contacts they can make. Further, we have the well known phenomenon that dilute and concentrated polymer solutions differ essentially in segment density. Finally, we acknowledge the effect back bending of chains may have, in particular in poor solvents like the system with which we are dealing.

That the theoretical treatments are still too rough to come up with quantitative answers is hardly surprising. The various contributions to ΔG, arising from these physically evident effects, are not seldom larger than ΔG itself. Even minute inaccuracies, unavoidable with approximate theories and negligible in each term separately, may seriously distort the resulting ΔG value. Large errors, in paticular in phase relations, may easily ensue. The prediction of such properties requires a high precision on ΔG which, at present, can only be had at the cost of a relatively large number of parameters.

In the end we retained a correction for the concentrated regime that so far seems to escape obvious molecular interpretation within the model. Free volume, left out of consideration here, must certainly be held responsible for part of the concentration dependence of g [60–63]. Inclusion will doubtless change the values of the parameters obtained so far, but allow excess volumes and the influence of pressure to be treated. This point is the subject of current research.

Acknowledgement

The authors are indebted to Mr. Ian Bowskill for the determination of critical points in the system diphenylether/polyisobutylene.

References

1. See eg: Prigogine I (1957) The Molecular Theory of Solutions, North-Holland Publ Co, Amsterdam
2. Kleintjens LA, Koningsveld R, Stockmayer WH (1976) Br Polym J 8:144
3. Onclin MH (1980) PhD Thesis, University of Antwerp, Belgium
4. Nies E (1983) PhD Thesis, University of Antwerp, Belgium
5. Scholte ThG (1970) Eur Polym J 6:1063
6. Scholte ThG (1971) J Polym Sci, Part A2, 9:1553
7. Scholte ThG (1972) J Polym Sci, Series C 39:281
8. Scholte ThG (1970) J Polym Sci, Part A2, 8:841
9. Rietveld BJ, Scholte ThG, Pijpers J (1972) Br Polym J 4:109
10. Hashizume I, Teramoto A, Fujita H (1981) J Polym Sci, Polym Phys Ed 19:1405
11. Derham KW, Goldsbrough J, Gordon M (1974) Pure Appl Chem 38:97
12. Irvine P, Gordon M (1980) Macromol 13:761
13. Irvine P, Kennedy JW (1982) Macromol 15:473
14. Koningsveld R (1970) Disc Farad Soc, No 49:144
15. Koningsveld R, Kleintjens LA, Shultz AR (1970) J Polym Sci, Part A2, 8:1261
16. Kuwahara N, Nakata M, Kaneko M (1973) Polymer 14:415
17. Nakata M, Kuwahara N, Kaneko M (1975) J Chem Phys 62:4278
16. Nakata M, Dobashi N, Kuwahara N, Kaneko M, Chu B (1978) Phys Rev, A 18:2683
19. Krigbaum WR (1954) J Am Chem Soc 76:3758
20. Flory PJ (1941) J Chem Phys 9:660
21. Flory PJ (1942) J Chem Phys 10:51
22. Huggins ML (1941) J Chem Phys 9:440
23. Huggins ML (1942) Ann NY Acad Sci 43:1
24. Staverman AJ, van Santen JH (1941) Rec Trav Chim 60:76
25. Staverman AJ (1941) Rec Trav Chim 60:640
26. Koningsveld R, Staverman AJ (1968) J Polym Sci, A2, 6:325
27. Flory PJ (1953) Principles of Polymer Chemistry, Cornell Univ Press
28. Guggenheim EA (1952) Mixtures, Clarendon Press, Oxford
29. Rehage G (1963) Kunststoffe 53:605
30. Koningsveld R (1968) Adv Interf Coll Sci 2:151
31. Shultz AR, Flory PJ (1952) J Am Chem Soc 74:4760
32. Derham KW, Goldsbrough J, Gordon M, Koningsveld R, Kleintjens LA (1975) Makromol Chem Suppl 1:401
33. Staverman AJ (1937) Rec Trav Chim 56:885
34. Langmuir I (1925) Coll Symp Monogr 3:48
35. Butler JAV (1933) J Chem Soc 19:681
36. See eg: Huggins ML (1948) J Phys Coll Chem 52:248
37. See eg: Kanig G (1963) Kolloid Z u Z Polym 190:1
38. Bondi A (1968) J Phys Chem 68:441
39. Rowlinson JS (1959) Liquids and Liquid Mixtures, Butterworth, London
40. Koningsveld R, Kleintjens LA (1971) Macromol 4:637
41. Staverman AJ (1938) PhD Thesis, University of Leiden
42. Silberberg A (1968) J Chem Phys 48:2835
43. Koningsveld R (1985) Proc Intern Disc Meeting Polym Sci Technol, Rolduc, Netherlands, April
44. Staverman AJ (1985) Proc Intern Disc Meeting Polym Sci Technol, Rolduc, Netherlands, April
45. Flory PJ (1949) J Chem Phys 17:1347
46. Yamakawa H, (1971) Modern Theory of Polymer Solutions, Harper & Row, New York
47. See eg: Olabisi O, Robeson LM, Shaw MT (1979) Polymer-Polymer Miscibility, Academic Press
48. See eg: Kurata M (1982) Thermodynamics of Polymer Solutions, Harwood Acad Publ
49. Koningsveld R, Stockmayer WH, Kennedy JW, Kleintjens LA (1974) Macromol 7:73
50. Chu SG, Munk P (1977) J Polym Sci, Polym Phys Ed 15:1163
51. Tong Z, Ohashi S, Einaga Y, Fujita H (1983) Polymer J 15:835
52. Staverman AJ (1950) Rec Trav Chim 69:163
53. Delmas G, Patterson D, Somcynski T (1962) J Polym Sci 57:79
54. Flory PJ, Daoust H (1957) J Polym Sci 25:429
55. Dušek K (1969) Coll Czechoslov Chem Comm 34:3309
56. Kennedy JW (1970) J Polym Sci, Part C, No 39:71
57. Koningsveld R, Kleintjens LA (1973) Pure Appl Chem, Macromol Chem 8:197
58. Kleintjens LA, Koningsveld R, Gordon M (1980) Macromol 13:303
59. Stockmayer WH (1949) J Chem Phys 17:588
60. Flory PJ (1965) J Am Chem Soc 87:1833
61. Orwoll RA, Flory PJ (1967) J Am Chem Soc 89:6814, 6822
62. Kleintjens LA, Koningsveld R (1980) Coll & Polym Sci 258:711
63. Koningsveld R, Kleintjens LA, Leblans-Vinck AM, Ber Bunsenges, in press
64. Lal M, Swinton FL (1968) Physica 40:446
65. Singh J, Pflug HD, Benson GC (1968) J Phys Chem 72:1939

Received July 29, 1985;
accepted August 15, 1985

Authors' address:

E. Nies
Polymer Technology
University of Technology
Eindhoven, Netherlands

Interaction between block copolymer micelles in solution*)

Č. Koňák, Z. Tuzar, P. Štěpánek, B. Sedláček, and P. Kratochvíl

Institute of Macromolecular Chemistry, Czechoslovak Academy of Sciences, Prague, Czechoslovakia

Abstract: Interactions between micelles of the three-block copolymer polystyrene-block-(hydrogenated polybutadiene)-block-polystyrene in selective solvent mixtures 1,4-dioxane/0–30 vol. % n-heptane were investigated by light scattering methods. Experimental values of the second virial coefficient, A_2, from the time-averaged light scattering data agree with theoretical values for the hard sphere model. The diffusion virial-coefficient values, k_D, from quasielastic light scattering measurements agree within the limits of experimental error with the experimental k_D value determined by Kops-Werkhoven and coworkers for spherical silica particles. Virtually monodisperse block-copolymer micelles appeared to be suitable for testing theoretical results obtained for hard sphere interactions.

Key words: Quasielastic light scattering, time-averaged light scttering, block copolymer micelles, interparticle interactions, hard sphere model.

Introduction

From the quasielastic light scattering (QELS) experiment, i. e., by measuring the relaxation times of particle fluctuations in solution, τ_c, one may obtain information on the dynamic properties of polymer and colloid systems [1]. From the relaxation times, values of the so-called mutual or collective diffusion coefficient $D_c(q)$ can be obtained, q being the scattering vector. The τ_c and D_c values are generally concentration-dependent due to direct, thermodynamical, and indirect, hydrodynamical, interactions. In sufficiently diluted dispersions (solutions) of particles, $D_c(O)$ may be expressed through a virial series expansion in the particle volume fraction ϕ or in the particle mass concentration c $(\mathrm{g\,cm^{-3}})$

$$D_c(O)/D_o = 1 + k_D \phi = 1 + k'_D c, \qquad (1)$$

where D_o is the diffusion coefficient at infinite dilution k_D and k'_D are the diffusion virial coefficients; $k_D = k'_D/\bar{v}$ and $\bar{v}$ is the partial specific volume of the par-

ticles. From the sign and magnitude of these coefficients one may draw conclusions as to the magnitude and character of interparticle interactions in dispersions (solutions).

Experimental studies in this field stimulated considerable theoretical interest in the diffusion of interacting particles (e. g., [2–14]). In the calculations of k_D, two types of interactions are considered independently in most theories: (i) direct thermodynamical interactions caused by interparticle forces (static part of the problem), (ii) indirect hydrodynamical interactions ensuing from the fact that the movement of one particle through the liquid generates a velocity field, which affects the movement of neighbouring particles. As regards the direct part, agreement between the theoretical result is satisfactory. In the case of hydrodynamical interactions, however, large differences exist, especially in results obtained for hard sphere interactions at $q \to 0$. The situation is illustrated in table 1, which summarizes the results of calculations of k_D for dispersions of hard spheres.

The experimental verification of theoretical results should be carried out with particles which approach the ideal hard spheres as closely as possible. Unfortunately, the majority of colloidal dispersions (latices,

---

*) Dedicated to Prof. Dr. H.-G. Kilian on the occasion of his 60th birthday.

Table 1. Theoretical values of the k_D constant in equation (1)

k_D	Ref.	Year	k_D	Ref.	Year
1.12	2	1941	1.00	9	1977
0.84	3	1964	1.56	10	1978
1.45	4	1972	2.50	11	1979
2.00	5	1973	1.45	12	1979
3.00	6	1975	1.45	13	1981
−1.83	7	1976	1.56	14	1982
−6.00	8	1976			

microemulsions, and the like) appeared to be unsuitable for such tests, especially because of non-negligible long-range (e. g., electrostatic) interactions between the particles. So far, the material approaching most the concept of an ideal hard sphere is represented by modified silica particles consisting of a nucleus (SiO_2), surrounded by a shell of chemically attached short hydrocarbon chains [15, 16]. Experimentally, $k_D = 1.3 \pm 0.2$ was obtained for such dispersions in cyclohexane [17]; with respect to the results of their own sedimentation measurements, the authors state that there is a good fit between these and the theoretical results of Batchelor [4] or Felderhof [10].

Very similar to these silica particles in their structure are block-copolymer micelles which arise by closed association of block copolymers in selective solvents (i. e., solvents which are good solvents form one block and precipitants for the other). These compact spherical particles with a narrow mass and size distribution consist of tens to hundreds of copolymer molecules, arranged so that the insoluble blocks for a compact core surrounded by a shell of soluble blocks [18]. In our earlier studies [19–21], we investigated the properties and behaviour of micelle of the three-block copolymer polystyrene-block-(hydrogenated polybutadiene)-block-polystyrene (SHBS) in various selective solvents and under various experimental conditions. Micelles of this copolymer in 1,4-dioxane, where the association equilibrium is distinctly shifted towards the micelles, behave as hard spheres, which is reflected, among other things, also in the coincidence of their geometrical and hydrodynamical dimensions [19]. These results have aroused a justified hope that micelles, although much less compact than silica particles, might still be suitable for use as hard-sphere-model particles for verification of the existing experimental and theoretical results.

This contribution is a study of direct thermodynamical and indirect hydrodynamical interactions of

SHBS micelles in mixtures of 1,4-dioxane/n-heptane. These systems were used in QELS and time-averaged light scattering (TALS) experiments with the aim of obtaining values of k_D and of the second virial coefficient A_2, and thus also an experimental basis for the evaluation of interactions between the micelles. Extension of QELS experiments by including TALS seems to be very promising, because it allows an independent determination and comparison of contributions of direct interactions, and thus helps to interpret the diffusion (QELS) experiments.

Experimental

Copolymer

The three-block copolymer polystyrene-block-(hydrogenated polybutadiene)-block-polystyrene (Kraton G-1650, Shell product) containing c. 1,5 mass % of homopolystyrene was fractionated in the system cyclohexane/n-propanol [22]. The middle fraction (c. 20 mass % of the sample), denoted below as $K - 1$, was free of homopolystyrene, its GPC diagram had a single sharp symmetrical peak, and the analysis of quasielastically scattered light gave the polydispersity [23] $M_w/M_n = 1.25$, comparable with that of the polystyrene standard. The molar mass (74×10^3 g mol^{-1}) and chemical composition (28 mass % of styrene units) of the fraction $K - 1$ do not differ within the limits of experimental error from values for unfractionated Kraton G-1650 [19].

In selective solvents consisting of 1,4-dioxane and 0–30 vol. % heptane the fraction $K - 1$, similarly to the unfractionated Kraton G-1650, forms multimolecular spherical micelles, the polydispersity of which lies below the resolution limit of the methods employed (GPC, QELS [23]). We therefore denote their molar mass by M, without an index specifying the type of averaging.

Polystyrene was a Pressure Chemical Co. standard, molar mass $M_w = 34 \times 10^3$ g mol^{-1}.

Solvents and sample preparation

1,4-dioxane and n-heptane, reagent grade (Lachema, Czechoslovakia), were rectified on a laboratory column. Solution used in the light scattering measurements were filtered through glass bacterial filters G 5, Jena, into cylindrical cells which could be sealed.

Time-averaged light scattering

The TALS measurements were carried out with a Sofica apparatus fitted with a He-Ne laser (vertically polarized light, $\lambda_0 = 632.8$ nm) and a digital voltmeter, in the angular range 30°–150°. The molar mass of the micelles, M, and the second virial coefficient A_2 were determined from the relation

$$Kc/R_o = 1/M + 2A_2c \qquad (2)$$

where K is the optical constant involving the squared refractive index increment, R_o in the Rayleigh ratio (proportional to the intensity of light scattered from micelles) extrapolated to the zero scattering angle θ, c is the polymer mass concentration.

The refractive index increment, dn/dc, of copolymer solutions was measured with a BP-2000 V Brice-Phoenix differential refractometer ($\lambda_o = 632.8$ nm). $dn/dc = 0.090$ cm^3 g^{-1} in 1,4-dioxane of 25°C, $(dn/dc)_\mu = 0.102$ cm^3 g^{-1} in the mixture 1,4-dioxane/30 vol.% n-heptane, also at 25°C; $(dn/dc)_\mu$ was measured under the condition of osmotic equilibrium between the copolymer solution and the mixed solvent [24].

Quasielastic light scattering

Quasielastic light scattering (QELS) was analyzed with a homodyne photon spectrometer fitted with a thermostated sample holder. The solution in a concentration range of $1 \times 10^{-3} - 2 \times 10^{-2}$ g cm^{-3} were thermostated to 25 ± 0.05°C. The light source was a Spectra Physics He-Ne laser 125 A ($\lambda_o = 632.8$ nm). The scattered light was detected at the angle $\theta = 45°$ with an RCA C 31034 photomultiplier and a photon-counting system PC-1, manufactured by Spex. The photopulse signal was treated with a 96-channel correlator.

The autocorrelation functions $G(\tau)$ obtained with micellar solutions were perfectly single-exponential, which suggests that within the measured concentration range the micelles are virtually monodisperse and that the amount of unimer in solution is negligibly small. The $D_c(q)$ values were obtained from experimental autocorrelation functions by a forced single-exponential fit in the form

$$G(\tau) = A \exp\left[- 2 D_c(q) q^2 \tau\right] + B, \qquad (3)$$

where τ is the time delay, A and B are constants.

With respect to the small size of the micelles, it is evident that at the angle of measurement of quasielastically scattered light $\theta = 45°$, $qR_G < 1$, where R_G is the radius of gyration of the micelles. The values of $D_c(q)$ thus measured are, consequently, virtually identical with those of $D_c(O)$ in equation (1). The hydrodynamical radius, R_H, was therefore calculated from the Stokes-Einstein formula $R_H = kT/6\pi\eta D_o$, where η is the solvent viscosity. D_o was obtained by extrapolating D_c ($\equiv D_c(O)$) to $c \to 0$.

Results and discussion

It can be seen from the results of the quasielastic and time-averaged light scattering experiments (fig. 1) that the concentration dependences of D_c and Kc/R_o are linear. This means that, within the concentration range studied, neither coagulation nor dissociation of the micelles takes place, and that by extrapolating the measured values to infinite dilution we obtain values of the diffusion coefficient of the micelles, D_o, and of

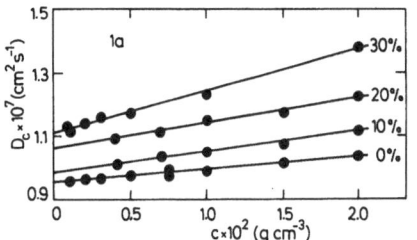

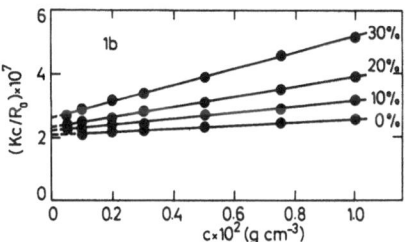

Fig. 1. Concentration dependence of D_c (1a) and Kc/R_o (1b) for micelles of the sample $K-1$ in 1,4-dioxane/n-heptane solvent mixtures at 25°C. Curves are labelled by vol-% of n-heptane

their molar mass, M. Hence, the slope of linear dependences can be interpreted as the coefficient k'_D (fig. 1a) and the second virial coefficient A_2 (fig. 1b) in the given solvent, respectively.

The experimental values of M, A_2, R_H and k'_D are summarized in table 2. By combining results of different methods, it has been demonstrated [19], that in 1,4-dioxane Kraton-1650 forms micelles with an aliphatic core and a polystyrene shell. 1,4-dioxane is a thermodynamically better solvent for polystyrene, while n-heptane is a better solvent of the central aliphatic block. An increase in the n-heptane content improves the thermodynamical quality of the mixture with respect to the micellar core and impairs it with respect to the polystyrene shell. The decrease in the M value and the increase in the A_2, R_H and k'_D values with increasing volume fraction of n-heptane in the solvent mixture

Table 2. Light scattering data for micelles of the copolymer sample $K-1$ in mixtures 1,4-dioxane/n-heptane at 25°C. A_2^{HS} and k_D are calculated quantities (for the definition see text)

n-heptane vol.-%	$M \times 10^{-6}$	$A_2 \times 10^6$ cm^3 g^{-2} mol	R_H nm	k'_D cm^3 g^{-1}	$A_2^{HS} \times 10^6$ cm^3 g^{-2} mol	k_D
0	4.80	2.1	19.2	4.02	3.07	1.09 ± 0.2
10	4.44	4.2	22.3	6.63	5.66	1.05 ± 0.2
20	4.20	7.5	23.2	7.54	7.19	1.01 ± 0.2
30	3.77	12.7	24.9	11.95	10.97	1.15 ± 0.2

(table 2) mean that the effect of improvement of the mixed solvent with respect to the micellar core predominates over the effect of impairment of the quality with respect to the micellar shell. This is also suggested by the fact that the decrease in the A_2 values for polystyrene is not pronounced (table 3), and the mixture 1,4-dioxane/30 vol. % n-heptane is still a good solvent for polystyrene. On the contrary, for the middle aliphatic block, 1,4-dioxane is a distinct precipitant, and only the mixture 1,4-dioxane/33 vol. % n-heptane, in which the copolymer dissolves molecularly is a thermodynamically good solvent.

The applicability of the hard sphere model to Kraton micelles in 1,4-dioxane has been proved earlier [19]. How far this model is satisfied also with slightly swollen micelles in mixtures with up to 30 vol. % n-heptane was proved by a comparison between theoretical A_2^{HS} values for hard spheres and exprimental A_2 values. In dispersions of ideal hard spheres, we have (e. g., [25]),

$$A_2^{HS} = 4 N_A V / M^2, \tag{4}$$

where N_A is the Avogadro number and V is the excluded volume of a spherical particle; $V = 4\pi R^3/3$, where R is the geometrical radius of the particle which, within the framework of the model, can be replaced by R_H. The relatively good agreement between the experimental A_2 values and the calculated A_2^{HS} values (table 2, fig. 2) justify a statement that direct thermodynamical interactions between the micelles are very close to those between the hard spheres.

For the comparison of the diffusion virial coefficients experimentally determined by us and other authors with theoretical values, k'_D has to be converted to k_D $(= k'_D / \bar{v})$. Since $\bar{v}$ is unknown for micelles swollen by solvents, we use as an approximation the volume of particles in solution per unit of M defined as $N_A V/M$. If we again use the R_H values instead of R (hard sphere model), then by using the M values we can calculate the k_D $(= k'_D M/N_A V)$ values given in table 2. Hence, it can be seen that the quality of the solvent is not reflected systematically in these results. Our k_D values agree, within the limits of experimental error, with $k_D = 1.3 \pm 0.2$ determined for silica particles [17], even though they are systematically lower. Perfect agreement would be reached on the assumption that $R_H > R$, for which there are no evident reasons. If, moreover, one considers the results of the time-averaged light scattering (A_2), it can be concluded that also indirect hydrodynamical interactions of micelles are very close to those between hard spheres. Compared with earlier data, our k_D values are closest to the theoretical value obtained by Hess and Klein ($k_D = 1.00$) [9].

It may be said, in conclusion, that micelles of the SHBS copolymer in selective solvents 1,4-dioxane/0–30 vol. % n-heptane behave as hard spheres and may be used as model particles for verifying theoretical results regarding the hard sphere interactions. Compared with dispersions of silica particles [17], the system used is advantageous, mainly due to the fact that it allows us, even if to a limited extent only, to vary the dimensions and compactness of the particles under study by varying the composition of the solvent.

Table 3. A_2 values for polystyrene in mixtures 1,4-dioaxane/n-heptane obtained from the time-averaged light scattering data at 25 °C

vol.-% n-heptane	0	10	20	30
$A_2 \times 10^4$, cm^3 g^{-2} mol	7.7	6.9	6.4	5.8

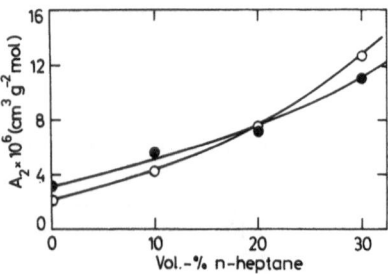

Fig. 2. Dependence of the second virial coefficient, A_2, of micelles of copolymer $K - 1$ on the compositon of the mixture 1,4-dioxane/n-heptane at 25 °C from the time-averaged light scattering data (O), and of the same quantity calculated using the hard spheres model (●)

Acknowledgement

The authors thank Dr. P. W. Glockner, Shell Development Co., Houston, Texas, USA, for the kind supply of the Kraton sample.

References

1. Berne BJ, Pecora R (1976) Dynamic Light Scattering, John Wiley and Sons, New York
2. Burgers JM (1941) Proc K Ned Akad Wet Amsterdam 44:1047, 1177; (1942) Proc K Ned Akad Wet Amsterdam 45:9, 126
3. Pyun CW, Fixman M (1964) J Chem Phys 41:937
4. Batchelor GK (1972) J Fluid Mech 52:245; (1976) J Fluid Mech 74:1
5. Altenberger AR, Deutch JM (1973) J Chem Phys 59:894
6. Harris S (1975) J Phys A: Math Gen 9:1895; (1976) J Chem Phys 65:5408

7. Anderson JL, Reed CC (1976) J Chem Phys 64:3240, 4336
8. Ackerson BJ (1976) J Chem Phys 64:242; (1978) J Chem Phys 69:684
9. Hess W, Klein R (1976) Physica 85:509
10. Felderhof BU (1978) J Phys A: Math Gen 11:929
11. Altenberger AR (1976) Chem Phys 15:242; (1978) Physica A 92:391; (1979) J Chem Phys 70:1994
12. Wills PR (1979) J Chem Phys 70:5865
13. van de Broeck C, Lostak F, Lekkerkerker HNW (1981) J Chem Phys 74:2006
14. Ohtsuki T, Okana K (1982) J Chem Phys 77:1443
15. van Helden AK, Jansen JW, Vrij A (1981) J Coll Interf Sci 81:354
16. Kops-Werkhoven MM, Fijnaut HM (eds) (1980) Degiorgio V, Corti M, Giglio M, Light Scattering in Liquids and Macromolecular Solutions, Plenum Press, New York, p 81
17. Kops-Werkhoven MM, Fijnaut HM (1981) J Chem Phys 74:1618
18. Tuzar Z, Kratochvíl P (1976) Adv Coll Interf Sci 6:201
19. Tuzar Z, Pleštil J, Koňák Č, Hlavatá D, Sikora A (1983) Makromol Chem 184:2111
20. Tuzar Z, Štěpánek P, Koňák Č (ed) (1985) Sedláček B, Physical Optics of Dynamic Phenomena and Processes in Macromolecular Systems, Walter de Gruyter, Berlin–New York, p 405

21. Tuzar Z, Bednář B, Koňák Č, Kubín M, Svobodová Š, Procházka K (1982) Makromol Chem 183:399
22. Tuzar Z, Sikora A, Straková D, Podešva J, Stejskal J, Kratochvíl P (1985) Coll Czech Chem Commun, in press
23. Štěpánek P, Tuzar Z, Koňák Č (ed) (1985) Sedláček B Physical Optics of Dynamic Phenomena and Processes in Macromolecular Systems, Walter de Gruyter, Berlin–New York, p 462
24. Tuzar Z, Kratochvíl P (1967) Coll Czech Chem Commun 32:3358
25. Yamakawa H (1971) Modern Theory of Polymer Solutions, Harper and Row, New York, Ch 6

Received May 30, 1985;
accepted July 8, 1985

Authors' address:

Č. Koňák
Institute of Macromolecular Chemistry
Czechoslovak Academy of Sciences
162 06 Prague 6, Czechoslovakia

Progress in Colloid & Polymer Science

Progr Colloid & Polymer Sci 71:20–25 (1985)

Temperature — concentration dependence of polyvinyl acetate gels with respect to the collapse phenomenon*)

M. Zrinyi and E. Wolfram †

Department of Colloid Science, Loránd Eötvös University, Budapest, Hungary

Abstract: The temperature dependence of equilibrium concentration of swollen polyvinyl acetate gels was studied below and above the θ temperature. Collapse was not observed. An analysis of the classical gel theories as well as the experimental results has shown that the James-Guth theory is adequate to describe the experimentally found universal dependence of ϕ_θ/ϕ on τ/ϕ_θ (where ϕ is the equilibrium concentration of the gel, τ is the reduced temperature and the θ subscript refers the θ state). It was also found that upon cooling below the θ point a continuous non-phase transition occurs.

Key words: Polyvinyl acetat gel, collapse, θ temperature, James-Guth theory.

Introduction

The excluded volume effect of isolated macromolecules, polymer solutions as well as gels, has aroused a great deal of interest in the past few years [1]. In particular the transition of a flexible chain from an extended coil to a collapsed, dense globula, effected by changing the quality of the solvent has been the subject of many discussions [2–5].

Since the theory of gels with chemicall cross-links is — in spirit — a single chain theory, consequently the possibility of collapse in swollen networks was soon realised [6,7]. According to Dusek and Patterson a first order transition can occur and two gel phases can appear or disappear by changing the temperature or the solvent composition. The necessary condition of the appearance of such transition is considerable of high value of the cross-linking density as well as of low value of the isotropic deformation factor. However, these two requirements, to some extent, are mutually exclusive.

A slightly different approach was proposed by Tanaka [8]. He concluded that a first order phase transition occurs when the cross-linking density exceeds a certain critical value.

Based on the idea given in [5] Khoklov derived an equation to describe the temperature dependence of equilibrium concentration of gels [9]. It was shown that phase transition from the coil to the globular state occurs. This transition is of first order in the case of a stiff chain network and of second order in the case of a flexible chain network. The basic parameter which controls the order of the phase transition was found to be the third virial coefficient of monomer interactions.

The first experimental evidence of coil-globula transition in solution was reported in 1979 by Nishio et al. [10, 11].

For neutral gels this sharp transition has not been observed experimentally as yet. Although polyacrylamide gels were found to have a sudden change in swelling degree within a narrow range of polymer-solvent interaction [8, 12–14], it turned out that the observed collapse was due to either electrostatic interactions [12,13], or formation of heterogeneous structures [14, 15]. These two effects were not considered in theories given in [6–8].

Experimental studies on swelling behaviour of neutral gels with respect to the collapse phenomenon are rather few [16–19]. The main purpose of the present work has, therefore, been to investigate the temperature — concentration dependence of polyvinyl acetate gels in the vicinity of the θ temperature. A theoretical consideration is also presented in which we use the classical network theories together with the traditional

---

*) Herrn Prof. Dr. H.-G. Kilian, der die Disziplin „Kolloidik" im breitesten Sinne auf deutschem Boden wieder ins Leben gerufen hat, zum Anlaß seines 60. Geburtstages herzlichst gewidmet.

assumption that the network chains deform affinely with the volume of gel. A scaling analysis of our achievement can be found elsewhere [19].

Theoretical background

There is no unique, generally accepted theory to describe the swellling and elastic properties of swollen networks [7, 31]. Even the classical works differ considerably. According to these [7] the condition of equilibrium with pure diluent can be given as follows:

$$v^* q_0^{-2/3} V \phi^{1/3} - B v^* V \phi + \ln(1 - \phi) + \phi + \chi \phi^2 = 0 \qquad (1)$$

where ϕ is the volume fraction of polymer in the gel, v^* is the moles of network chains per unit dry polymer volume, V is the partial molar volume of the diluent, χ is the temperature dependent interaction parameter and q_0 is the so-called memory term through which the system "remembers" those sets of states at which the cross-links were introduced.

The value of the coefficient B is rather contraversial. According to Flory [20] and Wall [21] $B = 2/f$ where f is the functionality of the cross-links. On the basis of the James-Guth theory [22] $B = 0$, while Kuhn and Hermans write $B = 1$ [23, 24].

There are different experimental results available from the literature. Van der Kraats et al. [25] and Froelich et al. [26] found that $B = 1/2$, Pennings et al. [27] and Rijke [28] obtained $B = 1$, and Horkay et al. [29] concluded that $B = 0$.

From now on we focus on the swelling behaviour of gels in the vicinity of the θ temperature. What we are interested in, is the dependence of the swelling degree (or the concentration) on temperature. First we will discuss the predictions of equation (1) and then compare them with the experimental results.

For weakly cross-linked networks the volume fraction of the polymer in the gel is usually much smaller than unity, and consequently the substitution for the logarithmic term in equation (1) of its third order series does not introduce appreciable error. From equation (1) one obtains

$$v^* q_0^{-2/3} V \phi^{1/3} - B v^* V \phi - u \phi^2 - w \phi^3 = 0 \qquad (2)$$

where $u = 1/2 - \chi$ and $w = 1/3$. It seems to be useful to generalize the above equation in order to take into account not only the binary, but also the ternery interactions between the monomers. Thus u and w can be

considered as successive (second and third) virial coefficients. u describes the interaction energy between solvent molecules and monomers in terms of the temperature dependent Huggins interaction parameter. It changes sign at the θ temperature.

The third virial coefficient, w is supposed to be temperature independent and dominated by the chain flexibility. The more stiff the chain is, the higher the value of w is. Let us introduce the expansion factor, α, of the network chains. If the swelling is considered to be affine with respect to the dimension of the network chains one can write that

$$\alpha^3 = \frac{R^3}{R_\theta^3} = \frac{\phi_\theta}{\phi} \qquad (3)$$

where R and ϕ are the radius of gyration of network chains and the volume fraction, respectively. The subscript θ refers to the θ state.

Equation (2) together with equation (3) results in

$$v^* q_0^{-2/3} V \phi_\theta^{1/3} \frac{1}{\alpha} - B v^* V \phi_\theta \frac{1}{\alpha^3} - u \phi_\theta^2 \frac{1}{\alpha^6} - w \phi_\theta^3 \frac{1}{\alpha^9} = 0 \qquad (4)$$

It is useful to introduce the following notations:

$$v^* q_0^{-2/3} V \phi_\theta^{1/3} = g \qquad B v^* V \phi_\theta = b$$

$$\frac{u \phi_\theta^2}{g} = U(T) \qquad w \phi_\theta^3 = W \qquad (5)$$

These quantities are not independent of each other. The correspondence between them can be immediately seen if one considers the θ state when $\alpha = 1$ and $u = 0$.

$$g \frac{V_\theta}{V} - b \frac{V_\theta}{V} - W = 0 \qquad (6)$$

Combining equations (4), (5) and (6) one can obtain how α is related to the temperature dependent U. (We note that all other quantitites in equation (4), except V, are independent of the temperature.)

$$\alpha^5 - \frac{b}{g} \alpha^3 + \frac{V_\theta}{V} \left(\frac{b}{g} - 1 \right) \frac{1}{\alpha^3} = U(T) \qquad (7)$$

It can be seen that the only quantity which controls the $\alpha - U$ dependence is b/g. (The effect of temperature on the value of V_θ/V, can be neglected.)

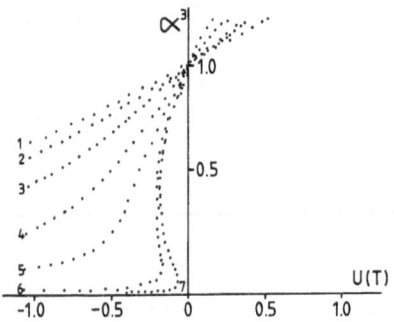

Fig. 1. Dependence of α^3 on $U(T)$ as given by equation (7). Curves denoted by numbers belong to different values of b/g.
Number: b/g; 1 : 0; 2 : 0.2; 3 : 0.5; 4 : 0.75; 5 : 0.9; 6 : 0.99; 7 : 0.999

In figure 1 the $\alpha^3 - U$ dependence is presented at different b/g values. It can be seen if b/g exceeds a certain value, a first order transition occurs. Above this critical value any alteration of U induced by either the temperature or the composition of the swelling liquid may result in a sudden change in the swelling degree.

b/g can be expressed by quantities which characterize the swollen gel. It is often assumed that $q_0 = \phi_c^{-1}$, where ϕ_c is the concentration at which the cross-links are introduced [7]. Thus

$$\frac{b}{g} = B q_0^{2/3} \phi_\theta^{2/3} \simeq B \left(\frac{\phi_\theta}{\phi_c}\right)^{2/3} \propto B \phi_c^{-2/3} (v^*)^{1/3} \quad (8)$$

If we accept that $B = 2/f$ or $B = 1$ then our analysis gives similar results to those obtained by Dusek and Patterson [6]: to realise a first order transition the product $q_0^{2/3}(v^*)^{1/3}$ must exceed its critical value.

So far we have considered the Flory-Wall and the Kuhn-Hermans theories where $B \neq 0$. If we accept the James-Guth result ($B = 0$ and also $b = 0$) equation (7) can be written in a more simple form:

$$\alpha^5 - \frac{V_\theta}{V} \frac{1}{\alpha^3} = U = \frac{u \phi_\theta^2}{g} \quad (9)$$

and in equation (2), g can be given as:

$$g = u \phi^2 + w \phi^3 \quad (10)$$

It follows from equation (10) that at θ condition $g(V_\theta/V) = w \phi_\theta^3$. Since g/V does not depend on the temperature it is possible to substitute for g/V of $(w/$

$V_\theta) \phi_\theta^3$ in equation (9). After the substitution one obtains:

$$\alpha^5 - \frac{V_\theta}{V} \frac{1}{\alpha^3} = \frac{V_\theta}{V} \frac{u}{w} \frac{1}{\phi_\theta} \quad (11)$$

(We note that w can be determined from the concentration dependence of the elastic modulus, see equations (14) and (15).)

We can express u by the reduced temperature, defined by $\tau = (T - \theta)/T$, as follows:

$$u = \frac{c}{\theta} \tau \quad (12)$$

where c is a constant which characterizes the polymer-solvent system. Combination of equations (11–12) gives

$$\alpha^5 - \frac{V_\theta}{V} \frac{1}{\alpha^3} = \frac{V_\theta}{V} \frac{c}{\theta w} \frac{\tau}{\phi_\theta} \quad (13)$$

For network homologous ($c/\theta w$ is constant) equation (13) predicts that $\alpha^3 = \phi_\theta/\phi$ is an unique function of τ/ϕ_θ.

Experimental section

Materials

Polyvinyl acetate (PVAc) gels were prepared by total acetylation of polyvinyl alcohol (PVA) gels according to a method described elsewhere [30].

Purification and characterization of materials as well as the network formation procedure were also described previously. A hydrolised Poval 420 (Kuraray Co., Japan) PVA sample was fractionated in a mixture of n-propyl alcohol-water in order to get a narrow molecular mass distribution. Some characteristics of fractionated sample are:

$$\bar{M}_n = 100\,000, \quad \bar{M}_w = 110\,000, \quad |\eta| = 0.68 \; dl \cdot g^{-1}.$$

PVA was cross-linked with glutaraldehyde (Merck, F.R.G.) at pH = 1.5 and 298 ± 0.1 K in water as solvent. Networks characterized by different degree of crosslinking were prepared at four different polymer concentrations (C_0 = 3.0, 6.0, 9.0 and 12.0 m%). The cross-linking density, DC, which means the moles of monomer units per mole of cross-linking agent, was varied from 50 to 400. (The gel samples are identified by the C_0/DC symbols, e. g. PVAc gel 6/100 means C_0 = 6 m% and DC = 100.)

Cylindrical gel specimens 1 cm in diameter and 1 cm in height were prepared in containers of a suitable frame. After the cross-linking reaction, the PVA gels were removed from the frame and the traces of the catalyst (2 n HCl solution) were washed by distilled water.

Then the media of the gel were replaced by a mixture of acetic anhydride (40 vol%) − acetic acid (10 vol%) − pyridine (50 vol%). The acetylation reaction was continued at 363 K for 8 hours before it was terminated. The acetylation mixture was renewed in each hour. After the reaction had been completed several solvent exchanges, each taking 48 hours or more, were made, in order to get rid of the reaction mixture. The complete wash cycle involved not less that ten solvent exchanges and took over 1 month. After the solvent exchanges the gels were carefully dried.

The dry networks were swollen to equilibrium in iso-propyl alcohol. The gels were stored not less than 1 month at each temperature before testing.

Methods

In order to determine the swelling equilibrium concentration, the gels were weighed, then evaporated to dryness and weighed again. Density of the dry networks and that of the iso-propyl alcohol were also determined as a function of the temperature. The volume fraction of the polymer in the gel was calculated from the mass and density measurements, supposing the additivity of the specific volumes.

The error of concentration measurements did not exceed 0.2 %. The equilibrium concentrations were determined at 8 different temperatures (25, 30, 37, 45, 50, 55, 60, 70 °C).

Results and discussion

In order to study the equilibrium concentration of gels in the vicinity of the θ temperature for the PVAc networks, iso-propyl alcohol was chosen as a solvent. The θ temperature of PVAc/iso-propyl alcohol systems was found to be 54.7 °C [18, 19]. In our previous papers we reported that the cloud point temperatures of these gels are — within the experimental accuracy — the same as those of the corresponding solutions at the same polymer concentration. We concluded that network elasticity associated with the presence of cross-links does not play any role, and consequently the θ temperature of the PVAc solutions corresponds to that of the gels. (For loosely cross-linked networks the effect of chemical modification, due to the presence of

Table 1. Symbols, equilibrium concentrations belonging to the θ temperature and b/g values calculated by equation (8) for the PVAc/iso-propyl alcohol gels

Symbol	ϕ_θ	b/g
12/50	0.350	1.02
12/200	0.191	0.68
9/50	0.312	1.14
9/100	0.209	0.87
9/200	0.168	0.76
9/400	0.117	0.59
6/50	0.235	1.24
6/200	0.127	0.82
3/50	0.140	1.39

cross-links, on the interaction parameter as well as the θ temperature can be neglected.)

Thus we considered 55 °C as the θ temperature of PVAc/iso-propyl alcohol gels. Mechanical measurements performed on these samples supported this consideration [19].

In figure 2 we plot the equilibrium concentration versus temperature for the studied systems. It can be seen that none of the gels have shown the macroscopic collapse phenomenon. For all the gels a continuous curve was obtained.

It has to be emphasised that upon cooling PVAc/iso-propyl alcohol gels exhibit optical and structural changes that can be visible to the naked eye. The originally transparent gels become turbid as the temperature decreases below a certain value. This cloud point temperature was found to depend only on the concentration of the gels. A more detailed description of the appearance of turbidity and structural changes can be found in two earlier papers [17, 18].

In table 1 we give the symbol of the gels, the equilibrium concentrations belonging to the θ temperature as well as b/g values calculated with the help of equation (8).

In order to show how different the $\alpha^3 - U$ curves are a model calculation based on equation (7) was made with the b/g values taken from table 1. The results referring to 5 selected samples can be seen in figure 3. It is obvious that the difference between the curves is rather significant. Moreover the gels denoted by numbers 4 and 5 should be separated — according to equation (7) — into two phases even in the range $U > 0$ which corresponds to $\chi < 1/2$. This unreal conclusion is probably due to either the failure of equation (7) or the assumption given by equation (8).

In figure 4 the measured $\alpha^3 = \phi_\theta/\phi$ quantities are plotted against τ/ϕ_θ. It can be seen that within the ex-

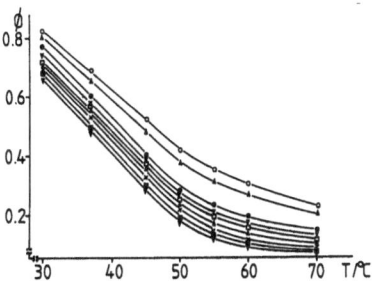

Fig. 2. Concentrations vs. temperature diagram for PVAc/iso-propyl alcohol gels [19]. Solid lines are guide for eyes. Symbols: 12/500 ○, 12/200 □, 9/50 △, 9/100 ▽, 9/200 ▲, 9/400 ▼, 6/50 ●, 6/200 ■, 3/50 ×

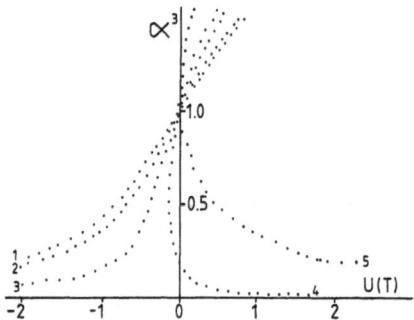

Fig. 3. Calculated $\alpha^3 - U$ dependence for different PVAc/iso-propyl alcohol gels. For the calculations equations (7), (9) and data in table 1 were used. Number: symbol; 1: 9/400; 2: 12/200; 3: 9/100; 4: 12/50; 5: 3/50

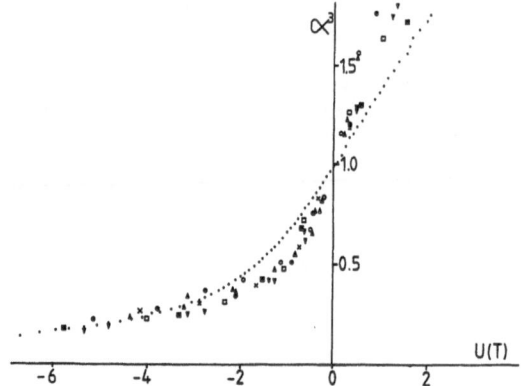

Fig. 5. Comparison between theory and experiments. Dotted line was calculated by equations (9) and (16). Symbols are the same as used in figure 2

perimental accuracy all points (72 points) are on the same curve. This result qualitativelly supports the validity of equations (9) and (13) since ϕ_θ/ϕ seems to depend only on one variablle of τ/ϕ_θ. This means that the James-Guth theory is adequate to predict the swelling degree — temperature dependence of PVAc/iso-propyl alcohol gels.

In order to compare quantitatively the experimental results with equations (9), one needs to know the quantity $c/w\theta$. The third virial coefficient (w) can be determined by mechanical measurements of gel homologous, at θ condition. Since we are convinced that $B = 0$ (or $b = 0$) it follows from equation (10) that

$$g\frac{V_\theta}{V} = w\phi_\theta^3. \tag{14}$$

On the other hand [7]:

$$g\frac{V_\theta}{V} = \frac{G_\theta V_\theta}{R\theta}, \tag{15}$$

where G_θ is the elastic (shear) modulus of the gel measured at θ temperature.

In a previous paper [19] the concentration dependence of the elastic (shear) modulus of network homologous was studied at different temperatures. It was found at 55 °C that $G_\theta(\phi_\theta) = 4492.6\ \phi_\theta^{3.04}\ kPa$. Since $V_\theta/R\theta = 28.88 \cdot 10^{-6}\ (kPa)^{-1}$ was obtained, w equals to 0.13. Now the quantity c remains the only adjustable parameter needed for the direct comparison between theory and experiments. The best fitting between U and τ/ϕ_θ was achieved by

$$U = 0.81\frac{\tau}{\phi_\theta}\quad\text{or}\quad\frac{c}{\theta w} = 0.81 \tag{16}$$

for our gel systems. In figure 5 the theoretical curve calculated by equations (9) and (16) and the experimental points are shown. The agreement is quite satisfactory, but it must be mentioned that in the close vicinity of the θ temperature equation (9) predicts a less increasing dependence than that of the experimental points.

Conclusions

The comparison of the different classical network theories with the experimental results leads to the following results:

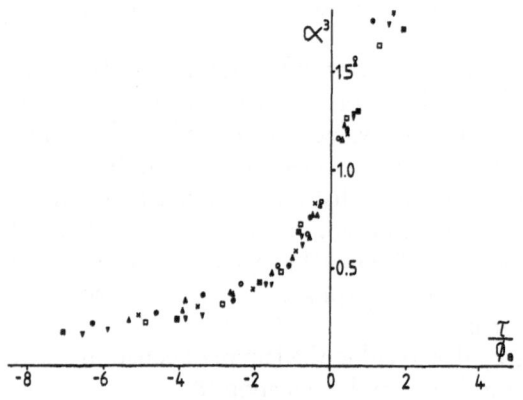

Fig. 4. Dependence of $\alpha^3 = \phi_\theta/\phi$ on τ/ϕ_θ for PVAc/iso-propyl alcohol gel homologous. Symbols are the same as used in figure 2

— The equilibrium concentration — temperature curves show a continuous non-phase transition below the θ temperature.

— The James-Guth theory is adequate to describe the observed dependence of ϕ_θ/ϕ on τ/ϕ_θ.

Acknowledgements

The authors thank Dr. F. Horkay for useful discussions.

References

1. de Gennes PG (1979) Scaling Concepts in Polymer Physics, Cornell University Press, Ithaca and London
2. Flory PJ (1949) J Chem Phys 17:303
3. Stockmayer WA (1960) Makromol Chemie 35:54
4. Lifshitz IM, Grosberg AY, Khoklov AR (1978) Rev Mod Phys 50:683
5. de Gennes PG (1975) J Phys Paris 36:L55
6. Dusek K, Patterson D (1968) Journal of Polymer Sci, Part A2 6:1209
7. Dusek K, Prins W (1969) Adv Polym Sci 6:1
8. Tanaka T (1978) Phys Rev Let 40:820
9. Khoklov AR (1980) Polymer 21:376
10. Nishio U, Sun ST, Swislow G, Tanaka T (1979) Nature 281:208
11. Williams C, Brochard F, Frisch HL (1981) Ann Rev Phys Chem 32:433
12. Ilavsky M (1981) Polymer 22:1687
13. Ilavsky M (1982) Macromolecules 15:782
14. Janas VF, Rodriguez F, Cohen C (1980) Macromolecules 13:977
15. Geissler E, Hecht Am, in press
16. Aharoni SM, Wertz DH (1983) J Macromol Sci Phys B22:129
17. Zrinyi M, Molnár T, Horváth E (1981) Polymer 22:429
18. Zrinyi M, Wolfram E (1982) Colloid and Interface Sci Vol 90 1:34
19. Zrinyi M, Horkay F (1984) Macromolecules 17:2805
20. Flory PJ (1953) Principles of Polymer Chemistry, Cornell Univ Press, Ithaca
21. Wall FT, Flory PJ (1951) J Chem Phys 19:1435
22. James HH, Guth J (1947) J Chem Phys 15:669
23. Kuhn W (1936) Kolloid Z 76:258
24. Hermans JJ (1962) J Polym Sci 59:197
25. van der Kraats EJ, Winkeler MAM, Patters JM, Prins W (1969) Rec Trav Chim 88:449
26. Froelich D, Crawford D, Rozek T, Prins W (1972) Macromolecules 5:100
27. Pennings AJ, Prins W (1961) J Polym Sci 49:507
28. Rijke AM (1966) J Polym Sci 4:131
29. Horkay F, Nagy M (1984) Acta Chim Hung 115:305
30. Horkay F, Nagy M, Zrinyi M (1981) Acta Chim Acad Sci Hung 108:287
31. Candau S, Bastide J, Delsanti M (1982) Adv in Polym Sci 44:27

Received May 30, 1985;
accepted August 15, 1985

Authors' address:

Dr. M. Zrinyi
Department of Colloid Science
Loránd Eötvös University
H-1088 Budapest, Puskin u. 11–13
Hungary

Progress in Colloid & Polymer Science   Progr Colloid & Polymer Sci 71:26–31 (1985)

Polymorphic phase transition and monomolecular spreading of synthetic phospholipids*)

T. Handa[1]), C. Ichihashi, and M. Nakagaki

Faculty of Pharmaceutical Sciences, Kyoto University, Kyoto, Japan, and
[1]) Gifu Pharmaceutical University, Gifu, Japan

Abstract: The lyotropic and thermotropic polymorphisms of synthetic phosphatidyl-choline (DMPC, DPPC, DSPC) and phosphatidylethanolamine (DPPE) are investigated by means of differential scanning calorimetry. The thermal phase transitions of the anhy-drated phase and the hydrated lamellar phase of PCs are reversible, while the transition of anhydrated DPPE is irreversible. The spreading pressures of the anhydrated and hydrat-ed lamellar phases are measured on water. The spreading pressure from the lamellar phase of DMPC shows a steep increment at the gel-liquid crystal transition temperature. The anhydrated phase of the phospholipid has a higher spreading pressure than that from the hydrated lamellar phase.

Key words: Phospholipid, lyotropic polymorphism, thermotropic polymorphism, spreading of monolayer.

Introduction

Phospholipids have been known to have various lyotropic and thermotropic polymorphisms. Much work has been done with both phosphatidylcholine (PC) and phosphatidylethanolamine (PE). As for the lyotropic polymorphism of synthetic PC with saturat-ed fatty acids, anhydrated, monohydrated and ten hy-drated states have been identified [1–3]. The ten hy-drated state has a lamellar structure of lipid bilayer. Further addition of water to this phase gives a two phase system, consisting of the lamellar phase dispersed in the phase of excess water [1, 3]. Further-more, every lyotropic form has a thermotropic poly-morphism. For example, the lamellar phase has a phase transition temperature, T_c, below and above which the lamellar phase is in the gel and liquid crystal-line states, respectively [1–3]. In the case of synthetic PE, more complicated lyotropic and thermotropic polymorphisms have been elucidated [4, 5]. The poly-morphisms of mixed lipids have been closely correlat-ed to the structures and functions of biological mem-branes. In this connection, the phase diagrams of *n*-alkane mixtures are very suggestive [6]. In these very large studies, however, the interrelationships between various polymorphic forms of phospholipid (espe-cially between anhydrated forms) have not been clearly explained.

On the other hand, insoluble monolayers of phos-pholipid spread on an aqueous surface have been stud-ied comprehensively [7–9]. Surface pressure of an insoluble monolayer in equilibrium with a definite bulk phase is called equilibrium spreading pressure, F_s. The equilibrium spreading pressure of phospholipid was measured by Phillips and Hauser [10] and by Gershfeld and Tajima [11]. It was reported that the lamellar phase of L-α-dimyristoyl phosphatidylchol-ine (DMPC) showed very low spreading pressure, F_s, (0.01–0.005 mN/m), below the gel-liquid crystal tran-sition temperature, T_c, but gave rise to a steep increase in the F_s value by the increase in temperature above T_c [11].

In this study, in addition to the equilibrium spread-ing pressures from the hydrated lamellar phase, the values from the anhydrated state of phospholipid were also measured and examined in relation to the bulk property of phospholipid.

---

*) Dedicated to Prof. Dr. H.-G. Kilian on the occasion of his 60th birthday.

Experimental

Materials

L-α-Dimyristoyl, -dipalmitoyl, -distearoyl phosphatidylcholines (DMPC, DPPC, DSPC, respectively) and L-α-dipalmitoyl phosphatidylethanolamine (DPPE) were purchased from Carbiochem. CO. All lipids were examined with thin layer chromatography to show single spots. Lamellar phases were prepared by addition of excess amount of water to anhydrous lipids and by swelling at the temperature 10 °C above T_c. Water was twice distilled before use.

Measurements

The differential scanning calorimeter used was a Shimadz DSC-30. The rate of temperature change was 5 °C min^{-1}. The details of DSC measurements are represented elsewhere [3, 5].

For the measurement of spreading pressure, Wilhelmy's plate method was employed. The lipid in anhydrated or lamellar states was sprinkled on the surface of double distilled water in a teflon coated duralmin trough [9, 12]. PC and PE in the lamellar state tend to sink from the surface and the anhydrous phase on the surface was slowly hydrated. This resulted in the cessation of monolayer spreading. Therefore, occasional addition of the lipids in anhydrous or lamellar states was necessary to obtain the final value of F_s. The particulars in surface pressure measurement can be obtained from other studies [12].

Resluts

Polymorphism of bulk lipid

DSC measurements were carried out for phospholipids in different hydrated states between various temperature ranges [3, 5]. The results were summarized as follows:

1. Three lyotropic polymorphic states were observed for PCs. Each lyotropic form had a thermotropic polymorphism, and the thermal phase transitions were reversible (i. e. the endothermic and exothermic transition temperatures of the phase change were of the same value). Anhydrous PC heated to 110 °C and then cooled to room temperature was the S_1 form which was identified to be the C form of Luzzati's nomenclature [2]. The results are shown in figure 1 and table 1.

2. Swelling of anhydrated PCs was rapid above the gel-liquid crystal transition temperature. Below the temperature, the swelling was very slow, however, after a few days the clear gel-liquid crystal phase transition peak was detected on the DSC recorder. The swelling of DPPE below the transition temperature (64.5 °C) was even slower.

3. Three lyotropic polymorphic states were also observed for DPPE. Each lyotropic form showed a thermotropic polymorphism. The thermal phase transition of anhydrated form was not reversible. In the

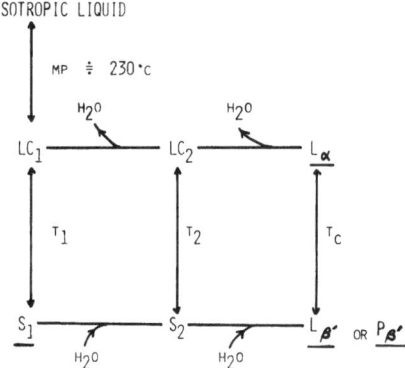

Fig. 1. Relation between polymorphic phases of phosphatidylcholine (PC). S_1 = anhydrated solid (Luzzati's C form [2]; LC_1 = anhydrated liquid crystal; S_2 = monohydrated solid; LC_2 = monohydrated liquid crystal; $L_{\beta'}$ and $P_{\beta'}$ = lamellar phase of gel; L_α = lamellar phase of liquid crystal. The underlined phases were used in the spreading experiment

Table 1. Phase transition temperatures of phosphatidylcholine (see fig. 1)

Lipid	T_1/°C	T_2/°C	T_c/°C
DMPC	87.5	61.0	23.0
DPPC	97.5	75.0	41.0
DSPC	101.0	86.0	54.0

cooling of anhydrated liquid crystal 1 (LC_1), an exothermic transition into the S_1 phase was observed at 51.5 °C. The S_1 phase was heated and transformed to the LC_1 phase with an endothermic transition at 51.5 °C. The polymorph C can be obtained only by the annealing of the S_1 form at 65–70 °C for 15 hours. The heating of C form shows an endothermic transition into LC_1 at 95 °C. The C form was identified to be an A form in Chapman's work [1]. The results obtained are shown in figure 2.

Monolayer spreading from polymorphic forms

The spreading of the monolayer on the water surface was carried out using anhydrated PC (S_1 form), hydrated lamellar PC [gel ($L_{\beta'}$ or $P_{\beta'}$) and liquid crystal (L_α)], anhydrated DPPE (C form), and hydrated lamellar phase of DPPE (L_β).

In figure 3, the variations of surface pressure with time are shown. The arrows in this figure indicate the additions of DMPC on the surface. Final values of surface pressure (equilibrium spreading pressure), were obtained from the plateau values. At 17 °C, the DMPC

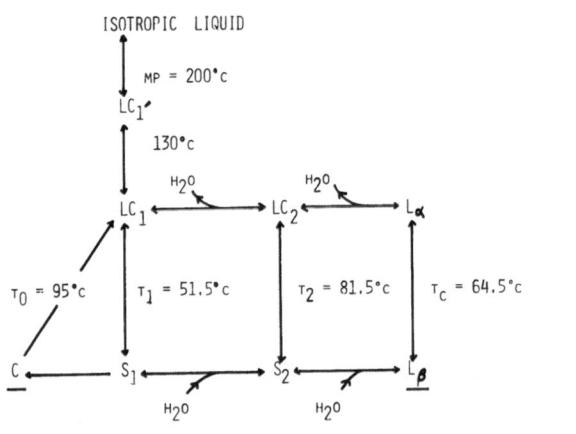

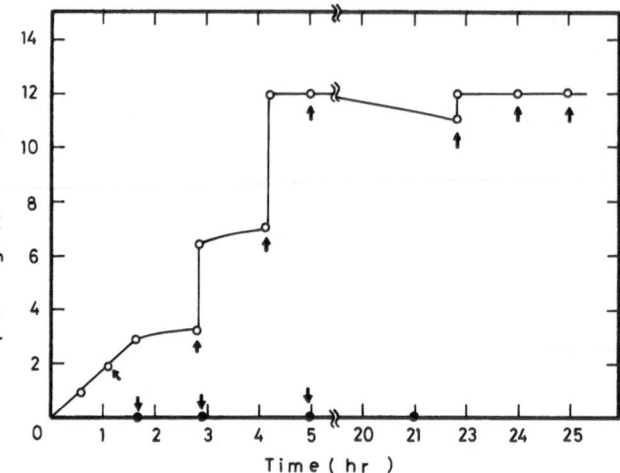

Fig. 2. Relation between polymorphic phases of DPPE. C = anhydrated crystal (Chapman's A state [1], obtained by annealing of S_1); S_1 = anhydrated solid; LC_1 = anhydrated liquid crystal 1; $LC_{1'}$ = anhydrated liquid crystal 1'; S_2 = monohydrated solid; LC_2 = monohydrated liquid crystal; L_β = lamellar phase of gel; L_α = lamellar phase of liquid crystal. The underlined phases were used in the spreading experiment

Fig. 4. Spreading of DSPC at 25 °C. $\bigcirc = S_1$ phase; $\bullet = L_{\beta'}$ phase. Arrows indicate additions of bulk DSPC (S_1 or $L_{\beta'}$) on water surface

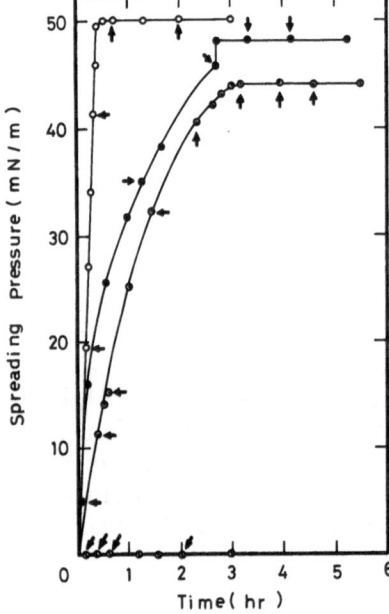

Fig. 3. Spreading of DMPC. $\bigcirc = S_1$ phase at 25 °C; $\odot = S_1$ phase at 17 °C; $\bullet = L_\alpha$ phase at 25 °C; $\oslash = P_{\beta'}$ phase at 17 °C. Arrows indicate additions of bulk phase (S_1, L_α or $P_{\beta'}$) on water surface

lamellar phase ($T_c = 23$ °C) showed undetectably (with our apparatus) a low value of F_s (lower than 0.1 mN/m), and at 25 °C the value was considerably high ($F_s = 44.2$ mN/m). These results were in good agree-

ment with those of Gershfeld and Tajima [11]. On the other hand, the anhydrous phase S_1 (the transition temperature of $S_1 \longleftrightarrow LC_1$ of DMPC was 87.5 °C as seen in table 1) gave high values of F_s (17 °C : 48.2 mN/m, 25 °C : 50.1 mN/m). Similar measurements were performed for DPPC and DSPC. The results obtained for DSPC at 25 °C are shown in figure 4. The anhydrated S_1 form gave the equilibrium spreading pressure of 12.0 mN/m, while the F value of the monolayer in equilibrium with the lamellar phase ($L_{\beta'}$) was an undetectably low value. The results obtained for PCs here, did not agree with the discussion done by Phillips and Hauser, in which they concluded that the anhydrated phase would not give an appreciable value of spreading pressure [10]. It was considered that they were measuring F_s with a hydrated lamellar form of PC, because their value of F_s was obtained above T_c and their temperature coefficient of F_s was negative above T_c. Above T_c, the lamellar phase is in a liquid crystalline state.

For DPPE, the spreading pressures of the anhydrated form (C form, $T_o = 95$ °C) and the hydrated lamellar phase (L_β, $T_c = 64.5$ °C) were measured at 25 °C, and the results are shown in figure 5. It is found that F_s from the lamellar phase, L_β, is an appreciably high value (32.0 mN/m) and is close to the value from the C form (33.5 mN/m).

The values of spreading pressure of anhydrous and lamellar phases of phospholipids are summarized in table 2.

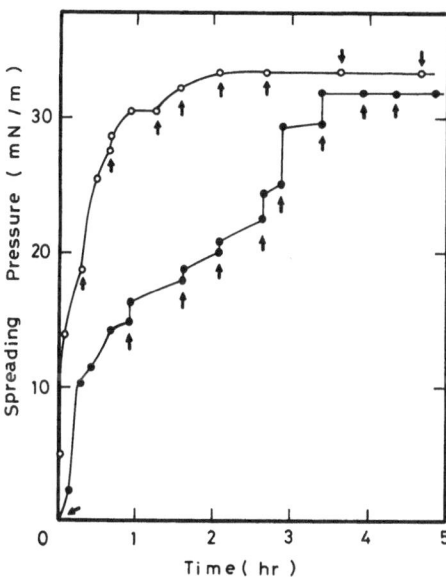

Fig. 5. Spreading of DPPE at 25 °C. ○ = C phase; ● = L_β phase. Arrows indicate additions of bulk DPPE (C or L_β) on water surface

Table 2. Spreading pressures, F_s, of PCs and DPPE

Lipid	Phase	Transition Temp./°C	Temp./°C	F_s/mN/m
DMPC	S_1	87.5	25	50.1
DMPC	S_1	87.5	17	48.2
DMPC	L_α	23.0	25	44.2
DMPC	$P_{\beta'}$	23.0	17	< 0.1
DPPC	S_1	97.5	25	44.0
DPPC	$L_{\beta'}$	41.0	25	< 0.1
DSPC	S_1	101.0	25	12.0
DSPC	$L_{\beta'}$	54.0	25	< 0.1
DPPE	C	95.0	25	33.5
DPPE	L_β	64.5	25	32.0

Discussion

From figures 3–5 and table 2, it is found that the lamellar phase of PC at a temperature below gel-liquid crystal transition temperature, T_c, does not give the appreciable value of spreading pressure, F_s, ($F_s < 0.1$ mN/m), while the anhydrated phase below its transition temperature, T_1 ($T_1 > T_c$), shows the considerably high value of F_s. These results are explained as follows: according to the DSC studies of the PC and water system, even at a lower temperature than T_c, the anhydrated state, S_1, is unstable and spontaneously

undergoes hydration to slowly form a lamellar phase in gel state, $L_{\beta'}$ (DPPC, DSPC) or $P_{\beta'}$ (DMPC) [3]. This phenomenon is represented in a comparison of the chemical potentials as

$$\mu(S_1) + 10\,\mu(H_2O) > \mu(L_{\beta'}) \tag{1}$$

where $\mu(S_1)$ and $\mu(L_{\beta'})$ are the chemical potential of PC of anhydrated phase S_1, and hydrated PC of lamellar $L_{\beta'}$, respectively, and $\mu(H_2O)$ is that of water. The hydration number of PC is 10 [1–3]. When the PC monolayer on water is in equilibrium with the bulk phase, $L_{\beta'}$,

$$\mu(L_{\beta'}) + (m - 10)\,\mu(H_2O) = \mu(M, L_{\beta'}). \tag{2}$$

Here, m is the hydration number of the PC molecule in the monolayer and is considered as $m \geqq 10$. $\mu(M, L_{\beta'})$ is the chemical potential of the PC in monolayer in equilibrium with the $L_{\beta'}$ phase. The equilibrium between the anhydrated phase, S_1, and the monolayer is represented as

$$\mu(S_1) + m\mu(H_2O) = \mu(M, S_1)$$
$$= \mu(M, L_{\beta'}) + \int_{F_s(L_{\beta'})}^{F_S(S_1)} A\,dF \tag{3}$$

where $F_s(S_1)$ and $F_s(L_{\beta'})$ are the equilibrium spreading pressures of the S_1 and $L_{\beta'}$ phases, respectively, $\mu(M, S_1)$ is the chemical potential of the PC in monolayer in equilibrium with the S_1 phase, and A is the area per a lipid molecule in monolayer. From equations (1)–(3), the following relation is obtained

$$\int_{F_s(L_{\beta'})}^{F_S(S_1)} A\,dF > 0. \tag{4}$$

Because $A > 0$, it is concluded from equation (4) that $F_s(S_1) > F_s(L_{\beta'})$. This means that the anhydrated state, S_1, exhibits a higher value of spreading pressure, F_s, than that of the lamellar phases, $L_{\beta'}$, and $P_{\beta'}$. Also for DPPE, the value of $F_s(C)$ is considered to be larger than that of $F_s(L_\beta)$. Similar results are obtained in the spreading from the L_α phase and $F_s(S_1) > F_s(L_\alpha)$. These conclusions agree with the experimental facts shown in figures 3–4.

Figure 6 shows the schematic relations between the temperature and spreading pressure of phospholipids. Anhydrated PC has appreciably high values of F_s in solid (S_1) and liquid crystalline (LC_1) states (T_1 is the transition temperature between S_1 and LC_1). Anhydrated DPPE also has appreciably high values of $F_s(T_o$

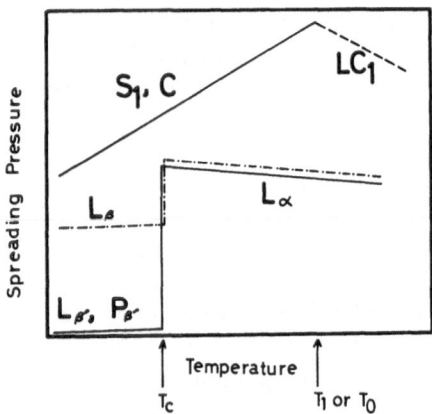

Fig. 6. Schematic presentation of molecular spreading from polymorphic phase of phospholipid. The $L_{\beta'}$ or $P_{\beta'}$ phases of PC give very low spreading pressure ($F < 0.1$ mN/m, solid line), but the L_β phase of DPPE shows the appreciably high F_s value (chained line)

is the transition temperature between C and LC_1). In hydrated lamellar states, PC has a very low spreading pressure in the gel state ($L_{\beta'}$, $P_{\beta'}$, $T < T_c$) and a very high spreading pressure in the liquid crystalline state (L_α, $T > T_c$), while DPPE gives very high spreading pressures in both gel (L_β, $T < T_c$) and liquid crystalline (L_α, $T > T_c$) states as shown by the chained line.

The gel phases of PC ($L_{\beta'}$ or $P_{\beta'}$) have the hydrocarbon chains tilted with respect to the normal to the lamellar plane, and the gel phase of DPPE (L_β) has the hydrocarbon chains oriented perpendicularly to the plane of the lamellar and packed with rotational disorder in a two dimensional square lattice [1, 2]. This rotational disorder is considered to be the reason why the gel phase of DPPE (L_β) has a higher spreading pressure than that of PC ($L_{\beta'}$ or $P_{\beta'}$).

Alternatively, monolayers of various lipids are spread with the aid of spreading solvent. The monolayer of DMPC thus formed at 25 °C underwent collapse at the surface pressure (46.0 mN/m) [9] close to the value of $F_s(L_\alpha)$ as shown in table 2. Therefore, it is concluded that the DMPC monolayer collapses and the lamellar phase, L_α, is separated at 25 °C. On the other hand, monolayers of DPPC and DSPC are compressed to give surface pressure of a higher value than $F_s(L_{\beta'})$, the latter being practically zero at the temperature lower than T_c. The monolayers are therefore thermodynamically unstable states. The DPPC monolayer at room temperature shows the phase transition between expanded and condensed monolayers [13], but this is the transition in the supercompressed monolayer.

For the DPPE system, the spreading pressure, $F_s(L_\beta)$ is 32 mN/m and indicates that the monolayer of DPPE at 25 °C is thermodynamically stable up to $F = 32$ mN/m.

To sum up, the DMPC monolayer at 25 °C collapses into the hydrated lamellar phase, L_α, and the collapse pressure [9] is equal to the spreading pressure, F_s. The DPPE monolayer at 25 °C is in a supercompressed state above the surface pressure of 32 mN/m. The DPPC and DSPC monolayers at 25 °C are also in supercompressed states at the appreciable surface pressure ($F > 0.1$ mN/m). These results observed in these phospholipid systems are quite similar to those obtained for triglyceride monolayers, for which the collapse pressure is much higher than the spreading pressure because of supercompression [14]. These can be compared with the systems of fatty acid, where the monolayers are easily compressed to surface pressure higher than the spreading pressure from anhydrated bulk fatty acid [15]. Further investigations on the spreading from lyotropic states of fatty acid are required to discuss the monolayer equilibrium.

On the basis of these studies, it may be concluded that the spreading of a monolayer from an anhydrated bulk phase is promoted by the hydration of spreading molecules, and also that the spread monolayers of most phospholipids are easily supercompressed and are thermodynamically metastable even though the monolayers perform phase transition with each other.

References

1. Keith AD, Snipes W, Chapman D (1977) Biochemistry 16:644; Williams RM, Chapman D (1970) Phospholipids, Lipid Crystal and Cell Membranes, Prog Chem of Fats & Other Lipids, vol 11, Oxford–London
2. Tardieu A, Luzzati V, Reman FC (1973) J Mol Biol 75:711
3. Nakagaki M, Handa T, Ohashi K (1979) Yakugaku Zasshi 99:1
4. Chapman D, Byrne P, Shipley GG (1966) Pro Roy Sco London A290:1115
5. Nakagaki M, Handa T, Ohashi K (1979) Yakugaku Zasshi 99:564
6. Asbach GI, Kilian HG, Stracke Fr (1982) Coll & Polym Sci 260:151
7. Shah DO, Schulman JH (1966) J Lipid Res 8:215
8. Joos P, Demel RA (1969) Biochim Biophys Acta 183:447
9. Handa T, Nakagaki M (1979) Coll & Polym Sci 257:374
10. Phillips MC, Hauser H (1974) J Coll Interf Sci 49:31
11. Gershfeld NL, Tajima K (1977) J Coll Interf Sci 59:597
12. Nakagaki M, Handa T (1976) Bull Chem Soc Jpn 49:880
13. Phillips MC, Chapman D (1966) Biochim Biophys Acta 163:301
14. Nakagaki M, Funasaki N, Fujita K (1972) Nippon Kagaku Kaishi :243
15. Jalal IM, Zografi G (1979) J Coll Interf Sci 68:196; Jalal IM, Zografi G, Rakshit AK, Gunstone F (1980) J Coll Interf Sci 76:146

Received May 6, 1985;
accepted May 15, 1985

Author's address:

Professor M. Nakagaki
Faculty of Pharmaceutical Sciences
Kyoto University
Sakyo-ku
Kyoto, 606, Japan

Progress in Colloid & Polymer Science   Progr Colloid & Polymer Sci 71:32–43 (1985)

Adsorption-entanglement layers in flowing high-molecular weight polymer solutions.

I. Direct observation of layer formation*)

R. A. M. Hikmet, K. A. Narh, P. J. Barham, and A. Keller

H. H. Wills Physics Laboratory, University of Bristol, Bristol, England

Abstract: This four-part series is concerned with thick multimolecular layers, termed "adsorption-entanglement layers", which according to our observations develop during flow of high molecular weight polymer solutions along the solid surfaces defining the flow channels. It is concerned with the formation and nature of these layers and with the effect they have on fluid transport in general and viscosity measurements in particular. In the present first part the existence of such layers is established through a flow visualization method and the layer formation diagnosed in situ is correlated with macroscopic flow effects. Two flow geometries are used, Couette and channel flow. In the Couette flow, where the shear rate is kept constant, the adsorption-entanglement layers first form, then thicken and eventually partially break off until a steady layer thickness is reached. These effects are reflected by a concommitant increase and subsequent decrease and final levelling off of the apparent viscosity as a function of shearing time, there being a near quantitative correlation between the measured changes in apparent viscosity and layer thickness. In channel flow, where it is the pressure head which is kept constant, the formation and thickening of layers leads to a decrease in flow rate with flow time, amongst others giving the impression of shear thickening. It is demonstrated that owing to the fact that adsorption entanglement layers alter the geometry of the flow channel the time dependence of flow effects, per se need not indicate shear thickening (or thinning), in fact that there is any departure from Newtonian flow behaviour. The above is of obvious consequence for the rheological interpretation of flow behaviour of solutions of very high molecular weight polymers.

Key words: Flow, adsorption, entanglement, apparent, viscosity.

Introduction

This paper is the first in a series of four in which we shall describe our recent experiences on flow anomalies in polymer solutions which we ascribe to the formation of thick adsorption-entanglement layers along the walls of the flow channels. (In this context we mean by 'thick' that the layers extend into solution over distances considerably larger than the radius of gyration of the molecule in the particular solvent). Following preliminary work already reported [1–3] these paper will describe the flow observations themselves, the verification of the existence of the adsorption entangle-

ment layers postulated previously, some properties of these layers, some factors influencing their formation, together with the presentation of a model of how they may arise with all the consequences for existing practice and interpretations in measuring solution viscosity. First the past history will be briefly presented.

Our own interest in adsorption-entanglement layers stems from our efforts to produce polyethylene fibres having ultra-high modulus and strength via solution processing. One such method termed the 'surface growth' method designed by Zwijnenburg and Pennings [4–7] consists of pulling polyethylene fibres out from the surface of a glass or Teflon cylinder while rotating in a solution of polyethylene at a temperature usually somewhat higher than the normal crystallization temperature of this polymer. It was the success of

---

*) Dedicated to Prof. Dr. H.-G. Kilian on the occasion of his 60th birthday.

this experiment which prompted Zwijnenburg and Pennings to postulate the existence of an adsorption entanglement layer of polyethylene along the cylinder surface, in the first instance as a source of fibrous crystal growth. For our part [2, 3] we have shown that this fibre production by 'surface growth' consists of the actual stretching of this adsorption-entanglement layer itself, a kind of gel network adhering to the cylinder surface, being constantly replenished from the solution interior during the rotation of the cylinder. We verified the existence of such a layer by fixing thin tapes or foils to the rotating cylinder and examining these subsequently by infrared spectroscopy or by simple weighing, which revealed a substantial accumulation of polyethylene on the tapes during cylinder rotation. Nevertheless this mode of verification could only be made after cylinder rotation had stopped and the foils were removed from the cylinder. One of the objectives of the present work (Part I) is to record the layer formation during rotation in situ.

First, however, a different kind of background relying on rheological measurements needs to be invoked. Starting with our own experiences [1] we measured the solution viscosity of polyethylene ('apparent' viscosity as we shall term it) in the same kind of apparatus which was used in the 'surface growth' method of fibre production, i. e. in rotating cylinder viscometers. The measurements were carried out at temperatures above those at which crystallization normally occurs, or even above temperatures used for fibre production. A pronounced time dependence of the apparent viscosity was observed. This consisted of the viscosity value passing through a peak before reaching a constant plateau. One such curve for polyethylene is contained in figure 5. We noted further, that the nature of the whole solution was affected by the treatment involved in the recording of such a viscosity-time curve, even if only in a temporary manner. Amongst others, when the solution was cooled after the rotation had been stopped it gelled, a behaviour not displayed by the unstirred control. The sum total of the observations led us to the following model [1]. An adsorption-entanglement layer builds up along the walls of the cylinders (at that stage we believed along both cylinders — see later) gradually constricting the flow channel and thus increasing the apparent viscosity. Eventually portions of these layers break off (if they did not, the flow channel would block and flow or rotation would stop), hence the peak in viscosity. The breaking off and reformation eventually reaches a steady state value which produces the final plateau in curves as in figure 5. The broken-off layer portions are molecular aggregates,

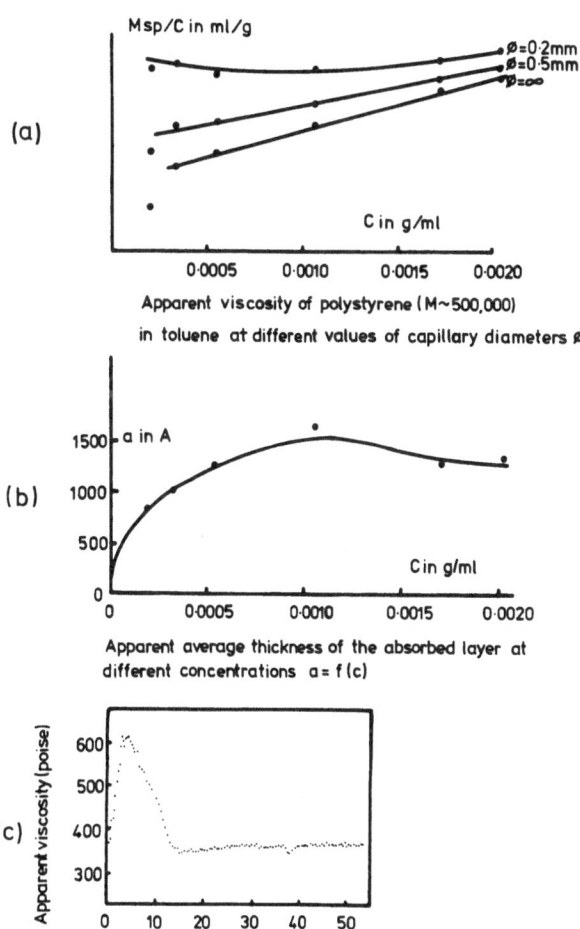

Fig. 1. (a) Apparent viscosity ($\eta_{sp/c}$) as a function of concentration for 500,000 molecular weight polystyrene from the work of Öhrn [10]; (b) Thickness of the adsorbed layer in capillaries deduced by Öhrn [10] from the data in (a) above; (c) Apparent viscosity during continuous shearing of a high molecular weight PMMA solution using a cone and plate viscometer. From Peterlin et al. [22]

kinds of gel fragment, hence the profound change in solution properties even after rotation has stopped. The gelation itself, which sets in the quiescent state, is envisaged to be due to the coalescence of these aggregates.

Initially we believed that the above decribed phenomena were associated with incipient crystallization of polyethylene. However, we soon became aware of the fact that similar effects have been observed in earlier literature on intrinsically uncrystallizable polymers (see e. g. fig. 1c). In fact, we found later that curves such as in figure 1c are only specific manifestations of a much larger body of material on anomalies observed during viscosity determinations using a variety of polymers, solvents and flow methods. Some of these were left unexplained, while in other cases the authors explicitly invoked adsorption layers of appreciable thickness to account for their effects. However, irrespective of whether such effectts were explained and how, all these works appear to have been forgotten. Only this way could it happen that in most recent times, in fact concurrently with our work, Cohen and Metzner [8] announced that viscosities of high molecular weight polystyrenes were affected by the capillary surfaces, leading them to postulate adsorption layers sufficiently thick to modify appreciably the flow properties of their solutions.

In summary, in the light of our own experiences and of a variety of earlier, largely bypassed observations and the rediscovery of effects already recognized long ago, we are now in a position to state that the subject of flow behaviour, viscosity measurements in particular, especially in the case of high molecular weight polymer solutions, abound with anomalies and that these anomalies can be attributed to thick adsorption-entanglement layers forming during flow. This assertion is quite general, since it applies to polymers of all kinds, crystallizable, polar, ionic etc. There are certainly differences between the above classes of polymers, regarding the readiness of forming such layers and the magnitude of the final effects, but not the occurrence of the effect itself.

In what follows we shall first briefly review the earlier literature, as this to our knowledge has not been done before from the present point of view.

First, there are a number of papers describing anomalies in the flow of high molecular weight solutions through capillaries, mostly relating to polystyrene. As early as 1955 Öhrn [9] attributed the rises (or in some cases peaks) in the $\eta_{sp/c}$ against c curves observed by himself [10] — as illustrated in figure 1a taken from his own works — and in several others [11–13], in terms of the formation of thick layers. For example in his own work with a polystyrene of molecular weight 500,000 in toluene (for which the radius of gyration is ca. 300 Å), Öhrn calculates from the increased flow times that a layer ca. 1500 Å thick must have formed at the capillary walls (fig. 1b). The recent work by Cohen and Metzner referred to above [8], done without knowledge of Öhrn's earlier work, reached a similar conclusion in a similar experiment.

In this context we must also mention the researches of Silberberg and his coworkers who have been using changes in flow times through capillaries to monitor adsorption [14, 15]. They often find adsorption layers whose hydrodynamic thickness is somewhat greater

than the radius of gyration of the adsorbing molecules, nevertheless they argue that these layers are in fact monomolecular. One argument against multimolecular layers comes from Silberberg's analysis [16] of the possibility of forming such thick layers during adsorption from quiescent solutions under equilibrium conditions. He shows that multimolecular layers will only form under these conditions as a kind of localised phase separation in poor solvents close to θ conditions. Nevertheless, as we shall show, multimolecular layers can arise in good solvents under *non-equilibrium* conditions in *flowing* solutions. We suggest that any differences between the work by ourselves [17–19] and by others [8–15] such as relate to capillary flow with the suggestion of thick multimolecular layers on the one hand, and those by Silberberg which are interpreted as monomolecular layers on the other, may lie in the chemical nature of the capillary surface. We shall show in the second paper of this series [17] that the build-up of layers depends strongly on the chemical nature of the surface. Indeed Cohen and Metzner [8] have already reported that silanization of their capillaries prevented, or reduced, the build-up of thick layers. The work of Silberberg [14–16] is particularly informative for our purposes in that it sets a baseline for multimolecular layers. Layers whose hydrodynamic thickness is much greater than those he observed must be of a multimolecular nature.

Secondly, a more direct observation of thick layers is to be found in the work of Hand and Williams [20]. They observed flow of PMMA dissolved in chloroform through a cell in an infrared spectrometer and found that the infrared absorbance due to PMMA increased at the same time as the flow rate decreased, indicating thick layers (≥ 0.1 mm) forming at the flow cell windows. In fact they even observed complete blockage of flow caused by such layers.

Finally, there are the examples of anomalous viscosity-time curves, recorded during continuous shearing in a Couette, or cone and plate viscometer which predate our more recent studies [1]. We illustrate this effect by figure 1c, reproduced from the work of Peterlin [21–25] already mentioned above. Peterlin and his co-workers obtained such viscosity-time curves for polystyrene, polyethylene oxide and polymethylmethacrylate. However, they did not invoke the formation of adsorption entanglement layers, but attempted to explain their results by postulating intrinsic changes in the structure of the solution itself, a line which was never taken further.

The principal purpose of the present series of papers is to demonstrate that adsorption-entanglement layers commonly occcur when high molecular weight polymer solutions flow past a foreign surface, and to draw attention to some of the consequences.

In this first publication we seek first to demonstrate the existence of the layers by direct observation, and further, to show how the presence of such layers, in the firm knowledge that they exist, may be deduced by indirect means, such as the variation in apparent viscosity during continuous shearing. We shall then draw upon the numerous examples of similar behaviour previously reported in the literature (where it is not usually attributed to layer formation).

In the present paper we deliberately restrict ourselves to a single polymer-solvent system. We have chosen as our example PMMA in dimethylphthalate (DMP) for reasons to be detailed later.

We have used two distinct flow systems in the present, as in the subsequent papers; Couette flow and flow through a rectangular tube (channel flow as we have called it). It is essential to appreciate the differences between these flow systems.

In the Couette apparatus the solution is continuously sheared at a constant shear rate (which for our purposes means that the build-up of a layer manifests itself as an increase in shear stress and hence in apparent viscosity). In contrast, in the channel flow apparatus the solution passes through the flow cell only once and, since the pressure is held constant, any layer formation simply reduces the flow rate. We should note that we are in the laminar flow regime throughout; in no instance have we used flow rates, or shear rates such as would lead to turbulent flow.

In the case of channel flow in particular, adsorption entanglement layers will cause spurious dependence of the viscosity on the flow rate which can lead to erroneous conclusions regarding the onset of non-Newtonian behaviour to be demonstrated in the present paper.

Adsorption entanglement layers have many further interesting and challenging consequences. Some of these will form the subject of the subsequent papers of this series. In specific detail, the second paper to follow [17] will be primarily concerned with the chemical nature of the surface at which the layers are formed. We shall show that the thickness of the layers is strongly dependent upon the nature of the surface and that in some cases layer formation can be prevented by suitable surface treatment. The third paper [18] will be concerned with the effect of solution concentration and solvent power. We shall show that in concentrated solutions thicker and stronger layers are formed, and that for a given concentration the layers form faster from better solvents. In the fourth and final paper [19]

we examine the rates at which the layers form, and once formed, decay on cessation of flow. We shall also present in that paper a qualitative model to account for the totality of our observations and point out the wider consequences for experimental practice on flowing macromolecular solutions.

Experimental

Materials and solution preparation

The polymer for the present experiments was PMMA, chiefly because by past reports it displays some of the most pronounced viscosity anomalies [20, 22, 25]. The solvent chosen was dimethylphthalate (DMP) which owing to its low vapour pressure allowed comparatively easy handling and storage compared with the more commonly used volatile solvents such as chloroform and toluene. Also this was one of the solvents used earlier by Matsuo et al. [25] displaying peaks in viscosity v. time curves (as shown in fig. 1c). Furthermore, the rate of layer build-up from DMP proved to be convenient for it to be followed in the present experiments. DMP is a moderate solvent for PMMA at room temperature; we infer that the θ temperature must be around $-10\,^\circ$C.

We intentionally used very high molecular weight PMMA for which the effects in question were most pronounced. Our base material had $M_w \sim 7 \times 10^6$ as supplied by I.C.I. Plastics and Petrochemicals Division in the form of cast sheets.

As the high molecular weight PMMA did not dissolve readily in DMP it was first dissolved in chloroform. DMP was then added to the chloroform solutions after which the chloroform was evaporated at 60 °C. The solutions were filtered before use.

The flow visualisation apparatus

The purpose of these experiments has been to observe the build-up of layers at a surface in situ during flow. We have used the two flow geometries indicated in the Introduction: Couette flow and channel flow, though the latter in a tube with a rectangular cross section. In order to observe the layers directly we have taken advantage of the fact that under the stress created by the flow the layers become birefringent, no doubt due to the creation of molecular orientation within them. Thus, such layers will become directly visible as birefringent zones along the surfaces of the Couette or channel. In order to register in situ we moved a fine laser beam across the flow field and monitored the intensity of light transmitted between crossed polaroids, with the polarisers placed either side of the flow system, as a function of the position of the light probe. The methods of realisation are illustrated in figure 2. The laser beam first passes through a lens system to reduce its size to ca. 300 μm then through a rotating glass cube. This glass cube is rotated typically at 40 Hz and displaces the beam along a preselected direction in a periodic manner. The beam is then polarised and passed through the flow cell itself. In the case of the Couette apparatus the movement of the beam is along a radial direction between the inner and outer cylinders, and in the case of the channel system it is normal between the adjacent sides. Finally, the beam passes through an analyser, and a focussing lens is set to collect all transmitted light at a photodiode, the output of which is sent to a storage oscilloscope. A second, trigger signal from the motor driving the rotating glass cube is also sent to the oscilloscope. The osscilloscope thus displays continuously the intensity of transmitted light as a function of position across the

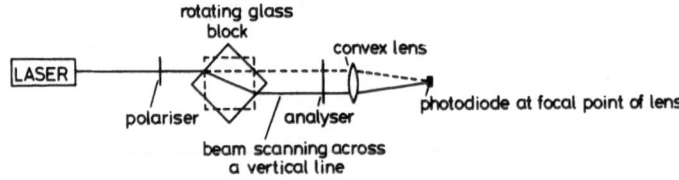

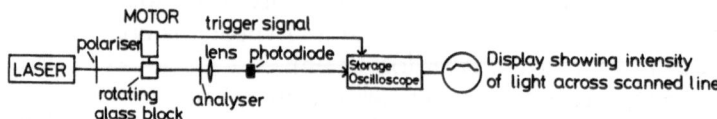

Fig. 2. Schematic diagram of the optical and electronic arrangements used in our flow visualisation technique

flow field. In addition, we can plot such intensity profiles at regular intervals on a chart recorder. Each scan of the flow field takes ca. 5 ms.

The Couette viscometer

A modified Ferranti Couette viscometer was used. Two principal alterations were made to the viscometer. First, a rotational differential variable transformer was fitted to the torque spring so that an electrical signal proportional to the torque is measured. Secondly, the cylinders were modified by providing them with optically flat glass top surfaces so that light can pass through them. In addition an extra pair of cylinders were constructed so that we could rotate the inner cylinder while monitoring the torque on the outer one in certain special experiments. The essential details of the viscometer and the optical path through the viscometer, together with the orientation of the polariser and analyser are shown in figure 3. The cylinders used in this work had diameters 2.15 cm (inner) and 2.39 (outer) in experiments in which the outer cylinder was rotated; in experiments when rotating the inner cylinder the corresponding cylinder diameters were 2.5 and 2.75 cm respectively. The cylinders were all made either from stainless steel or brass.

The "channel flow" apparatus

The channel flow cell was constructed, as sketched in figure 4, from aluminium plates forming a slit with glass plates glued to the top and bottom to form a rectangular tube. This cell was connected

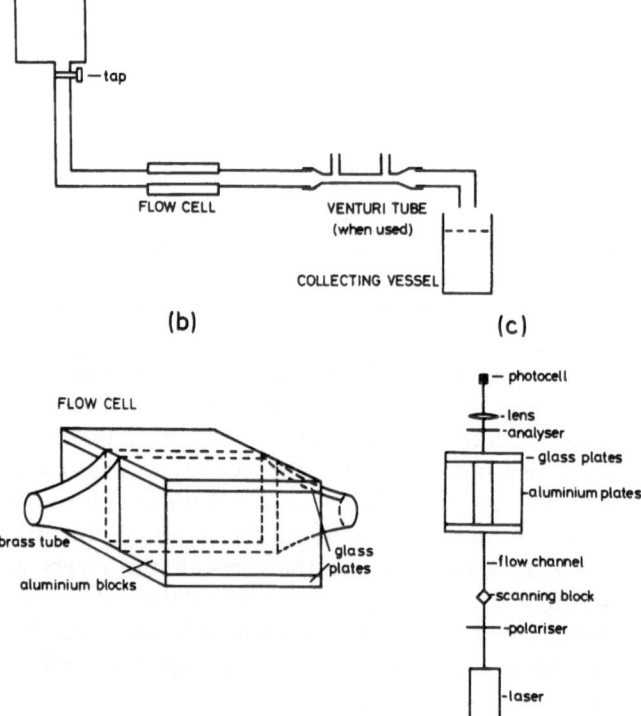

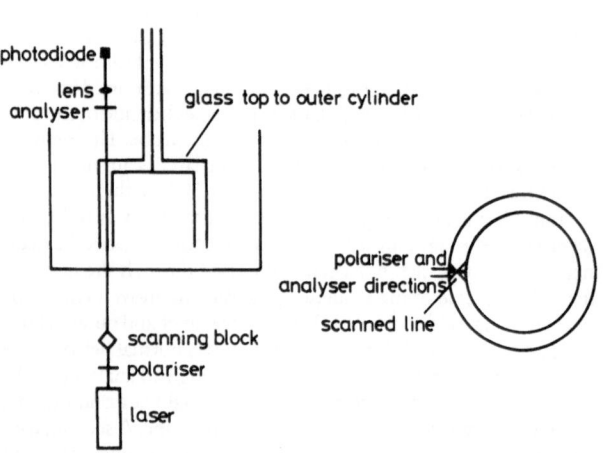

Fig. 3. A diagram showing the optical path through our Couette apparatus

Fig. 4. The channel flow equipment: (a) an overall diagram; (b) Detail of the flow cell; (c) a diagram indicating the optical path through the flow cell.

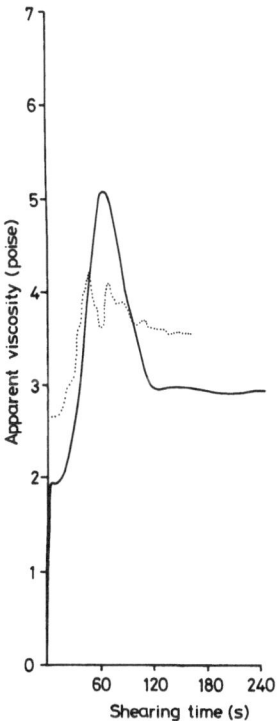

Fig. 5. Apparent viscosity during continuous shearing in a Couette viscometer of two different solutions. The solid line is from a 0.6% (w/w) solution of polyethylene ($M_w \sim 1.2 \times 10^6$) in decalin at 120 °C, and the dotted line is from a 1.3% (w/w) solution of PMMA ($M_w \sim 7 \times 10^6$) in DMP at room temperature. In both cases the rotation speed of the outer cylinders was 64 r.p.m.

to a supply tank filled with the solution which was pressurised with nitrogen. The overall mass flow rate of the solution was measured and monitored directly by collecting the outflowing liquid while the instantaneous flow rate was estimated from the pressure drop across a Venturi tube on the exit side of the flow cell. This Venturi tube was calibrated by using a simple liquid (glycerol) and, in some cases, the solution itself. The optical path through the flow cell and the orientation of the polariser and analyser are shown in figure 4.

Results

Couette viscometer experiments

First we shall briefly recapitulate our own previous work involving polyethylene solutions. We reproduce in figure 5 some typical results obtained from couette viscometry of a solution of a high molecular weight polyethylene in decalin [1]. We have previously argued that this behaviour is due to a layer of polymer built up at the viscometer cylinder surfaces, effectively reducing the gap and increasing the shear stress (or in this case the "apparent viscosity"). As the layer grows

thicker so the shear stress across it increases and eventually exceeds the failure stress of the adsorption-entanglement layer. Now the layer starts to break up and the shear-stress (apparent-viscosity) falls. Eventually a steady state between layer formation and breakdown is achieved leading to the final plateau value of apparent viscosity shown in figure 5. In the case of PMMA we see very similar behaviour (fig. 5; see also earlier experiments on PMMA by Peterlin [22] reproduced in fig. 1). With PMMA, in our experiments at least, in place of a single peak we see multiple peaks of decreasing magnitude. We would argue (and shall shortly show) that this is due to the repeated build-up and subsequent breakdown of the adsorption-entanglement layers. Thus the principal difference between PMMA and polyethylene is that the steady-state is reached more readily and with fewer fluctuations in polyethylene. We shall now proceed to apply our new flow visualisation technique to the case of PMMA. In the first instance we shall aim to make the postulated adsorption-entanglement layers visible in situ, and this accomplished, to correlate directly changes in apparent viscosity with layer formation.

Figure 6 shows some results obtained using a 1.3% PMMA solution at 20 °C at a shear rate of 70.67 s⁻¹ with outer cylinder rotating. This figure displays (a) the variation in apparent viscosity with shearing time, and (b) the optical scans across the Couette gap recorded simultaneously with a. These latter curves represent the transmitted light intensity across the Couette gap at consecutive times.[1])

It is clear that there is an increase in transmitted intensity (corresponding to the building up of a layer at the inner cylinder). We can measure both the hight of this peak and the total area above the base-line. Both of these should be a measure of the build-up of the layer, both in terms of the amount of material in the layer and its orientation. In parts (c) and (d) of figure 6 we have plotted both these quantities against the shearing time. There is a very clear, in fact remarkable correlation between each of these curves and the apparent viscosity given in (a). It is also most noteworthy that the adsorption-entanglement layer is only seen at the inner rotor surface. In order to check whether this is associated with the fact that the inner surface was the one which was in motion we performed a similar experiment where the outer cylinder was rotated. The result of this

---

[1]) A correction to allow for the reduction in beam size near the walls of the cylinders has been applied to these data.

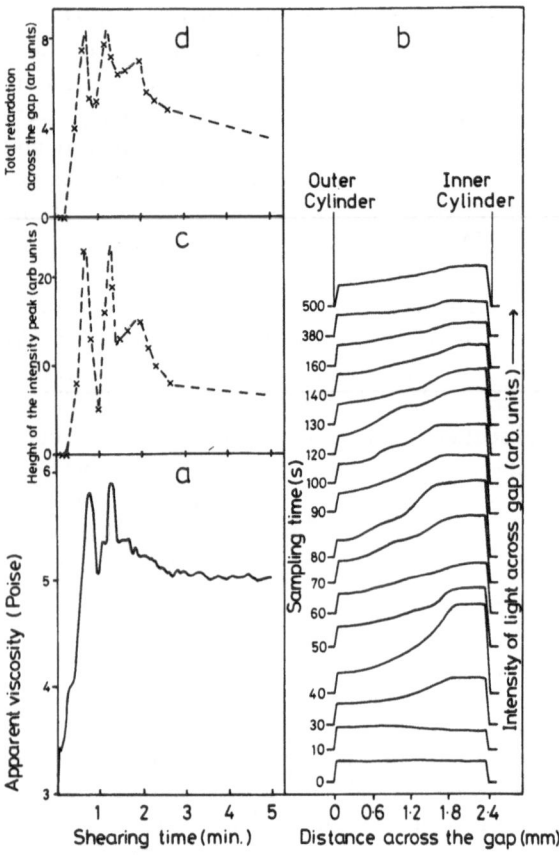

Fig. 6. Data obtained during continuous shearing at 70.67 s^{-1} of a 1.3 % PMMA ($M_w \sim 7 \times 10^6$) solution in DMP at room temperature: (a) the apparent viscosity; (b) a series of recordings of the optical retardation across the gap, the sequence of traces from bottom upwards corresponding to sampling at consecutive times; (c) the height of the peak in optical retardation near the inner cylinder wall; (d) the total area (above the base line at zero time) of the retardation curves

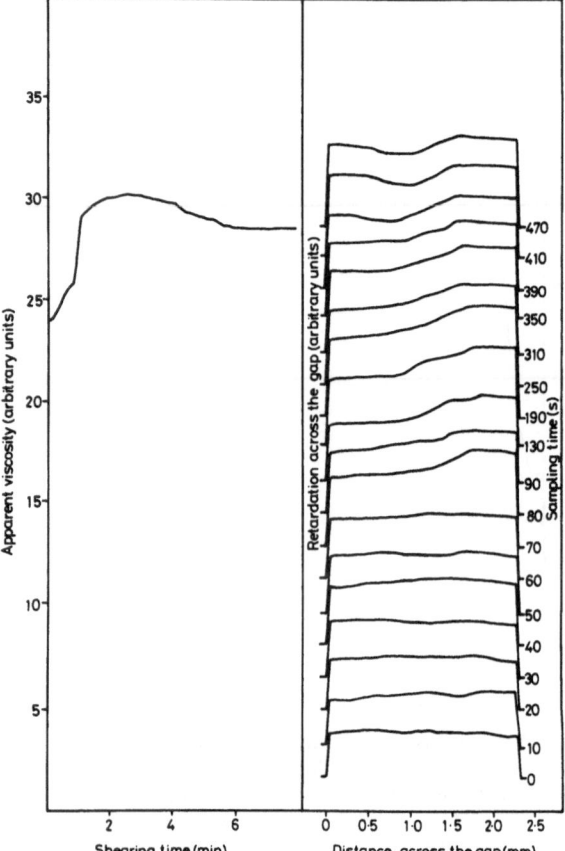

Fig. 7. The variation in apparent viscosity during continuous shearing at approx. 7 ls^{-1} of a 1.0 % PMMA solution in DMP at 20 °C. In this experiment the inner, rather than the outer, cylinder of the Couette viscometer was rotated

experiment is shown in figure 7. The viscosity v. time behaviour again shows a peak, even if its detailed structure is different. Nevertheless, the layer as recorded by the light optical trace is again located at the inner rotor surface, and only there.

Nevertheless, layers can be seen to form at the outer rotor surface when the shear rate is increased — figure 8. Here, as before, a layer is first built up at the inner cylinder but then followed by the growth from the outer cylinder. Eventually the two layers nearly coalesce leading to a very narrow flow channel with consequent unstable flow.

Channel flow experiments

For this work we again used a 1.3 % DMP solution. Figure 9 shows how the optical retardation varies across the channel when the solution is flowing, for three different flow rates (as produced by different pressures in the supply tank). Several points should be noted. First, the retardation is everywhere much higher than that expected from conventional flow birefringence. Secondly, there is a zone of constant retardation near the walls, the width of this zone increasing with flow rate (pressure), and thirdly there is a finite retardation at the centre of the slit. We attribute the zone of constant retardation near the walls to an adsorption-entanglement layer on the aluminium side surfaces of the slit, and the finite minimum retardation at the centre as part of a uniform background due to adsorption layers also on the top and bottom glass surfaces seen flat on. In order to demonstrate that the light transmission is due to molecular orientation in the flow direction we conducted a series of experiments

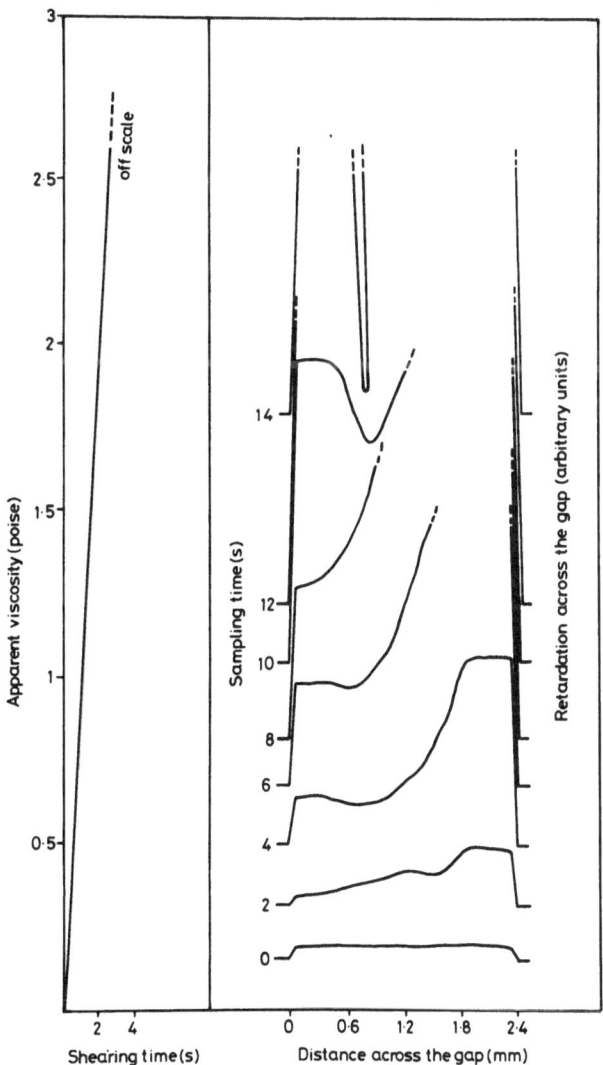

Fig. 8. Data for the 1.3 % PMMA solution as in figure 6, but now at the higher shear rate of 168 s⁻¹

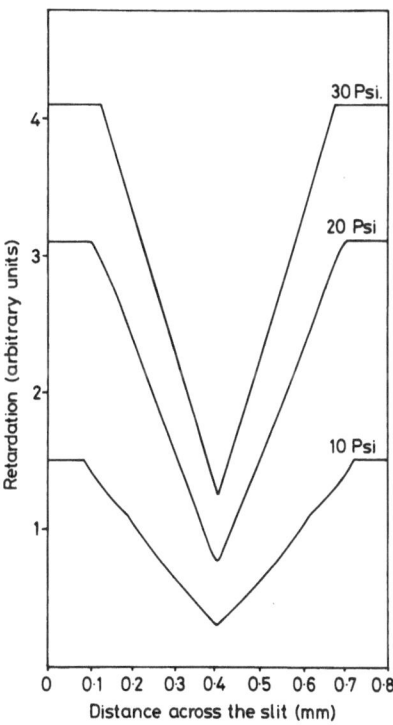

Fig. 9. Variation in retardation across the channel flow cell during flow at various pressures. 1.3 % solution of PMMA in DMP at 20 °C

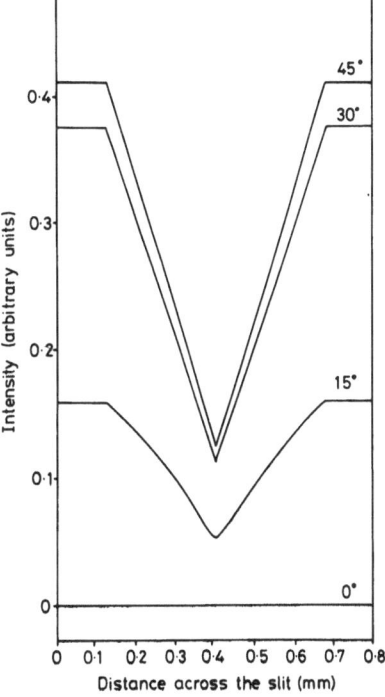

Fig. 10. Data as for figure 9 using a pressure of 30 psi. In this case the polarisers were rotated at various angles to the flow direction. (45 °C corresponds to the case in figure 9)

where the polariser and analyser were rotated. The results are shown in figure 10. It is apparent that the intensity is zero when the polariser is parallel to the flow direction and is a maximimum when it is at 45 °C.

The layer build-up in the channel flow experiments can also be observed indirectly by changes in the flow rate. When the flow is started the flow rate increases with time: we observe a fast rise in the flow rate (as mesured by the pressure drop across a Venturi tube), when in the case of a Newtonian fluid, a steady-state is achieved. However, when using the PMMA solutions we see a subsequent decrease in flow rate (see fig. 11).

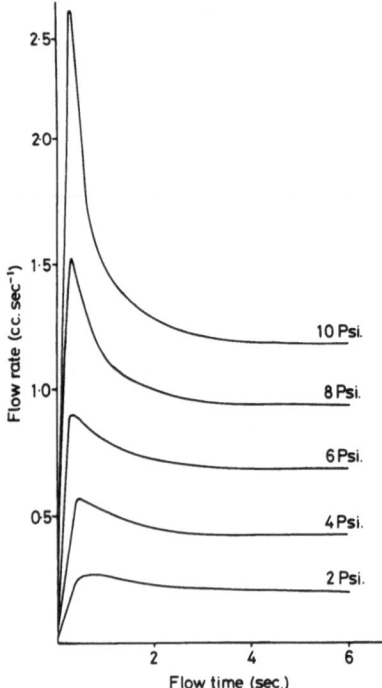

Fig. 11. The variation in instantaneous flow rate of a 1.3 % PMMA solution in DMP at 20 °C (obtained by measuring the pressure drop across a Venturi tube) at various flow pressures. The peaks do not occur when Newtonian fluids are used

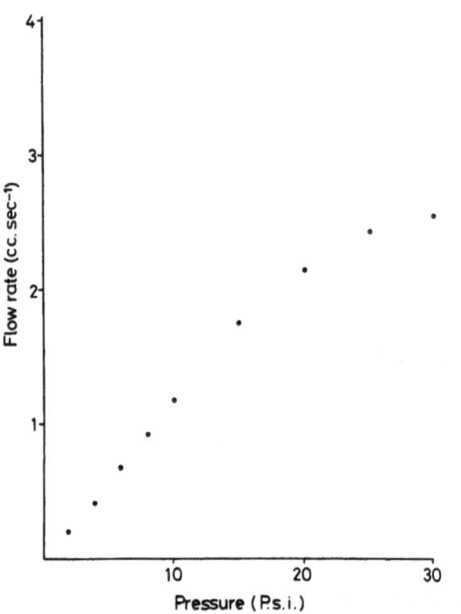

Fig. 12. The steady-state, mean flow rates of the 1.3 % PMMA solution as in figure 11, measured by collecting the outflowing solution, as a function of the flow pressure

We interpret this behaviour as follows. The initial rise in flow rate is due to the acceleration of the fluid as the pressure is applied. The subsequent drop in flow rate is due to the formation of the layers which reduce the width of the flow channel. We do not show any data at higher flow pressures since the venturi tube in figure 11 itself introduces flow instabilities – possibly due to the formation of adsorption entanglement layers at its surface.

We also measured the mean flow rate, with the venturi tube removed, by collecting the outflowing solution. We present these data in figure 12 in the form of a plot of flow rate as a function of pressure in the supply tank. These results, if interpreted by assuming no adsorption-entanglement layers, would suggest that the solution exhibits shear thickening (i. e. increasing viscosity with increasing shear rate). However, if we allow for the effective reduction in the size of the flow cell, due to the presence of the adsorption entanglement layers, then this alone can explain the observed reduced flow rates. To illustrate this we have calculated the "viscosity" making two (extreme) assumptions. In both cases we have taken η, the viscosity, as:

$$\eta = \frac{P \omega^3 h}{12 l Q}$$

where P is the pressure across the flow cell (i. e. in this case the pressure in the supply tank), Q is the volumetric flow rate and ω, h and l are the width, height and length of the flow channel.

We have in the first case taken ω and h as the dimensions of our flow cell, and in the second, we have reduced them by the thickness of the adsorbed layers as assessed by the width of the zone of constant retardation in figure 9. The results of these calculations are shown in figure 13 in the form of a plot of viscosity against flow rate. It is clear that the first assumption leads to the appearance of shear thickening, while the second would suggest shear thinning. It follows, that by merely observing departures from linearity in the macroscopically measured flow rate-pressure relation we cannot conclude that we have shear thickening; we cannot even assert that the flow is non-Newtonian, not to speak of the direction of departure from non-Newtonian behaviour if non-Newtonian flow indeed prevailed.

Discussion

The first, and most important issue which emerges is the existence and generality of the adsorption-entan-

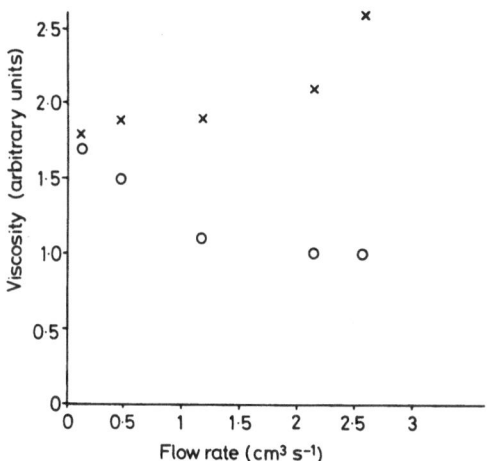

Fig. 13. The 'viscosity' (in arbitrary units) of the PMMA solution in figures 11 and 12 calculated using two different assumptions (see text) x — with no allowance for layers, o — allowing for a reduction in the size of the flow cell due to layer formation

glement layers. In our past work [1–3] we have, albeit indirectly, inferred that such layers must form in sheared solutions of high molecular weight polyethylene. In the present work we have been able, by using an optical technique, to observe the layers directly in PMMA solutions and to correlate their appearance, almost quantitatively, with changes in shear stress in the Couette viscometer and flow rate in our channel flow experiments. We therefore suggest that whenever rheological effects, such as presently described, are observed, they are most likely to be due to the formation of adsorption-entanglement layers. It may be appropriate at this stage to list some of the systems which, to our knowledge, exhibit one or other of these phenomena. Accordingly, in table 1 we have collected reports on systems which show either peaks in shear stress during continuous shearing, or reduced flow times in capillary viscometry, drawing both on the general literature and on our own work. We see from this table that a wide range of polymer-solvent systems can exhibit behaviour characteristic of adsorption-entanglement layer formation which should be sufficient to demonstrate the generality of such layers within flowing high molecular weight polymer solutions. All the reports we quote in table 1 refer to high (i. e. > 500,000) molecular weight polymers; they cover a very wide range of polymers and include both good and bad solvents.

The work we have reported here raises several other issues. These include the questions of why the layers form preferentially at the inner cylinder in Couette flow and of the causes of the fluctuations in the layer thickness before settling down to a more or less steady state as evidenced by figure 6. Further, there is the problem of the different shape of the 'apparent viscosity' — time curve when the inner, rather than the outer cylinder of the Couette is rotated. We do not at present have definitive answers to any of these questions, nor do we have more than a simple pictorial idea of the underlying mechanism of layer formation. Two issues however deserve some further discussion.

First, there is the difference in the rate of build-up of layers between the Couette and channel flow experiments. Although not well defined, the shear-rates used were very much higher in the channel flow work. We shall return to this in a subsequent publication [19] where we shall show that the rate of build-up of layers depends almost linearly on the shear rate.

Secondly, and most importantly, there is the issue of shear thickening. We have shown that, had we not recognized the existence of the adsorption-entanglement layers, we would have deduced that PMMA in DMP solutions exhibit shear thickening. However, as we have shown, there is insufficient evidence in this particular case to postulate any non-Newtonian behaviour. Further, since we would expect similar behaviour with any high molecular weight polymer solution, we suggest that the possibility of adsorption-entanglement layer formation should be carefully investigated where departures from non-Newtonian behaviour in general, and shear thickening in particular, are indicated by the flow rate dependence of the viscosity.

There are many other areas which may be influenced by adsorption-entanglement layers. We shall merely list some of these here. (Some others will be mentioned in the subsequent papers in this series [17–19]). It is widely known that drag reduction is, at least in part, caused by the adsorption of very long molecules at foreign surfaces and their subsequent modification of the flow field. The possibility that such layers may be of a multi-molecular adsorption-entanglement type described here has not, to our knowledge, been considered. Another area is the blocking of pores during filtration. The familiar experience that the GPC technique ceases to be operational at "high" concentrations clearly falls in the same category where the recognition of thick layer formation would have significant consequences. An understanding of the mechanisms of adsorption-entanglement layer formation will, we

Table 1

Polymer	Solvent	Minimum quoted concentration (%)	Method	References
a-PS	Toluene	0.1	Capillary flow	17–19
	Decalin	0.01	Capillary flow	17–19
	Chloroform	0.02	Capillary flow	9–13
	Aroclor	0.05	Shear stress	25
HDPE	Xylene	0.1	Shear stress	2,26
	Decalin	0.1	Shear stress	1
		0.08	Capillary	17–19
PP	Xylene	–	Shear stress	27, 28
PEO	Aroclor	0.3	Shear stress	25
PVC	Plasticisers	–	Shear stress	29
PMMA	DMP	2.5	Normal stress difference	30, 31
		1	Shear stress	25, 17–19
			Cappillary flow	17–19
			Direct observation	this work
	a-chloronapthlene	5	Shear stress	25
	chloroform		Shear stress	25, 17
			Direct observation	17
	Xylene	0.5	Shear stress	17
	Aroclor	0.1	Shear stress	21–24
Polyacrylamide	H$_2$O	5	Capillary flow	8
Polyisobutene	oligomers		Shear stress	
	Primol 355		Shear stress	

will, we expect, greatly aid the detailed comprehension of the surface growth technique [5] and of stirring induced gelation [2].

To summarise, we have shown that the formation of thick adsorption-entanglement layers is a common, although until recently largely unrecognised occurrence in flowing high molecular weight polymer solutions. The existence of such layers and their general recognition should have significant consequences in many areas of polymer science.

References

1. Narh KA, Barham PJ, Keller A (1982) Macromol 15:464
2. Barham PJ, Hill MJ, Keller A (1980) Coll & Polym Sci 258:899
3. Barham PJ (1979) Comment in Faraday Discussion Meeting No 68
4. Zwijnenburg A, Thesis PhD (1978) Groningen
5. Zwijnenburg A, Pennings AJ (1976) Coll & Polym Sci 254:868
6. Barham PJ, Keller A (1980) J Mats Sci 15:2229
7. Barham PJ (1982) Polymer 23:1112
8. Cohen Y, Metzner AB (1982) Macromol 15:1425
9. Öhrn OE (1956) J Polym Sci 19:199
10. Öhrn OE (1955) J Polym Sci 17:137
11. Streeter DJ, Boyer RF (1954) J Polym Sci 14:5
12. Umstätter H (1954) Makromol Chem 12:94
13. Batzer H (1954) Makromol Chem 12:145
14. Silberberg A (1976) ANTAS 275:2
15. Silberberg A (1968) J Chem Phys 48:2835
16. Silberberg A (1972) J Coll Int Sci, 38:217
17. Narh KA, Barham PJ, Himket RA, Keller A, Coll & Polym Sci, submitted
18. Barham PJ, Hikmet RA, Narh KA, Keller A, Coll & Polym Sci, submitted
19. Barham PJ, Coll & Polym Sci, submitted
20. Hand JH, Williams MC (1973) Chem Eng Sci 28:63
21. Peterlin A, Turner DT (1965) Polym Lett 3:517
22. Peterlin A, Quan C, Turner DT (1965) Polym Lett 3:521
23. Peterlin A, Phillipott W, Turner DT (1965) Kolloid Z 204:21
24. Burrow S, Peterlin A, Turner DT (1964) Polym Lett 2:67
25. Matsuo T, Pavan A, Peterlin A, Turner DT (1967) J Coll Interf Sci 24:241

26. Pennings AJ (1977) J Polym Sci Polym Symp 59:55
27. Vinogradov GV, Malkin AY (1969) Z Z Prik Mekh I Tek Fiz 5:66
28. Narh KA, Thesis PhD (1978) Bristol
29. Glukov EE, Vinogradov GV, Klaz S (1963) Vyskomol Soedin 5:1543
30. Adams N, Lodge AS (1964) Phil Trans Roy Soc A256:149
31. Lodge AS (1961) Polymer 2:195
32. Bartenev G, Vishnitskaya LA (1964) Makromol Verb Moskau 6:751–757

Received June 3, 1985;
accepted June 15, 1985

Authors' address:

R. A. M. Hikmet
H. H. Wills Physics Laboratory
University of Bristol
Tyndall Avenue
Bristol BS8 1TL, England

Influence of plug porosity on streaming potential measurements*)

J. Schurz and G. Erk

Institut für Physikalische Chemie der Universität Graz, Graz, Austria

Abstract: The Happel-Ciriack model for flow through a porous fiber plug is used for evaluating measurements of streaming potential. To this end a numerical solution is given and tested experimentally. Contrary to simpler models, the new method yields porosity independent ζ-potential values with synthetic fibers. With cellulose fibers, the values are no longer independent of plug density. An empirical correction is proposed which eliminates this short-coming.

Key words: Zetapotential, porosity, fibers.

1. Happel's streaming model for cylindrical fibers

The general description of streaming current and streaming potential phenomena in porous systems is based on flow in more or less curved capillaries (common hydraulic radius) and on the Gouy-Chapman theory for the diffuse part of the electric double layer. However, measurements of streaming current displayed significant differences between theoretical expectation and experimental results. Therefore Happel [1, 2] devised a new model for slow flow perpendicular to a mesh network of cylindrical fibers. He proposed the "model of free surface", as shown schematically in figure 1. The outer cylinder with radius b is the frictionless surface, where the shear rate is zero. It is assumed that both viscosity and density of the liquid phase between outer and inner cylinder are constant and have the same value as in the bulk phase. Contrary to the classsical capillary bundle model (irregularly curved liquid channels), no additional effect arises when the fiber plug is compressed to lower porosity, since thereby the fibers are neither kinked nor oriented. All the fibers lie in a plane perpendicular to the direction of flow, their orientation in this plane is random.

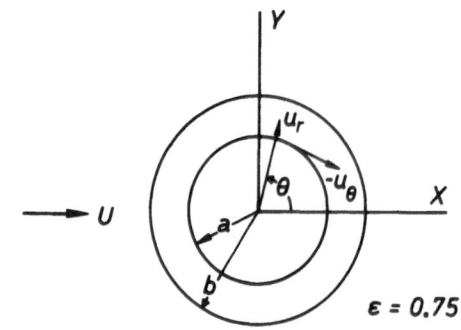

a : radius of cylinder
b : radius of liquid shell
U : velocity of liquid approaching the cylinder
ε : porosity $(b^2 - a^2)/b^2$

Fig. 1. "Free surface model" for flow vertical to a stationary cylinder

2. The streaming-current equation of Ciriack

Based on the analogies between the electric double layer model and the free surface model, Ciriack [3] derived a new equation for the streaming current. Since the arrangement of the cylindrical fibers will not be influenced by a compression of the fiber plug, is could be assumed that the discrepancies of the older

*) Dedicated to Prof. Dr. H.-G. Kilian with best congratulation for his 60th birthday.

capillary bundle model could be eliminated. Ciriack obtained the following equation:

$$\zeta = -4C_4 \pi a \chi^{-2} (I/U)/[A D_K (T_3 - 0.375 C T_1 - 0.5 D_1 T_2) - T_3 (E + 0.25 D_1) + F T_4].$$

The following abbreviations are used:

$$D_1 = -2/\ln[1/(1-\varepsilon)^{0.5}] + 4[(1-\varepsilon)^2 - 1]/[(1-\varepsilon)^2 + 1]$$

$$F = D_1 \cdot a^2/[4(1-\varepsilon)^2 + 4]$$

$$C = -a \cdot F (1-\varepsilon)^2/a^4$$

$$E = 1 + F/a^2 - 0.5 D_1 (\ln a + 0.5) - 0.375 C a^2$$

$$T_1 = a^{0.5} \int_a^{a+\tau} [\exp(\chi a - \chi r) r^{1.5}] dr$$

$$T_2 = a^{0.5} \int_a^{a+\tau} [\exp(\chi a - \chi r) \ln r/r^{0.5}] dr$$

$$T_3 = a^{0.5} \int_a^{a+\tau} [\exp(\chi a - \chi r)/r^{0.5}] dr$$

$$T_4 = a^{0.5} \int_a^{a+\tau} [\exp(\chi a - \chi r)/r^{2.5}] dr$$

$C_4 = 9.0$ transformation constant to obtain uniform dimensions

ζ : zeta potential
χ : Debye-Hückel parameter
I : streaming current
U : streaming velocity
D_K: dielectric constant
ε : porosity
a : fiber radius.

For practical application of this system of equations, a Fortran IV program was written, in which a numerical integration was performed according to the Romberg method. The program allowed a simultaneous evaluation according to the capillary model by Goring and Mason [4], and the improved model by Biefer and Mason [5]. The program and further details have been given be Erk [6].

3. Measurements and their evaluation

Measurements of streaming current were performed in an apparatus described some time ago [7].

The fiber plug was prepared according to a special method and in a special device. 1.0 g fibers were suspended in 100 ml electrolyte solution (10^{-4} n KCl). Then this 0.01 % fiber suspension is transferred by suction into the streaming current measuring cell by means of the device shown in figure 2, whereby a constant vacuum of 550 Torr is maintained. The fibers are sedimenting slowly into the measuring cell and will form, under the conditions described, a homogeneous plug free of gas bubbles, in which the fibers are oriented within the yz-plane, that is perpendicular to the direction of flow in compliance with the assumption of the model. Too strong suction (higher vacuum) or a more concentrated fiber suspension will impair the plug formation considerably, since in the first case the fibers will also be oriented parallel to the flow direction and in the second case flocculation may take place. Thereby no more single fibers, but fiber flocs will be introduced into the measuring cell, whereby correct orientation is no more guaranteed. In a few cases we checked the orientation within the fiber plug by means of inspection with a light microscope. We observed good and exclusive orientation within the yz-plane.

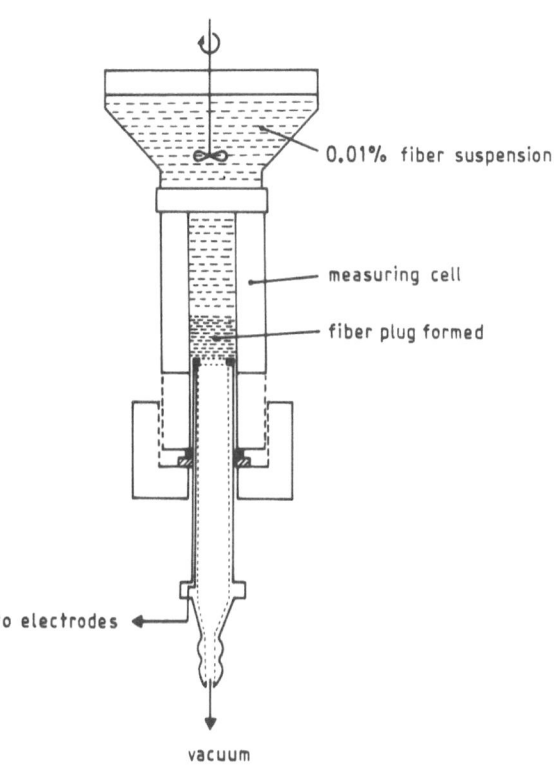

Fig. 2. Preparation of the fiber plug

4. Check of the homogeneity of the fiber plug

In order to check the homogeneity of the fiber plug, we used the Konzeny factor [8], which depends on the pressure drop along the plug and on other parameters. Furthermore, we used the empirical Davies factor [9], which describes the dependence on porosity of the Konzeny factor. If, for a given packing density or porosity both factors have the same value, then inhomogeneities can be excluded. Therefore, if accordingly both factors are compared for the range of porosities in which measurements are made, any inhomogeneity arising during compression of the plug will be detected.

The Konzeny factor reads:

$$k_K = \frac{p \cdot \varepsilon^3}{[UL(1-\varepsilon)^2 \eta S_v]}$$

with

p : pressure drop
U : average flow velocity
L : length of the plug
η : bulk phase viscosity
S_v : specific surface per unit volume of a fiber.

The porosity ε is given by:

$$\varepsilon = 1 - \alpha c$$

α : speicific volume of fibers in cm³/g
c : concenration of fibers in g/cm³.

The Davies factor, which does not depend on pressure drop p and flow velocity U, reads:

$$k_D = k_1 \frac{\varepsilon^3}{\sqrt{1-\varepsilon}} \cdot [1 + k_2(1-\varepsilon)^3] \quad \begin{array}{l} k_1 = 3.5 \\ k_2 = 57.0 \end{array}$$

Both equations hold for cylindrical fiber plugs made of cylindrical single fibers according to the method described before. For the calculation of both k_K and k_D, the unknown figures U, S_v and α must be obtained. For the determination of U, during measurement of the streaming current, besides the pressure drop p the volume flow per unit time V is measured for every porosity, and then divided by the constant cross section area of the measuring cell to obtain U. The specific surface S_v (in cm²/cm³) is calculated by the ratio of the surface F of a fiber to its volume V. For cylindrical fibers with radius a and length l we obtain:

$$S_v = \frac{F}{V} = \frac{2\pi a^2 + 2\pi a l}{2\pi a^2 l}.$$

Here the average fiber length l and the average fiber radius a are required. Both figures can be determined in a light microscope, if they are not provided by the producer of the fibers. The influence of an eventual swelling has to be taken care of. S_v is independent of porosity.

The specific volume α can be determined in a pyknometer, especially for swelling fibers. If the fiber diameter (titer) is known, the specific volume can also be calculated.

For the calculation of the zeta potential, the thickness of the diffuse layer must be known. Ciriack calculated a value of 0.6 µm in a solution of 10^{-4} n KCl. We have adopted this figure, since all our measurements were done in the same medium.

5. Results and discussion

5.1. Polyamide fibers (Nylon 66)

In a first set of experiments, the applicability of the Ciriack method was tested by means of polyamide (Nylon 66) fibers in 10^{-4} n KCl at 23 °C and pH = 4.55. Some characteristic of the fibers investigated are given in table 1. The results of the numerical calculations are compiled in table 2. In figure 3 the calculated zeta potentials are plotted against the porosity of the fiber plug. We note that the results based on the capillary bundle model (ζ_1 and ζ_2) decrease with porosity in the porosity range 0.6 to 0.79, while the Happel-Ciriack evaluation (ζ_3) yields porosity independent figures

Table 1

Characteri-zation of the fiber sample	Titer (dtex)	Average radius a (µm)	Average length l (mm)	S_v (cm²/cm³)	α (cm³/g)
Rhodiaceta-Nylon 66, dull	3.3913	10.015	2.8	2004	0.929

Table 2

L (cm)	k_K	k_D	ζ_1 (mV)	ζ_2 (mV)	ζ_3 (mV)	ε
4.0	5.9	5.8	−40.9	−28.8	−7.5	0.795
3.5	5.7	5.6	−34.6	−25.9	−7.5	0.765
3.0	5.7	5.6	−29.6	−23.9	−7.8	0.726
2.5	5.5	5.6	−23.7	−21.5	−8.1	0.671
2.0	5.4	5.5	−18.3	−20.3	−8.5	0.589

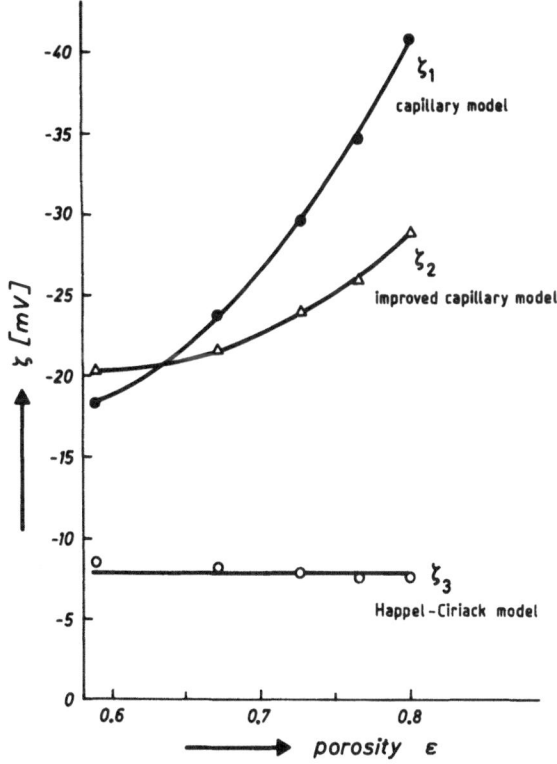

Fig. 3. Comparison of capillary models with the Happel-Ciriack model. System: Nylon 66 in 10^{-4} n KCl

Table 3

Characteri- zation of the fiber sample	Titer (den.)	Average radius a (μm)	Average length l (mm)	S_V (cm²/cm³)	α (cm³/g)
Cuprammoni- um rayon (Cuprama) glossy	1.4	7.985	3.3	2512	0.659

Table 4

L (cm)	ε	k_K	k_D	ζ_1 (mV)	ζ_2 (mV)	ζ_3 (mV)
3.0	0.806	6.45	5.89	−35.2	−24.4	−4.7
2.5	0.767	5.46	5.63	−29.4	−21.9	−4.9
2.0	0.709	9.49	5.56	−31.5	−27.2	−7.2
1.5	0.611	17.62	5.58	−30.4	−31.8	−9.9

find constant figures for the zeta potential. In the low porosity ranges however, significant deviations between k_K and k_D are found (compare the two lowest porosity values). Accordingly, in this range the zeta potential shows a strong rise with decreasing porosity.

within the limit of measuring errors. Both k_K and k_D agree well, which proves that the new method for making the fiber plug works satisfactory with Nylon 66 fibers.

5.2. Cellulose fibers (Cuprama)

A further test of the new equations was made with Cuprama fibers (regenerated cellulose rayon spun from cuprammonium), which are easily deformable and will swell in aqueous solutions. The data of these fibers are compiled in table 3. The measurements were made in 10^{-4} n KCl solution at 23 °C and pH = 5.33. The evaluation yielded the data shown in table 4 and plotted in figure 4. It is seen the neither the capillary bundle model nor the Happel-Ciriack model yield the expected constant values for the zeta potential. The reason may lie in the strong swelling tendency and the high deformability of the cellulose fibers, which could give rise to inhomogeneities in the fiber plug even with the new preparation technique. A closer inspection of the factors k_K and k_D as given in table 4 shows that they agree well in the range of high porosity, where we also

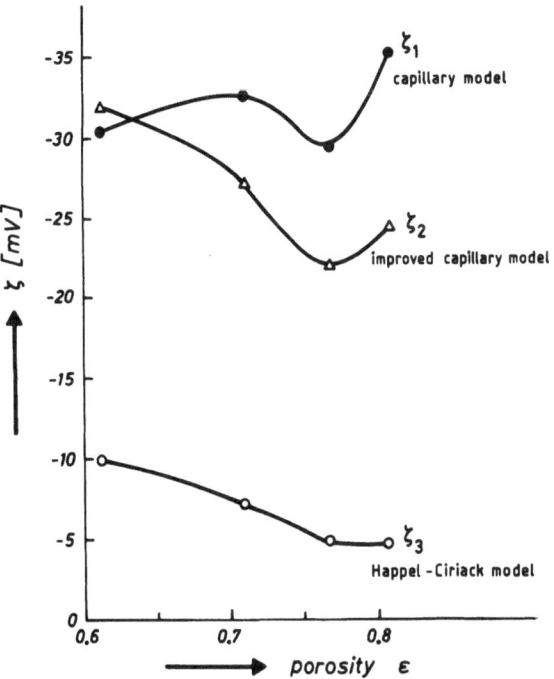

Fig. 4. Comparison of capillary models with the Happel-Ciriack model. System: Cuprammonium rayon (Cuprama) in 10^{-4} n KCl

The different values for both k_K and k_D indicate inhomogeneities in the fiber plug. While the empirical Davies factor k_D remains constant over the whole range, the Konzeny factor k_K, as obtained from the experimental figures p and U, shows a significant increase, especially for low porosities. Since the empirical Davies factor, contrary to the Konzeny factor, is independent of inhomogeneities of the fiber plug, it can be used to apply a correction to the measured slope $(I/U)_{exp}$. For instance, one measurement yielded for $\varepsilon = 0.611$ a k_K value of 17.62 and a $(I/U)_{exp}$ of 1.7 $\mu A \cdot s \cdot cm^{-1}$. The correctly packed fiber plug with $k_D = 5.58$ would correspond to a corrected slope $(I/U)_{corr}$, which can be calculated according to:

$$\frac{k_K}{k_D} = \frac{(I/U)_{exp}}{(I/U)_{corr}} \text{ resp. } (I/U)_{corr} = (I/U)_{exp} \cdot \frac{k_D}{k_K}.$$

Table 5

ε	$(I/U)_{exp}$ ($\mu A \cdot s \cdot cm^{-1}$)	$(I/U)_{corr}$ ($\mu A \cdot s \cdot cm^{-1}$)	$\zeta_{1, corr}$ (mV)	$\zeta_{2, corr}$ (mV)	$\zeta_{3, corr}$ (mV)
0.806	0.30	0.274	−32.2	−22.2	−4.3
0.767	0.38	0.392	−30.3	−22.5	−5.0
0.709	0.76	0.445	−19.0	−15.9	−4.2
0.611	1.70	0.537	− 9.6	−10.0	−3.2

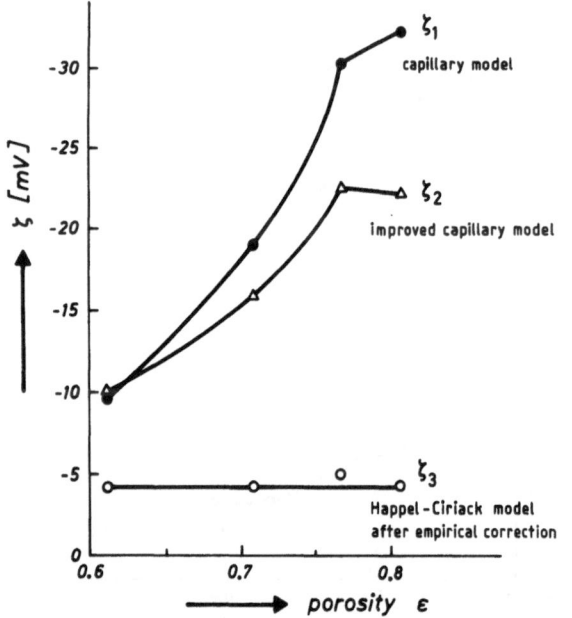

Fig. 5. Comparison of capillary models with the Happel-Ciriack model after empirical correction. System: Cuprammonium rayon (Cuprama) in 10^{-4} n KCl

In our case we find $(I/U)_{corr} = 5.58 \cdot 1.70/17.68 = 0.537$ $\mu A \cdot s \cdot cm^{-1}$. If this correction is applied for each porosity measured, the corrected values $(I/U)_{corr}$ yield the zeta potentials shown in table 5. In figure 5 these corrected zeta potentials are plotted against porosity. While the capillary bundle data are again unsatisfactory, the corrected Happel-Ciriack model yields porosity independent zeta potential and thus fulfills the theoretical expectation.

The reason for the inhomogeneities of the fiber plug is not known. An inspection of the dry fiber plug in the light microscope indicated that several single fibers were oriented parallel to the flow direction and that due to the high swelling an exact cylindrical cross section of the fibers was no longer guaranteed.

Both effects may be the reason why the ideal Happel model is no longer fulfilled in the real case. Therefore, the empirical correction must be applied, which leads to satisfactory results despite the shortcoming in the plug formation.

References

1. Happel J, Brenner H (1965) Low Reynolds Number Hydrodynamics, Prentice-Hall, Englewood-Cliffs, NJ, p 553
2. Happel J (1959) A I Ch E Journal 5:174–177
3. Ciriack JA (1967) An investigation of the streaming current method for determining the zetapotential of fibers, Ph D Thesis, Lawrence University
4. Goring DAI (1949) Ph D Thesis, McGill University; Goring DAI, Mason SG (1950) Can J Research 828:307
5. Biefer GJ, Mason SG (1959) Trans Faraday Soc 55:1239–1245; Biefer GJ (1952) Ph D Thesis, McGill University
6. Erk G (1974) Diss TH Darmstadt
7. Erk G, Schempp W, Schurz J (1975) GIT 19:772
8. Brown Jr JC (1949) Determination of the exposed surface area of pulp fibers from air permeability measurements, using a modified Konzeny-equation, Doctors Dissertation, The Insititute of Paper Chemistry, Appleton, Wisc, p 103–112; (1950) Tappi 33:130–137; Charman PC (1937) Trans Inst Chem Eng, London 15:150
9. Han ST (1965) The status of the sheet-forming process, a critical review, The Insititute of Paper Chemistry, Appleton, Wisc, Dec 31; Han ST (1969) Pulp and Paper Mag Can 70:65–77; Davies CN, (1952) Proc Inst Mech Eng, London B1:185

Received May 9,1985;
accepted July 26, 1985

Authors' address:

Professor J. Schurz
Institut für Physikalische Chemie
Universität Graz
Heinrichstr. 28
A-8010 Graz, Austria

Progress in Colloid & Polymer Science   Progr Colloid & Polymer Sci 71:49–58 (1985)

Effect of the initiator acetyl perchlorate upon the morphology, kinetics and mechanisms of crystal growth of poly(oxymethylene)*)

M. Rodríguez-Baeza and R. E. Catalán Saravia

Universidad de Concepción, Area de Polímeros, Departamento de Química, Facultad de Ciencias, Chile

Abstract: The kinetics of the cationic polymerization of trioxane initiated by different concentrations of acetyl perchlorate in dichlormethane solution is analyzed. Hexagonal single crystals of poly(oxymethylene) (POM), are obtained after an induction period. The kinetic of the crystal growth have been followed by electron and light microscopic techniques. The influencee of the initiator concentration on the morphology of the POM-crystals, together with the fact that the initiator is adsorbed on the crystal surface, enabled the discussion bf possible reaction mechanisms of polymerization and crystallization. Morphological observations show that extended chain crystals are obtained. The growth spirals of the crystals due to screw dislocations in the growing crystals can be quantitatively interpreted. This was conducted on the basis of the BCF-theory of crystal growth, with the particularity that in this case the initiator is adsorbed on the crystals surface, assuming a Langmuir-type adsorption behaviour. This heterogeneous reaction is an example of a simultaneous polymerization and crystallization. On the crystals surface cationic actives sites are generated, in which the polymerization and crystallization process occurred. We emphasize the role of the counterion, $OClO_3^-$, upon the polymerization reaction. The results were compared with those previously reported by using perchloric acid as initiator.

Key words: Trioxane, acetyl perchlorate, poly(oxymethylene), morphology, kinetic, mechanisms.

Introduction

Extensive studies have been carried out on the influence of crystallization on the overall kinetics of polymerization of trioxane [1, 2]. In addition, the morphology of POM-crystals have been also extensively studied [3–8].

The growth of hexagonal POM-crystals, which were formed by cationic polymerization of trioxane from nitrobenzene solution by using boron trifluoride etherate as initiator, was previously investigated. The purpose was to study the correlation between the mechanism of polymerization and crystal growth [6].

However, only recently the kinetic laws of the crystal growth of POM were described, on the basis of morphological observations [9]. They obtained nascent poly(oxymethylene) by cationic polymerization of trioxane, initiated by perchloric acid in dichloromethane solution. In this case, an adsorption equilibrium regarding the catalyst was established, in the presence of the rather large surface area of the growing polymer crystals. It was assumed a Langmuir-type adsorption behaviour and the observed growth spirals could be quantitatively interpreted, on the basis of the BCF-theory of crystal growth [10, 11].

According to these results, it seemed to be of interest to examine the effect of the initiator acetyl perchlorate on the overall kinetics of trioxane polymerization and on the kinetics of the crystals growth of POM, since both initiator $HClO_4$ and $CH_3COOClO_3$ have the same counterion.

Our goal was to study the effect of the initiator concentration upon the kinetics of polymerization, induction times, morphology and on the crystallization of poly(oxymethylene).

---

*) Dedicated to Prof. Dr. H.-G. Kilian on the occasion of his 60th birthday with our best wishes.

This was a good system to study both processes: the kinetics of polymerization and crystallization, which occurred simultaneously. Polymerization of trioxane is mentioned as one of the striking examples of simultaneous polymerization and crystallization [12, 13].

A chemical mechanism of polymerization and crystallization is proposed, on the basis that the initiator acetyl perchlorate is adsorbed on the crystalline surface. This can be explained as catalyzed crystal growth, similar to the case when perchloric acid was used as initiator. The proposed mechanisms corroborated that the adsorption of initiator on the crystalline surface is in good agreement with a Langmuir-type adsorption behaviour. Also, the growth features can be quantitatively interpreted, on the basis of the BCF-theory of crystal growth and therefore with the kinetic laws of the crystal growth of POM described by Wegner and co-workers [9].

In order to study the kinetics of crystal growth, it was necessary to find reaction conditions that would allow an optimum morphological observations of the crystal growth by electron and light microscopy. Acetyl perchlorate is an effective initiator for cationic polymerization [14, 15]. Besides, dichloromethane is not only a very suitable reaction medium for cationic polymerizations, but it also facilitates the preparation of in situ generated crystals for electron microscopy investigations.

The POM-crystals obtained in our laboratory are very useful as models in order to obtain a better understanding of the general crystallization behaviour and crystals properties of long chain molecules.

Experimental

Trioxane (m.p. 62 °C) was refluxed over sodium-potassium for 48 h and then distilled under dry nitrogen prior to polymerization. Dichloromethane was refluxed over LiAlH$_4$ for 48 h and then distilled under dry nitrogen before polymerization.

Acetyl perchlorate was prepared and purified according to Burton and Praill [16]. Acetyl perchlorate in dichloromethane solutions is stable for about 24 h. Their characterization was carried out in dichloromethane solution by IR-analysis (spectrophotometer, Perkin-Elmer model 237 and 577).

For polymerization, 25 ml of a solution of trioxane in CH$_2$Cl$_2$ was transferred under atmosphere of dry-nitrogen into a 50 ml polymerization flask, which was decontaminated from water by evacuating and heating with a torch. The flasks were set in a thermostat at 25 °C. Solutions of the initiator in dichloromethane were added by means of a syringe through a rubber stopper. Total concentration of the monomer was 2.5 mol/l and initiator concentrations ranged from 5 · 10^{-6} to 5 · 10^{-4} mol/l. The polymerizations were stopped by adding a solution of triethylamine in methanol, at different reaction times. The polymer crystals were carefully filtered and washed with CH$_3$OH and CH$_2$Cl$_2$ and then dried in vacuo at room

temperature. The yields of each polymerization were determined gravimetrically, on the basis of the initial quantity of trioxane.

The kinetic of lateral growth of the crystals was followed by measuring the diameter of the crystals (photomicroscope, Zeiss) in relation to the reactions times. The morphology of the single crystals and the kinetic of the thickness growth were determined by electronic microscopy. Surface replicas of the crystals and of the fracture surfaces for electron micrographs were obtained as reported in the literature [6]. A transmission electronic microscope, Philips EM-200, was used for those experiments.

Thermal data were measured with a Perkin-Elmer DSC-2, differential scanning calorimeter at 20 °C/min.

Results and discussion

1. Effect of the initiator concentration on the yields and on the induction times

The plots of conversion against times of polymerization obtained with different concentrations of initiator showed, at the beginning of the reactions, induction periods (fig. 1) typical of the cationic polymerization of trioxane [1, 2, 17–21]. These figures also show that the rate of polymerization increases with a higher concentration of the initiator, as well as the yields obtained. The reactions reached in each case a maximum yield. These upper values depended on the initiator concentration and then we could conclude that the system is kinetically hindering the attainment of their equilibrium condition. This behaviour of the acetyl perchlorate as initiator is analogous to perchloric acid [9], but different to the behaviour of boron trifluoride etherate, in which the Ceiling-equilibrium is reached by a conversion of 75 % [6]. Then, we established that the polymerization reactions are stopped when a maximum conversion has been reached, even though the expected thermodynamic conversion was not

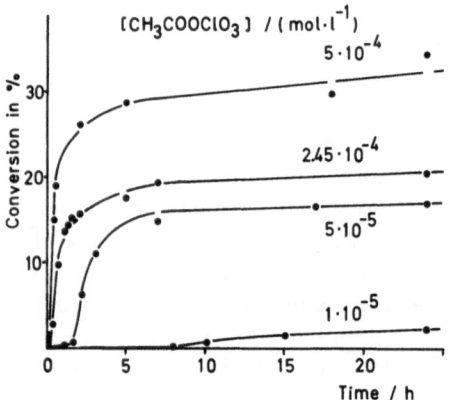

Fig. 1. Polymerization yield of trioxane as function of time, by using different concentrations of the initiator acetyl perchlorate, at 25 °C. Initial concentration of trioxane [TOX] = 2.5 mol/l. CH$_2$Cl$_2$ as solvent

attained. Several explanations have been reported [22, 23, 24].

Our investigations on the thickness growth of the polymer crystals (see section 4) show that the reaction rate does not decrease due to the physical inclusion of cationic actives polymer chain ends in the growing polymer crystals, but is caused for a special physical-chemical effect. This is closely connected with the crystal growth process, and therefore the reaction could not be reached their Ceiling-equilibrium, because the crystallization driving force became weaker.

After the addition of the initiator to the reaction solution, soluble oligomers are formed first and when the solution is supersaturated with oligomer molecules, beginning spontaneously the formation of a primary nucleuos in a short period.

A well defined population of crystals is formed during the nucleation event at the beginning of the polymerization.

The induction time (t_i) is defined as the time required for the appearance of the first crystals, visualized by the turbidity present in the heterogeneous system. A lineal relationship is obtained by plotting t_i against the reverse values of the concentration of the initiator (fig. 2), according to the following equation:

$$t_i = 0.07 \ [CH_3COOClO_3]^{-1}. \tag{1}$$

We established that a decrease in the initiator concentrations led to an increase in the induction times.

2. The lateral growth of the crystals

Immediately after the nucleation period, which depends on the initiator concentration (see fig. 2), small hexagonal crystals with a diameter of ca. 0.75 μm

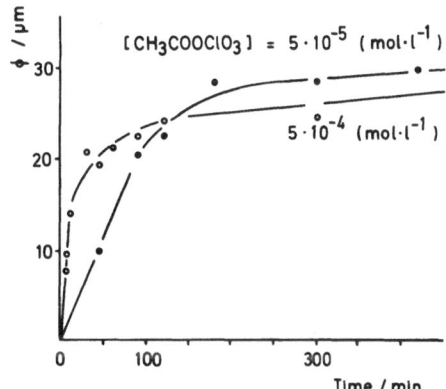

Fig. 3. Diameter of POM-crystals (ϕ) vs. polymerization times at two different initiator concentrations. (O): $[CH_3COOClO_3] = 5 \cdot 10^{-4}$ mol/l. (●): $[CH_3COOClO_3] = 5 \cdot 10^{-5}$ mol/l

are formed. These crystals grow rapidly (in diameter) in a lateral course with the increase of the polymerization time and conversion. One hundred crystals were measured, in order to obtain the average crystal diameter. These crystals were obtained at different polymerization times and concentrations of initiator, and were measured from edge to edge. Figure 3 shows the relation between polymerization times and diameter of the crystals at two different concentrations of acetyl perchlorate ($5 \cdot 10^{-5}$ mol/l and $5 \cdot 10^{-4}$ mol/l respectively). It can be observed that the rate of lateral crystals growth is higher with the increase of the initiator concentration. The crystals are larger at low concentrations of initiator due to a lower rate of polymerization and crystallization.

The crystals did not continue to grow in diameter after the maximum conversion had been reached. The results reported here are in slight disagreement with findings reported by Mateva et al. [6]. They have found that the crystals did not stop growing when the maximum conversion of 75% was reached. The crystal growth observed by them was explained as a special case of Ostwald ripening [25]. In this case, the solubility or vapor pressure over small particles depends on their size, and therefore smaller particles tend to redissolve in favor of further growth of already larger particles to minimize the free surface of the whole system. These authors performed essentially the same experiments as described by us, with the only difference that they used BF_3Et_2O and nitrobenzene as solvent.

We could observe that the rather small distribution in crystal sizes, mainly at the beginning of the reaction, indicated that the primary nucleation occurs as a bulk,

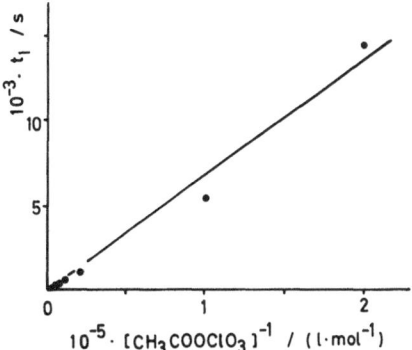

Fig. 2. Induction times (t_i) as a function of the reciprocal initiator concentrations

Fig. 4. Optical micrograph of hexagonal bipyramidal form of POM-crystals with small distribution in crystal sizes. $[CH_3COOClO_3] = 5 \cdot 10^{-5}$ mol/l; 1.5 h polymerization time; 0.53% conversion

simultaneously. Therefore the crystals have the same possibility to grow (see fig. 4).

The crystals show not only a thickness growth, but also a lateral growth at the [100] planes according to the morphological studies. We shall discuss separately both growth processes, since they depend on quite different chemical reactions.

The kinetic of the lateral growth assumes a $\sqrt{t}$ – law [9] given by

$$r = \sqrt{\frac{b_L\, v_t}{\pi}}\ \sqrt{t} \ . \tag{2}$$

Figure 5A shows a schematic respresentation of a poly(oxymethylene)-crystal indicating the meaning of b_L and v_t. The lateral growth of the hexagonal base lamella is due to the addition of monomer or oligomer to cationically active chains ends [1] (assuming chain folding), at translation steps, whose Burgers vector (b_L) is found normal to the [100] planes of the base lamella. Figure 6 indicates that the lateral crystal growth shows a linearity with the square root of the time, at least at the beginning of the polymerization. Corrections were applied by substracting the induction times to the polymerization times. This dependence was also found when $HClO_4$ as initiator was used [9].

3. Morphology of the crystals

The morphology of the crystal structure of nascent poly(oxymethylene), before reaching the equilibrium conditions of polymerization, is discussed on the basis of electron microscopy.

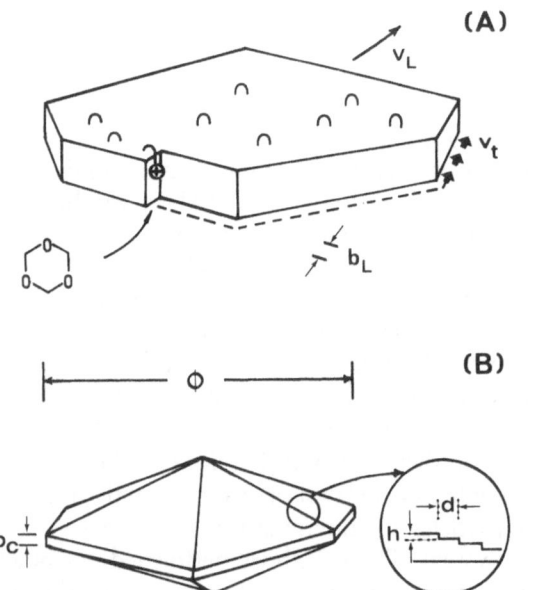

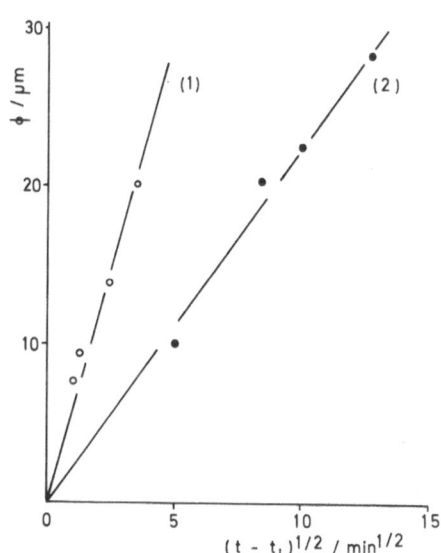

Fig. 5. (A) Schematic representation of a POM-crystal indicating the meaning of b_L and V_t in equation (2). b_L: Burgers-vector; v_t: tangential rate; v_L: rate of lateral growth. (B) Geometric model of a POM-crystal which grows on the basis of a screw dislocation to an ideal bipyramidal form. (Models as presented in literature [9])

Fig. 6. Variation of the diameter of the crystals (ϕ) with the square root of the polymerization times (difference between the polymerization times and induction times t_i), at two different initiator concentrations. (1): $[CH_3COOClO_3] = 5 \cdot 10^{-4}$ mol/l and (2): $[CH_3COOClO_3] = 5 \cdot 10^{-5}$ mol/l

Preliminary experiment carried out by us showed that the concentration of the initiator is a very relevant parameter which has a pronounced effect upon the morphology of the crystals.

The polymer crystals obtained in the course of polymerization possess the well known hexagonal structure of POM-crystals. The conditions of polymerization, in this case, allowed us to obtain single crystals of the polymer (figs. 7 and 8). Many of them show a smooth surface and they look like normal so-called polymer single crystals of POM, obtained by solution crystallization of high molecular-weight polymer, but they differ greatly in regard to morphology and physical properties. The melting point for the polymer with molecular weight of ca. 7.8×10^5 was 182 °C as determined by the peak maximum of the DSC-curve.

As conversion and time increase, these crystals also started to grow in thickness by development of growth spirals due to screw dislocation on the formerly smooth lamellar surfaces (fig. 9). The Burgers vector is found parallel to the plane [001], that means to be normal to the surface of the lamellar primary nucleous. We also observed growth spirals with regular rotation.

A twinned growth could occur at the center of the crystals when the polymerization time increases (fig. 10).

Higher polymerization times only produced an agglomerate of several crystals, not suitable for electron microscopy analysis.

It is remarkable that the thickness growth of the crystals is developed on the basis of only one or very few spiral growths in comparison to when BF$_3$-ether-

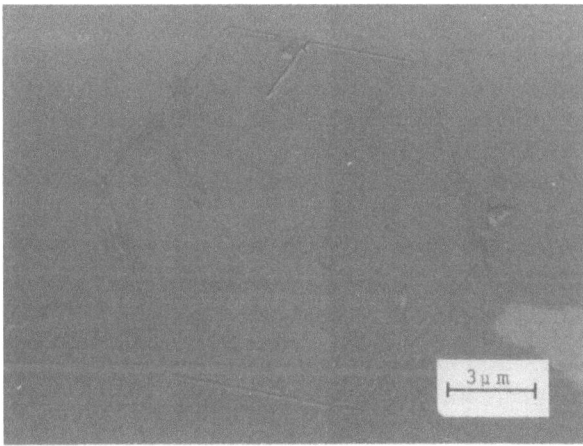

Fig. 8. POM-crystal with smooth surface. Note the incipient pyramid form and the regular spiral growth at the side of the crystal. $[CH_3COOClO_3] = 1 \cdot 10^{-5}$ mol/l; 24 h reaction time; 2.31 % conversion

ate was used, on the polymerization of trioxane [6, 8], in which a different growth process occurs.

The crystals finally grow, in the form of hexagonal bipyramids which can be well observed from the light microscope picture (see fig. 4). The pyramid's surfaces are formed on a hexagonal base lamella of b_c height. The height of the steps of the spiral growth and the

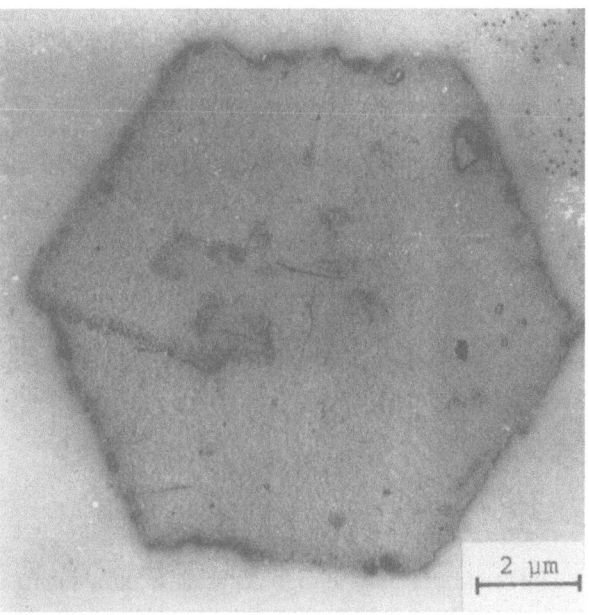

Fig. 9. Beginning of the thickness of POM-crystal via development of growth spiral due to screw dislocation on the formerly smooth lamellar surface. $[CH_3COOClO_3] = 5 \cdot 10^{-5}$ mol/l; 0.75 h polymerization time; 0.18 % conversion

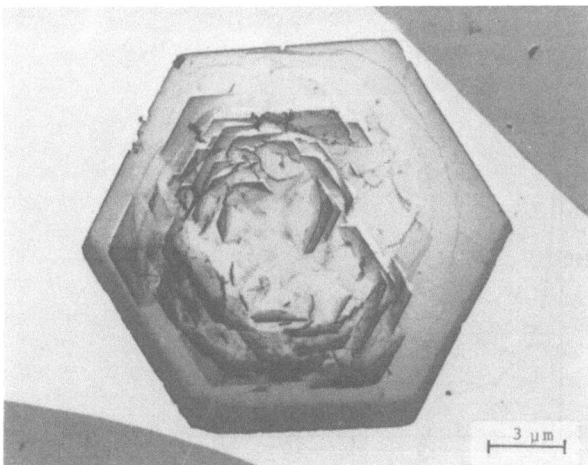

Fig. 7. Electron micrograph of a hexagonal POM-single crystal as obtained after 1.5 h reaction time. $[CH_3COOClO_3] = 5 \cdot 10^{-5}$ mol/l; 0.53 % conversion

Fig. 10. POM-crystal with irregular growth at the center, but with regular step width of the spiral growth at the periphery. $[CH_3COOClO_3] = 2.45 \cdot 10^{-4}$ mol/l; 0.58 h reaction time; 9.95% conversion

Fig. 11. Fracture surface of extended chain crystals of POM. $[CH_3COOClO_3] = 5 \cdot 10^{-5}$ mol/l; 3 h polymerization time; 11.01% conversion

width of the steps was designated by h and d respectively (compare fig. 5B). We think that the dimensions b_c and h at constant temperature does not depend on the concentration of the initiator, whereas the experimental results show that d depends strongly on the concentration of the initiator.

Figure 11 shows a micrograph of a fracture surface of an assembly of polymer crystals. The sample consists of lamellae with a rather large distribution in crystal thickness. Average crystal thickness is about 0.5 μm corresponding to a molecular weight of ca. 7.8×10^5, if one assumes that POM crystallizes as a 9/5 helix in a hexagonal unit cell [26, 27] with a period of 17.3 Å. Also, the fracture surface shows an extended-chain lamellae. The morphology of the polymer crystals thus obtained is very similar to the morphology of other extended-chain polymer crystals, e. g., the PE-crystals obtained by crystallization under high pressure [28, 29]. This extended-chain macroconformation is typical of a simultaneous polymerization and crystallization reaction.

4. The thickness growth of the crystals

The thickness growth of the crystals occurs due to a spiral growth as was shown on the electron microscopy pictures. The surface of the crystals at the beginning of the reaction shows regular steps of spiral growth, which is propagated from the center of the crystals.

The step width of the spiral growth, d, can be measured with high precision at the side of the crystals. d was determined in crystals obtained at about 1% conversion and to different concentration of the initiator. A reciprocal relationship between width of the steps d and concentration of $CH_3COOClO_3$ is shown in figure 12. In fact, the step width of the spirals increased with an increase of the polymerization times or the increase of the crystal surface, from the center to the side of the crystals.

The system studied shows that for small conversion, a large portion of the active initiator is fixed on the surface of the growing polymer crystals by adsorption and/or chemisorption, generating cationic active sites.

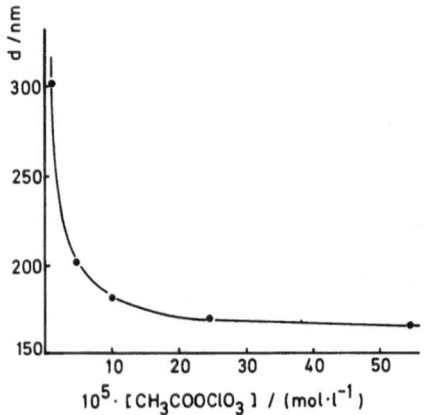

Fig. 12. Trend of the steps width (d) of the spiral growth of POM-crystals vs. initiator concentration

Fig. 13. POM-single crystal obtained by "sowing" to a trioxane solution, polymer crystals cationically actives

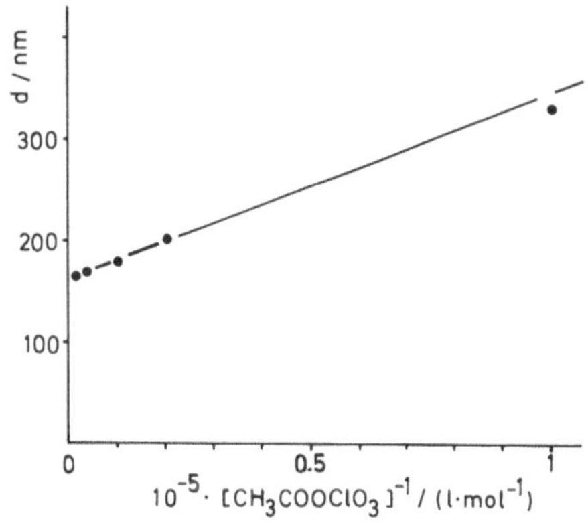

Fig. 14. Experimental verification of the equation (3) from the data of figure 12. (In equation (3): $d_\infty = 163$ nm and $b = 1.83 \cdot 10^{-3}$ nm mol/l)

This was demonstrated according to the following experiments:

a) a close system of trioxane solution with polymer already formed was filtered, and no further polymerization was detected in the filtrate until a new solution of the initiator was added again and

b) the polymerization was developed as usual when POM-crystals recently obtained, were added to a trioxane solution [30].

Figure 13 shows a polymer crystal obtained by "sowing" to a trioxane solution, POM-crystals cationically active. In this case, the polymerization reaction is carried out as normal, but without the presence of the initiator solution.

Immediately after the formation of the solid phase, an adsorption/desorption equilibrium occurred between the initiator that is in solution and the initiator adsorbed on the crystals surface.

The regular spiral growth of the crystals presents an interesting example of the theory of the crystal growth of Burton, Cabrera and Frank [10, 11].

A linear relationship between the step width of the spirals (d) and the inverse of the initiator concentration was found when $HClO_4$ as initiator was used [9]. This relation is also valid for acetyl perchlorate as initiator, according to the following equation:

$$d = d_\infty + b \, [CH_3COOClO_3]^{-1} \qquad (3)$$

in which the constant d_∞ is the intercept and b the slope of the straight line. This equation agrees with a Langmuir-type adsorption isotherm phenomenon of the initiator, and on the basis of BCF-theory [9]. The experimental results presented in figure 14 are in good agreement with this equation (3). The value of $d_\infty = 163$ nm indicates the minimum step width of the spirals growth at higher concentration of initiator of 5×10^{-4} mol/l. Several twinned growths are produced by a spontaneous nucleation on the crystal surface, whose morphology is not seen in detail by electron microscopy, with initiator concentrations higher than 5×10^{-4} mol/l. Therefore, the results discussed here are valid only for an initiator concentration $< 10^{-3}$ mol/l.

5. Chemical mechanisms of polymerization and crystallization

Several chemical mechanisms are available from literature which explain the lateral and thickness growth of POM-crystals, or the mechanism of polymerization of trioxane and crystals growth [1, 6, 9, 31, 32].

Before discussing the chemical mechanisms of polymerization and crystallization, it is necessary to analyze the nature of the initiator on the reaction media.

Acetyl perchlorate was prepared in situ by a metathetic reaction between equimolar quantities of acetyl chloride and silver perchlorate in dichloromethane as solvent. An absorption band at 2300 cm^{-1} obtained by IR-analysis indicated the existence of a cationic species corresponding to the oxocarbenium ion $CH_3\overset{+}{C}=O$. This cation is a resonance hybrid between the two canonical structures $CH_3-\overset{+}{C}=O$ and $CH_3-C\equiv\overset{+}{O}$ [33]. Also, a broad signal to $1775-1825$ cm^{-1} indicates the presence of the covalent ester form of molecular acetyl perchlorate. The existence of active polarised ester molecules (pseudocationic polymerization) was already recognised in 1966 [34]. We did not attempt a more exact characterization of the compound and we assumed that in CH_2Cl_2 solution existed partly ionized.

$$CH_3-\overset{\overset{O}{\|}}{C}-OClO_3 \rightleftharpoons CH_3-\overset{+}{C}=O \quad OClO_3^-.$$

$$(4)$$

We have not determined the residual concentration of water in the media used for polymerization, and thus cannot report anything about the degree of hydrolysis of the initiator. The degree of hydrolysis of the initiator could therefore have been quite high, which implies that in practice a mixture of acetyl perchlorate, acetic acid and perchloric acid are present in these polymerizing solutions [33]. We did not investigate the possible presence of free perchloric acid. It could well be that in fact these polymerizations are exact replicas of perchloric acid-trioxane ones [9]. All the features observed would agree entirely with such a possibility.

The induction times observed using $CH_3COOClO_3$ are approximately twice the quantity obtained when $HClO_4$ was used [9]. Thus, we can conclude that perchloric acid is more active than acetyl perchlorate. In both cases, the polymerization of trioxane was carried out in the same conditions. A possible explanation for such results was discussed by Gandini and Cheradame [33] for the system styrene-acetyl perchlorate and styrene-perchloric acid. Most probably, the real situation in these systems was that the initiator was a mixture of acetyl perchlorate and perchloric acid.

Again, the low initiator concentrations used in the presence of a possible wet medium cast serious doubts on the real nature of the initiator. It is disappointing to see that all the work so far carried out with acetyl perchlorate has failed to provide any real fundamental information about the way this catalyst operates.

The chemical mechanisms which are responsible for the POM-crystals growth, when perchloric acid as initiator was used, were particularly well conceived by Wegner et al. [9].

The discussion of the chemical elemental steps which are responsible for the crystal growth requests the assumption of a morphological model, that means,

the alternative, if a crystal surface contains a folded chain structure or chain ends corresponding to a macroconformation of fully extended-chain crystals.

The morphological studies, and the fact that the initiator is adsorbed on the crystal surface, together with the supposition that the real initiator is acetyl perchlorate since the polymerizations were carried out under severely dried conditions, indicate that other possible reaction mechanisms of polymerization and crystallization could be described.

Figure 15 shows that acetyl ester chain ends and perchlorate ester chain ends groups are found on the lamellar crystal surface. Folded chains are not present in this proposed mechanism. The crystal growth is carried out by insertion of a monomer between the labil perchlorate ester ends groups and the alkoxycarbenium ion of the POM-chain. This reaction could be considered as a pseudocationic polymerization. Also, it could play a role in the esterification reaction between the perchlorate ester end groups and the acetyl ester chain end groups. In this case, an inactive (CH_3-CO- end chain) site on the crystal surface is again activated and thus can be used in other crystal growth process. This reaction is favoured in a Kink position.

Figure 16 shows as alternative the participation of folded chains onto the thickness growth, either by an insertion reaction of perchlorate ester chain ends on a

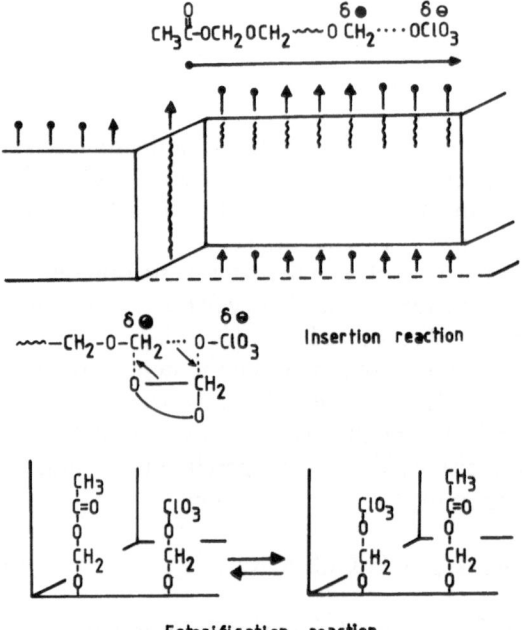

Fig. 15. Mechanism of the thickness growth. The crystal surface contained acetyl ester end groups and growing perchlorate ester end groups. Folded chains are not present

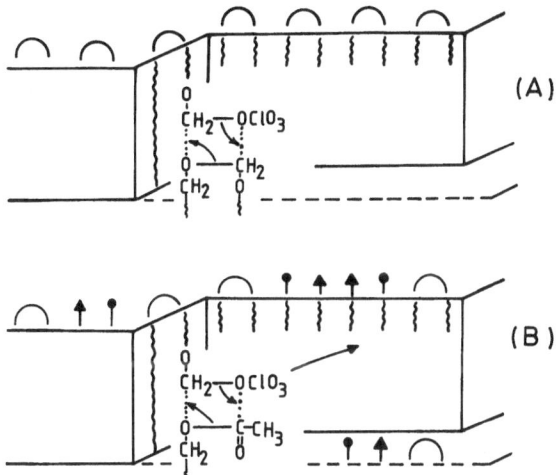

Fig. 16. Crystal growth by participation of folded chains: (A) insertion reaction, (B) liberation of the initiator. (Symbols as explained in fig. 15).

We found a kinetic behaviour entirely similar to that of the system trioxane-perchloric acid, and a striking resemblance of the phenomenology in both systems.

The above reported results emphasize the importance of the counterion $OClO_3^-$ of the initiator for the overall kinetics of trioxane polymerization. They also provide a pathway to the growth of extended-chain crystals. The mechanisms allowed to corroborate that polymerization and crystallization of the nascent polymer are not separated. However, they are interdependent stages of the same reaction, in which each chemical step of the polymerization is also a building step of the crystal.

Acknowledgement

The authors thank the "Dirección de Investigación" de la Universidad de Concepción (Chile) for financial support (Project N° 2.15.35).

folded chain or between perchlorate ester chain ends and acetyl ester end groups. In the latter case, on the basis of an esterification reaction between acetyl ester end groups and active chain ends, it free acetyl perchlorate is regenerated and simultaneously a single polymer chain is formed from these two different chain ends. This reaction is significant in relation to the discussion of the state of the initiator on the surface of the growing polymer crystals. The explanation of the growing phenomenon assumed that a portion of the initiator, on the basis of an adsorption isotherm, is reversibly bound to the crystal surfaces and from there it is released and another reaction site is available.

Polymerization and crystallization can also be carried out by free oxycarbenium ions of growing POM-chains and folded chains, by an intermolecular transacetalization reaction.

These mechanisms are in good agreement with the fact that extended-chain crystals were obtained. It was deduced that chain axes are normal to the surface and that the growth in thickness via growth spirals occurs regularly with respect to the geometry of the underlying mother lamella [35], from electron diffraction in those crystals.

Our results show that the reactions take place on the surface of single polymer crystals. They can also be characterized by the term simultaneous polymerization and crystallization whenever the reaction proceeds from the dissolved monomer directly to the crystalline polymer without forming intermediate, dissolved macromolecules [28].

References

1. Jaacks V (1969) Adv Chem Ser 91:371
2. Leese L, Baumber MW (1965) Polymer 6:269
3. Iguchi M, Kanetsuna H, Kawai T (1969) Makromol Chem 128:63
4. Iguchi M (1973) Br Polym J 5:195
5. Iguchi M, Murase I, Watanabe K (1974) Br Polym J 6:61
6. Mateva R, Wegner G, Lieser G (1973) J Polym Sci, Polym Lett Ed 11:369
7. Muñoz-Escalona A, Guerrero SJ (1976) Makromol Chem 177:2169
8. Wegner G, Fischer EW, Muñoz-Escalona A (1975) Makromol Chem Suppl 1:521
9. Wegner G, Rodríguez-Baeza M, Lücke A, Lieser G (1980) Makromol Chem 181:1763
10. Frank FC (1949) Discuss Faraday Soc 5:48
11. a) Burton WK, Cabrera N, Frank FC (1949) Nature, London 163:398
    b) Burton WK, Cabrera N, Frank FC (1951) Philos Trans R Soc London, Ser, A: 243:299
12. a) Wunderlich B (1968) Angew Chem 80:1009
    b) Wunderlich B (1968) Advan Polym Sci 5:568
13. Jaffe M, Wunderlich B (1967) Kolloid Z Z Polym 215-218:203
14. Masuda T, Higashimura T (1971) J Polym Sci, A-1, 9:1563
15. Masuda T, Higashimura T (1971) Polym Lett 9:783
16. Burton H, Praill PF (1953) J Chem Soc, 837
17. Kern W, Jaacks V (1967) Kolloid Z Z Polym 216-217:286
18. Weissermel K, Fischer E, Gutweiler K, Hermann HD, Cherdron H (1967) Angew Chem 79:512
19. Jaacks V (1967) Makromol Chem 101:33
20. Kern W, Cherdron H, Jaacks V (1961) Angew Chem 73:177
21. Okamura S, Higashimura T, Miki T (1972) Progr Polym Sci Japan 3:97
22. Iguchi M (1970) J Polym Sci, Part A-1, 8:1013
23. Boehlke K (1969) Dissertation, Mainz
24. Smirnov YuN, Volkov VP, Rozenberg BA, Yenikolopyan NS (1974) Vysokomol Soedin 16:283
25. Wegner C (1961) Ber Bunsenges Phys Chem 65:581

26. Uchida I, Tadokoro H (1967) J Polym Sci, (A-2) 5:63
27. Holdsworth PJ, Fischer EW (1974) Makromol Chem 175:2635
28. Wunderlich B (1973) Macromolecular Physics, Vol 1, Academic Press, New York
29. Wunderlich B, Melillo L (1968) Makromol Chem 118:250
30. Catalán S RE (1983) Ph D dissertation, Universidad de Concepción, Chile
31. Plesch PH, Westermann PH (1968) J Polym Sci C 16:3837
32. Mihailov M, Nedkov E, Terlemezyan L (1980) Polymer 21:66
33. Gandini A, Cheradame H (1980) Adv Polym Sci 34–35:211
34. Plesch PH (1966) Pure Appl Chem 12:117
35. Reneker DH, Geil PH (1960) J Appl Phys 31:1916

Received July 9, 1985;
accepted August 15, 1985

Authors' address:

Mario Rodríguez-Baeza
Universidad de Concepción,
Area de Polímeros,
Departamento de Química,
Facultad de Ciencias,
Casilla 3-C, Concepción, Chile

Progress in Colloid & Polymer Science   Progr Colloid & Polymer Sci 71:59–65 (1985)

Study on the copolymerization of phenyl ester-type monomers comprising mesogenic groups in the side chain*)**)

J. Horváth, F. Cser, and G. Hardy

Research Institute for Plastics, Budapest, Hungary

Abstract: Four copolymer systems from six acrylic monomers were prepared. The relative reactivity ratios in the copolymers were as follows:

for p-butoxyphenyl p'-acryloyloxybenzoate (BPAB) with p-phenyl p'-acryloyloxy ethoxybenzoate (PAEB), $r_{BPAB} = 0.39$ and $r_{PAEB} = 0.65$;

– for BPAB with p-butoxyphenyl p'-acryloyloxy ethoxybenzoate (BPAEB), $r_{BPAB} = 0.94$ and $r_{BPAEB} = 0.45$;

– for p-phenyl p'-acryloyloxybenzoate (PAB) with p-octyloxyphenyl p'-methacryloyloxy ethoxybenzoate (OPMEB), $r_{PAB} = 0.29$ and $r_{OPMEB} = 1.24$;

– for PAB with p-methoxyphenyl p'-acryloyloxy ethoxybenzoate (MPAEB), $r_{PAB} = 1.15$ and $r_{MPAEB} = 0.62$.

All of the copolymers are mesomorphic systems. Plots of their transition temperatures against the copolymer composition have a minimum for PAB/OPMEB and PAB/MPAEB systems and a break-point for BPAB/PAEB and BPAB/BPAEB systems.

Key words: Mesomorphic state of polymers, side chain mesomorphism, p-alkoxy-phenyl-benzoates, acrylate monomers, copolymerization.

Introduction

The synthesis and fundamental characteristics of nine monomeric phenyl esters of acryloyloxybenzoic acid and their polymers have been reported upon in a previous paper [1]. Based on the experimental results, two of these polymers were supposed to have a nematic structure. Results were obtained from contact preparates of the homopolymers with reference materials upon cooling. The complex investigation of miscibilities upon heating did not support our previous statements [2, 6].

Nematic and cholesteric states were unequivocally detected in copolymers [3] with different length of spacers in them. The spacers with different length force the hard mesogenic cores to be at a non uniform distance from the main chain which is necessary for the formation of the nematic state.

In order to study this effect, five of the nine aforementioned monomers were selected for preparation of four copolymer systems and characterization of regularities of their behaviour.

Pairs of comonomers were chosen by varying the length of spacer as well as of p-alkoxy chain.

Two of the copolymer systems comprised a common comonomer, p-butoxyphenyl p'-acryloyloxybenzoate (BPAB), copolymerized either with p-butoxyphenyl p'-acryloyloxy ethoxybenzoate (BPAEB) or with p-phenyl p'-acryloyloxy ethoxybenzoate (PAEB).

The homopolymer of BPAB was supposed [4, 5, 6] to have aperiodic helical structure while homopolymers of monomers comprising ethyloxy spacer show [2] a lamellar smectic structure.

The common monomer of the other two copolymer systems was p-phenyl p'acryloyloxybenzoate (PAB) copolymerized with p-methoxyphenyl p'-acryloyloxy ethoxybenzoate (MPAEB) or with p-octyloxyphenyl p'-methacryloyloxy ethoxybenzoate (OPMEB) having a long aliphatic side chain and a methacrylate main chain.

---

*) Dedicated to Prof. Dr. H.-G. Kilian on the occasion of his 60th birthday.

**) Polymerization in Liquid Crystal XX. Part XIX: J. Horváth, F. Cser, and G. Hardy: Acta Chim. Acad. Sci. Hung., in press.

Table 1. Characteristics of the monomers and homopolymers

$$CH_2 = C - COO - X\bigcirc - COO -\bigcirc - R$$
$$\quad | \quad$$
$$\quad Y$$

Marking	Y	R	x	Characteristics of monomers[a] (1)	Characteristics of polymers			
					$T_g/°C$	$T_f/°C$	$T_{lc}/°C$	$T_i/°C$
PAB	H	H	–	cr 77 i	80	180	–	228
BPAB	H	$-O(CH_2)_4H$	–	cr 84 lc 102 i	120	180	–	330
PAEB	H	H	$-(CH_2)_2O-$	cr 84 i	10	45	–	72
MPAEB	H	$-OCH_3$	$-(CH_2)_2O-$	cr 91 i	25	55	–	110
BPAEB	H	$-O(CH_2)_4H$	$-(CH_2)_2O-$	cr 81 i	30	64	132	154
OPMEB	CH_3	$-O(CH_2)_8H$	$-(CH_2)_2O-$	cr 54–58 i	90	109	151	192

[a]) cr = crystalline; lc = liquid crystalline; i = istoropic; T_g = glass transition temperature; T_f = fusion temperature; T_{lc} = liq. cryst. phase transition temperature; T_i = clearing temperature

An aperiodic helical structure was again supposed for the homopolymer of PAB [4, 5, 6] while these ethoxy-containing monomers also form a lamellar smectic homopolymer [2].

Experimental

Synthesis of monomers

The synthesis and characterization of the present monomers have been detailed in a previous paper [1]. Preparation of the lately introduced p-octyloxyphenyl p′-methacryloyloxy ethoxybenzoate (OPMEB) is reported as follows.

First of all, p-octyloxyphenol was prepared by condensation of hydroquinone with octyl bromide [7]. The second step of the synthesis lead to 4-/6-hydroxy ethoxy/benzoic acid as described previously [1]. This product was then esterified with methacrylic acid under the same conditions as in the acrylic esterification [1]. In the next step, p-methacryloyloxy ethoxybenzoic acid was transformed into acyl chloride with thionyl chloride in the presence of dimethyl formamide just as the corresponding acryloyl compound [1]. Finally, the acyl chloride was reacted with p-octyloxyphenol. The crude end-product was recrystallized from methanol three times. C, H microanalysis of the monomer (m.p. 54 to 58 °C) gave 71.2 % for C and 7.45 % for H (calculated 71.4 and 7.5, respectively) thus, purity of the produuct was about 99.7 %.

Synthesis and characterization of the homopolymers were detailed in the previous paper [1], the general formula and characteristics of the monomers and homopolymers are summarized in the table 1.

Copolymerization

Copolymerization was carried out in chloroform solution at 60 °C. Monomer mixture of 1.5 g was dissolved in 3 cm³ of chloroform. The reaction was initiated by 1 % of azobisisobutyronitrile (AIBN). Duration of the copolymerization reaction was adjusted on the basis of preliminary tests so that conversion was not higher than 30 %. Rate of copolymerization was highly dependent on the composition of the monomer mixture.

The copolymer was precipitated from the clear solution by methanol or ethanol. The precipitate was allowed to settle for 24 hours then it was filtered, washed several times by the precipitating agent, and dried up to constant mass.

Test methods

NMR spectroscopy:

The composition of the copolymers was determined by a 60-MHz Varion EM-360 instrument. Spectra were recorded in deuterochloroform solution and evaluated by using aromatic protons (δ = 7 to 8 ppm) as internal standards since their number is identical in all of the monomers. The extent of incorporation was determined in terms of protons in the ethoxy spacer (BPAB/BPAEB system) or the proton signal of the terminal methyl group of p-alkyl chain (BPAB/PAEB, PAB/OPMEB, and PAB/MPAEB systems). Relative reactivity ratios of comonomers were calculated by the Kelen-Tüdös-Turcsányi method [8].

Polarization microscopy

The anisotropy of copolymers was determined by a Zeiss-Polmi A polarization light microscope. The extent of birefringence was measured by a photometer coupled with the microscope. A Boetzius heated plate served for temperature control.

Thermal analysis (DSC):

Transition temperatures and the accompanying enthalpy changes of copolymers were determined by differential scanning calorimetry using Perkin-Elmer DSC 2 instrument at a heating rate of 10 °C/min.

Caloric evaluation of peaks was performed by calibration with high-purity ln standard.

Results and discussion

NMR spectra of the homopolymers and the copolymers from 1 : 1 monomer mixtures are shown in figures 1a and 1b. The reference bands of the internal standard are clearly separated from those related to the copolymer composition, so that the comonomer ratios can be calculated with relatively high accuracy (within ± 5 %).

Composition diagrams of copolymer systems containing BPAB monomer and those of PAB units are illustrated in figures 2 and 4, respectively.

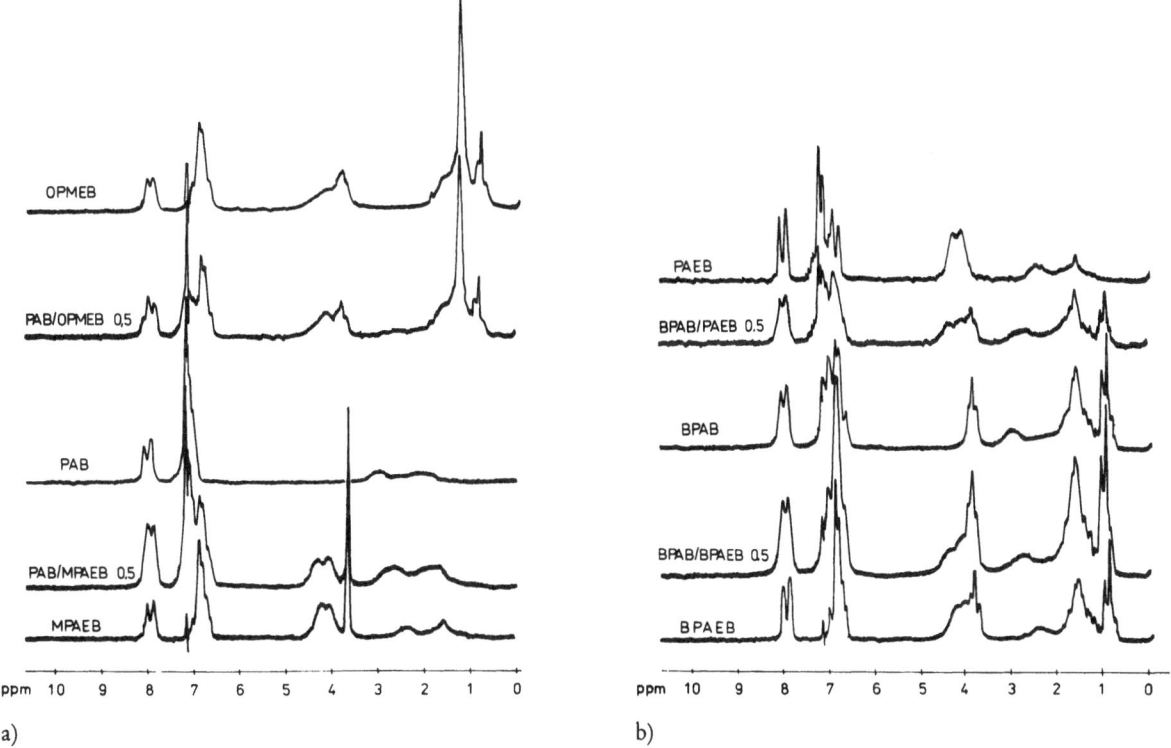

Fig. 1. NMR spectra of the homopolymers and of the copolymers from 1 : 1 monomer mixtures; a) NMR spectra of homopolymers and co-polymers of PAB; b) NMR spectra of homopolymers and copolymers of BPAB

The compostion diagram of the BPAB/PAEB system intersects the diagonal at $M_{BPAB} = 0.4$ to 0.5. Below the point of intersection, BPAB content of the copolymer is higher than that of the corresponding monomer mixture and crosswise above the intersection.

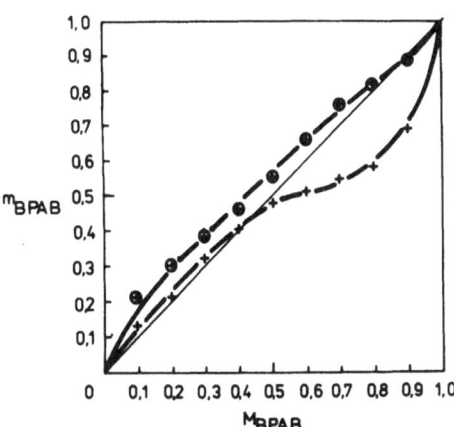

Fig. 2. Compostion diagram of BPAB/PAEB and BPAB/BPAEB copolymer systems; Notations: +: BPAB/PAEB system; ⊕: BPAB/BPAEB system

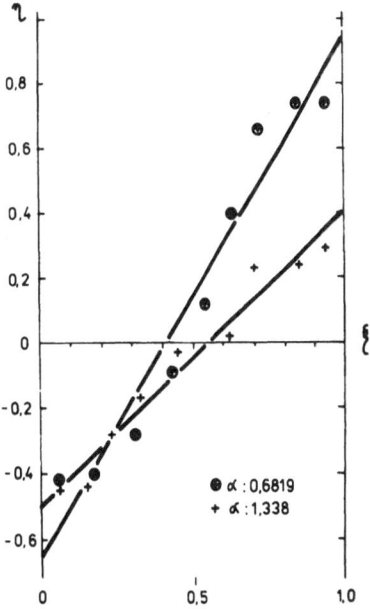

Fig. 3. Kelen-Tüdös-Turcsányi representation of the composition of BPAB/PAEB and BPAB/BPAEB co-polymers; Notations: +: BPAB/PAEB system; ⊕: BPAB/BPAEB system

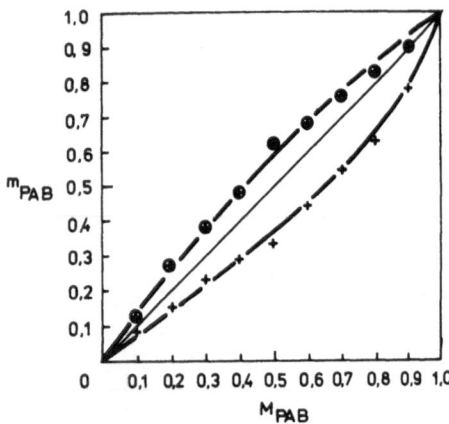

Fig. 4. Composition diagrams of PAB/OPMEB and PAB/MPAEB copolymer systems; Notations: +: PAB/OPMEB system; ⊕: PAB/MPAEB system

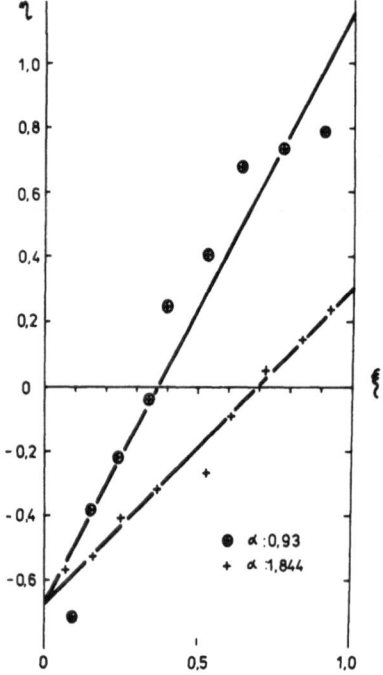

Fig. 5. Kelen-Tüdös-Turcsányi representation of the composition of PAB/OPMEB and PAB/MPAEB copolymers; Notations: +: PAB/OPMEB system; ⊕: PAB/MPAEB system

In the composition diagram of the BPAB/BPAEB system, this section point is at $M_{BPAB} = 0.8$ to 0.9 below which the incorporation of BPAB is relatively higher.

Kelen-Tüdös-Turcsányi diagrams for the copolymers of both systems are shown in figure 3. The experimental points fit the curve well in both cases.

Calculated reactivity ratios in the BPAB/PAEB system are as follows: $r_{BPAB} = 0.39$, $r_{PAEB} = 0.65$. The characteristic value of accuracy of curve fitting $\varrho^2 = 0.96$. As both r values are less than 1, i. e. both of the monomers react more readily with the other monomer, statistical copolymers tend to be formed. The Δe value from r_1 and r_2 values is 1.16 which is unreasonably high considering the similar polarity of the monomers. This discrepancy suggests a topochemical effect [9], inasmuch as surroundings of the molecule to be polymerized have a crucial effect on r_1 and r_2 i. e. on the copolymer structure thus, the formula of ideal radical reactivity ratios cannot be applied in the original form [9].

The reactivity ratios in the BPAB/BPAEB system are calculated as $r_{BPAB} = 0.94$ and $r_{BPAEB} = 0.45$ with a fitting accuracy of $\varrho^2 = 0.95$. A formation of statistical copolymer is expected again where BPAB is a more active monomer than its ethoxy analogue. Starting with a monomer mixture with a high concentration of BPAB monomer ($M_{BPAB} > 0.8$), composition of the copolymer is close to that of the initial monomer mixture.

Δe is 1.15 again as calculated from r values, which is still unreasonably high just as in the case of the BPAB/PAEB system.

Composition diagrams of copolymer system containing PAB units are presented in figure 4. Kelen-Tüdös-Turcsányi representations of copolymer compositions are shown in figure 5. Curve fitting is good in both cases. The diverse attitude of PAB to the two comonomers is clearly perceptible in the diagrams: incorporation of PAB is less into the PAB/OPMEB system in relation to its proportion in the initial monomer mixture while, in the PAB/MPAEB system, PAB is the more active comonomer.

The r values obtained by the linearized formula are as follows: $r_{PAB} = 0.29$ and $r_{OPMEB} = 1.24 / \varrho^2 = 0.98$ / for the PAB/OPMEB system; $r_{PAB} = 1.15$ and $r_{MPAEB} = 0.62 / \varrho^2 = 0.93$ / for the PAB/MPAEB system. The Δe value in the former case is 1.01, which is still too high as before, while that for the latter system is 0.58, being closer to the expected level concerning the theory of the ideal radical reactivity ratios. In terms of the r values, both copolymers are statistical.

It can be established by comparing the relative reactivity ratios of the chemically different monomers that incorporation of a spacer reduces the comonomer activities in the presence of the same p-substituent (cf. the BPAB/BPAEB system). On the other hand, if the spac-

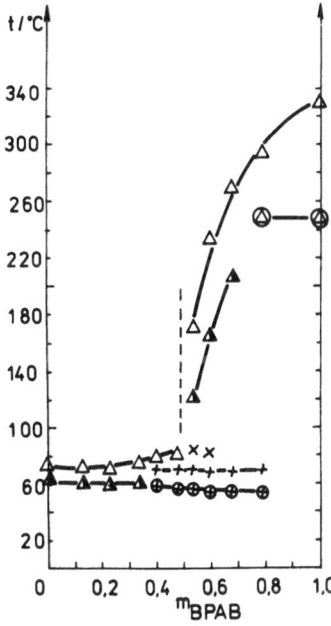
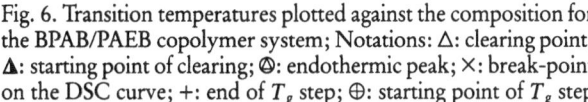
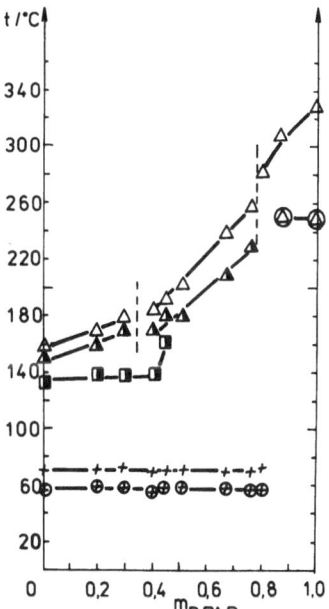

Fig. 6. Transition temperatures plotted against the composition for the BPAB/PAEB copolymer system; Notations: △: clearing point; ▲: starting point of clearing; ◎: endothermic peak; ×: break-point on the DSC curve; +: end of T_g step; ⊕: starting point of T_g step

Fig. 7. Transition temperatures plotted against the composition for the BPAB/BPAEB copolymer system; Notations: △: clearing point; ▲: starting point of clearing; ◎: endothermic peak; ▣: mesomorphic transition; +: end of T_g step; ⊕: starting point of T_g step

er chain is the same, lengthening the chain of the p-substituent also decreases the monomer activities in relation to an identical conomoner partner (cf. the BPAB/BPAEB and BPAB/PAEB systems).

Activities of methacrylic monomers are higher even in the presence of a spacer and a long p-alkyl substituent than those of the acrylics without a spacer or any p-alkyl substituent (cf. the PAB/OPMEB system) or from the other aspect, they are more active than an acryl comonomere with the same spacer but a much shorter p-alkyl chain (cf. the PAB/OPMEB and PAB/MPAEB systems).

Transition temperatures of the BPAB/PAEB copolymers are plotted against the copolymer composition in figure 6. Birefringence of the copolymers appears only at high BPAB contents, thus the melting point diagram was recorded on the basis of DSC curves. In the composition range of $m_{BPAB} \leq 0,5$, the feature of PAEB is dominant. Copolymers are optically isotropic even under shearing action. They appear to be isomorphic in this range, but more comprehensive investigations will be necessary to obtain further information on their structure.

In the other half of the diagram, for $m_{BPAB} > 0.5$, the phase state of copolymers undergoes abrupt changes, probably toward the helical structure of BPAB homo-

polymer. At compositions of $m_{BPAB} > 0.8$, DSC records indicate an additional endothermic signal before the melting peak referring to emergence of the helical structure. The formation of this configuration is assumed to originate from the helical blocks of BPAB linked together by more flexible copolymer chains. T_g of copolymers is detected as a separate step in the DSC trace only above BPAB content of 0.4 indicating that T_g at lower BPAB contents is close to the melting point of the polymers and the two transition processes are superimposed on each other.

A state diagram of the BPAB/BPAEB system is shown in figure 7. These copolymers are birefringent, regardless of the composition. The state diagram can be divided into three sections: at $m_{BPAB} < 0.4$, structure of the copolymers is similar to that of the BPAEB homopolymer, in fact, the mesophase of BPAEB is also present. At higher BPAB contents, $m_{BPAB} = 0.4$ to 0.8, this mesophase disappears and the structure of the copolymer is changed. On the basis of the available data, nothing further can be established about this structure, other than an enhanced order with increasing BPAB content, as indicated by the elevated melting enthalpy change. In the range of $m_{BPAB} > 0.8$, BPAB blocks appear, possibly tending to a structure like the high BPAB copolymers of the BPAB/PAEB system.

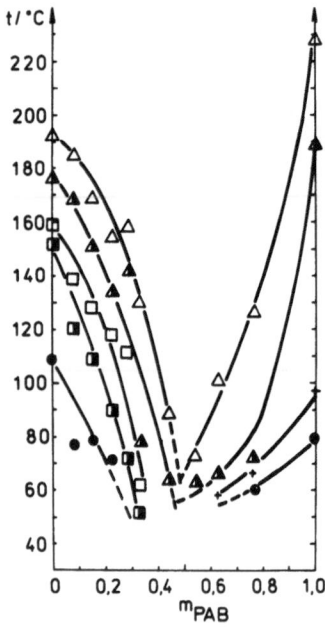

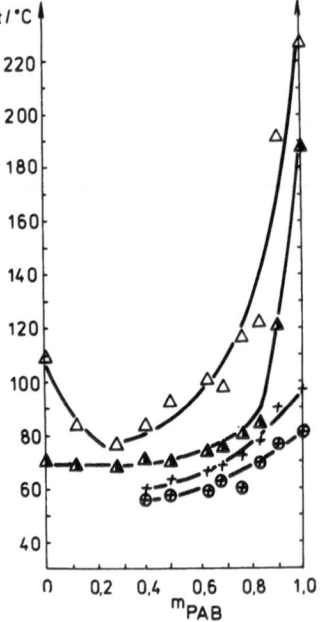

Fig. 8. Transition temperatures plotted against the composition for the PAB/OPMEB copolymer system; Notations: $\triangle$: end-point of clearing; $\blacktriangle$: starting point of clearing; $\square$: mesogenic transition; $\blacksquare$: starting point of mesogenic transition; $\bullet$: fusion point; +: end of T_g step; $\oplus$: starting point of T_g step

Fig. 9. Transition temperatures plotted against the composition for the PAB/MPAEB copolymer system; Notations: $\triangle$: end-point of clearing; $\blacktriangle$: starting point of clearing; +: end of T_g step; $\oplus$: starting point of T_g step

Melting point diagrams of PAB/OPMEB copolymers are presented in figure 8. At compositions of $m_{BPAB} \leq 0.4$, strong birefringence is observed by polarization microscopy. Before the clearing point, several transitions appear involving textural changes. Copolymers of higher PAB content but PAB homopolymer are not birefringent. Birefringence is, however, induced by shearing action. The temperature at which no birefringence was detected even under shearing stress was accepted as the clearing point. Because of the uncertainty of the polar optical observations, the state diagram was collected with respect to the transitions indicated in the DSC records. Transition temperatures have a minimum in the range of $m_{PAB} = 0.4$ to 0.5, referring to the boundary composition at a structural change where fundamental differences exist in the optical and caloric properties of the copolymers between the two sides of this composition. In the birefringent range, $m_{PAB} < 0.4$, the lamellar smectic structure characteristic of the OPMEB homopolymer is dominant (the sharp mesomorphic transition of these copolymers is similar to that of the OPMEB homopolymer). On the other side of the phase diagram, however, the rigid anisometric structure of the PAB homopolymer is attributed to the copolymers. Only the rigidity of the structure is reduced by the incorpo-

ration of the extraneous monomer as compared to that of the homopolymer.

A state diagram of the PAB/MPAEB copolymer system is shown in figure 9. Polar optical studies revealed that birefringence of all the copolymers was quite weak, in contrast to the homopolymers, and it disappeared above 50–60 °C. At higher temperatures, birefringence was observed only under shearing action. Because of the indistinct polar optical measurements, the state diagram was collected on the basis of the transitions detected by DSC curves again. The clearing point in this diagram has a minimum at $m_{PAB} = 0.2$ to 0.3 referring to the structural switch in the copolymer system. T_g steps are observed only at $m_{PAB} \geq 0.38$ since $T_g''s$ of other copolymers are below the measuring limit of 55 °C.

Conclusions

It can be established from the copolymer studies that an increase in flexibility of the side chain of the monomeric unit reduces the probability of its incorporation into the polymeric chain if the functional group is the same. This is in accord with the earlier observations supporting the hypothesis of the topochemical character of the copolymerization reaction [9].

Since homopolymers of the copolymerization partners had diverse structures, no isomorphic copolymer formation was expected over the whole composition range. Though the experimental data actually suggest an only partly isomorphic structure, the three different types of copolymer structure in the BPAB/ BPAEB system is surprising. Identification of the structure in the middle range of concentration requires further investigations. All the other copolymers did not show a nematic state.

Acknowledgement

The authors are indebted to Gábor Garzó (Technical University of Budapest) for the DSC measurements, to Attilla Csehi (Chemical Worsk of Gedeon Richter, Ltd.) for the NMR spectra, and to Sándor Czibor and Klára Golarits, graduating chemical engineers, for their contribution in the course of their diploma work.

References

1. Horváth J, Nyitrai K, Cser F, Hardy G (1985) Eur Polym J 21:251
2. Horváth J, Cser F. Hardy G, Magy Kém Folyóirat, in press
3. Finkelmann H, Koldehoff J, Ringsdorf H (1978) Angew Chem 90:992
4. Hardy G, Cser F, Nyitrai K, Samay G, Kalló A (1980) J Crystal Growth 48:191
5. Cser F, Nyitrai K, Hardy G. Pócsik L, Tompa K (1981) Magy Kém Folyóirat 87:337
6. Horváth J, Cser F, Hardy G, Acta Chim Acad Sci Hung, in press
7. Klormann E, Gotyos LW, Sretnov VA (1932) J Am Chem Soc 54:298
8. Tüdös F, Kelen T (1975) J Macromol Sci A9:1
9. Hardy G, Cser F, Nyitrai K, Bartha É (1982) Ind Eng Chem Prod Res Dev 21:321

Received March 11, 1985;
accepted July 29, 1985

Authors' address:

Professor G. Hardy
Research Institute for Plastics
H-1950 Budapest
Hungária krt. 114

Progress in Colloid & Polymer Science   Progr Colloid & Polymer Sci 71:66–70 (1985)

Double-diffusive fluctuations and the $v^{3/4}$-law of proton spin-lattice relaxation in biopolymers*)

R. Kimmich and F. Winter

Nuclear Magnetic Resonance Section, University of Ulm, Ulm, F.R.G.

Abstract: Field-cycling and conventional proton relaxation spectroscopy in the frequency range $10^4 \leq v \leq 3 \cdot 10^8$ Hz have been applied to diverse lyophilized and D_2O-hydrated globular proteins and poly-L-alanine. An overall T_1-dispersion $T_1 \sim v^{0.75 \pm 0.05}$ has been found. This $v^{3/4}$-law is tentatively interpreted by a double-diffusive fluctuation mechanism.

Key words: Proteins, fluctuations, spin-lattice relaxation, field-cycling, frequency dependence.

1. Introduction

Some years ago, when we started to investigate the nuclear magnetic relaxation behaviour of solid biopolymers [1] by the aid of the field-cycling technique (e. g. [2]) we were of the opinion that the proton spin-lattice relaxation dispersion was shaped by any ill-defined distribution of correlation times according to the complicated molecular structures. We therefore concentrated our efforts first on the study of the shape of the $^{14}N^1H$- or $^2H^1H$-quadrupole dips [1,2] which promised to yield information specific for the backbone dynamics.

The results obtained with many different biological systems [1,3–5], however, led to quite unexpected conclusions. First, the width and shape of the $^{14}N^1H$-quadrupole dips appear to be insensitive to the protein species, to the conformational properties and to the degree of hydration. Furthermore, the protons of solid proteins and homopolypeptides (lyophilized or weakly hydrated with D_2) show a uniform slope of the T_1-dispersion outside the dips. As a power law common to several proteins we find for the overall relaxation dispersion [4,5]

$$T_1 \sim v^{0.75 \pm 0.05} \tag{1}$$

where v is the proton Larmor frequency. This proportionality represents the experimental data outside the dip regions over up to more than four decades. The exponent is the average of the values found by least square fits to the data sets of figures 1 and 2. The limits are standard deviations.

The subject of this paper, relation 1, is a T_1-dispersion quite exceptional over such a wide range. To our knowledge there is no comparable result with other polymers. The generality with which this power law has been found even with a homopolypeptide contradicts any distribution of correlation times, so that we are probably confronted with an intrinsic property of biopolymeric backbones. A distribution of side group motions is expected to influence the T_1-dispersion merely at the highest frequencies. Frequently assumed theoretical models yielding special frequency dependences deviating from the standard BPP theory [6] are defect diffusion [7], translational diffusion [8], order fluctuations [9] and reptation in its modified form [10]. Exponents 0, 1/2, 3/2 or 2 appear. Most of these in a

---

*) Dedicated to Prof. Dr. H.-G. Kilian on the occasion of his 60th birthday.

more or less extended range and, hence, more or less conclusively have been verified in experiment.

With the reptation model in the version originally suggested by de Gennes [11], however, a $\nu^{3/4}$-power law fitting to relation 1 has been predicted. While this relation actually failed to describe the T_1-dispersion of synthetic polymer melts [10,12], the question arises whether it could account for the biopolymer data.

Clearly, reptation *per se* as a translational chain diffusion mechansim does not make any sense in context with biopolymers having well-defined secondary and tertiary structures. The essence of the $T_1 \sim \nu^{3/4}$ law, i. e. the special type of fluctuations leading to it, however, might provide the key for the derivation of models suitable for biopolymers. This is the aim of the present paper.

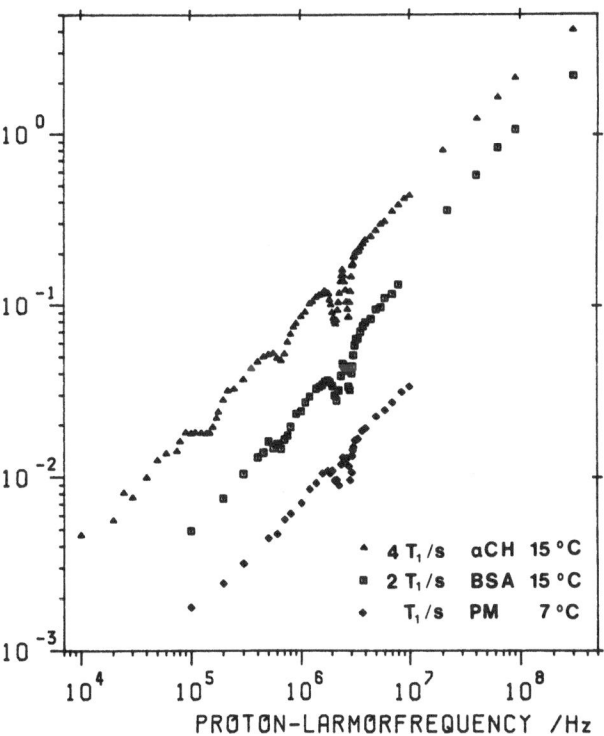

Fig. 2. Frequency dependence of the proton spin-lattice relaxation time of several deuterium exchanged and D_2O-hydrated proteins (see legend of fig. 1). In the case of α-chymotrypsin a $^2H^1H$ quadrupole dip has been resolved at 150 kHz in addition to the three $^{14}N^1H$ dips

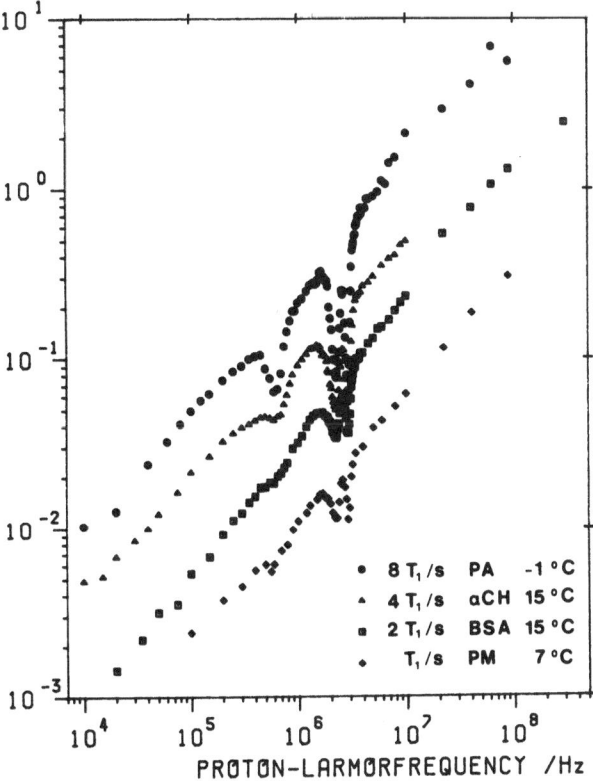

Fig. 1. Frequency dependence of the proton spin-lattice relaxation time of various lyophilized biopolymers (PA poly-L-alanine, α-CH α-chymotrypsin, BSA bovine serum albumin, PM purple membrane). The dips indicate the $^{14}N^1H$ resonance crossing [1]. Only the overall T_1-dispersion is considered in this paper. The data have been shifted by the indicated factors in order to avoid excessive overlapping. The polyalanine data refer to the backbone protons only. The relaxation rate of the PA methyl protons has been subtracted after analyzing this contribution from raw data in the range up to 300 MHz. Details of the procedure will be published elsewhere

2. Experimental

The data have been recorded mainly with a home-built field-cycling relaxation spectrometer. A detailed description of the improved version which is now in use will be published elsewhere. The high-frequency data above 20 MHz have been recorded with conventional Bruker pulse spectrometers using the saturation recovery technique.

Samples of α-chymotrypsin (from bovine pancreas; Fluka, Buchs, Switzerland), bovine serum albumin (electrophoretic purity 100%; Behring-Werke, Marburg/Lahn, F.R.G.), poly-L-alanine (molecular weight 10 000–25 000; Sigma, München, F.R.G.) and purple membranes (prepared in this lab [13]) have been investigated either in lyophilized form (> 1 day in high vacuum) or as deuterium exchanged and D_2O-hydrated complexes (several dissolving/lyophilizing cycles over a period in the order of weeks). The degree of hydration was 16%.

3. Formalism

Proton spin-lattice relaxation in biopolymers is normally governed by dipolar interaction to neighbouring nuclei. This coupling can be analyzed into certain transition operator terms multiplied by functions $F^{(j)}(\theta, \phi, r)$ of the spherical coordinates of the internuclear vectors [6, p. 289]. The fluctuations of these functions are characterized by their autocorrelation functions

$$G^{(j)}(t) = \langle F^{(j)}(\theta_o, \phi_o, r_o)\, F^{(j)*}(\theta_t, \phi_t, r_t)\rangle \qquad (2)$$

where the subscripts refer to the time at which the internuclear vector is regarded. With the Fourier transform of the functions equation (2), i.e. the intensity functions

$$I^{(j)}(\omega) = 2 \int\limits_o^\infty G^{(j)}(t)\, \cos \omega t\, dt, \qquad (3)$$

one obtains the spin-lattice relaxation rate

$$1/T_1 = C_1[I^{(1)}(\omega) + I^{(2)}(2\omega)] \qquad (4)$$

where C_1 is a constant. Thus the T_1-dispersion directly mirrors the type of fluctuation, provided that no distribution of correlation functions is valid.

The correlation functions are *average* quantities and, hence, are virtually determined by the probability W that the initial interaction state is still or again existent. Omitting the superscript we thus have

$$G(t) = C_2\, W(t) \qquad (5)$$

where C_2 is again a constant.

We now assume that the fluctuations of the internuclear vector r are generally governed by a random walk formalism [14]. The displacement of the internuclear vector in a time t is characterized by the mean square displacement $\langle(\Delta r)^2\rangle$. We thus can define an *ad hoc* diffusion coefficient D. The displacements Δr are considered to be of a general nature and we assume that they can arise both from translations and/or rotations. The probability that the actual displacement Δr has been reached in a time t is assumed to correspond to the solution of the conventional diffusion equation

$$P(\Delta r, t) = [2\pi \langle(\Delta r)^2\rangle]^{-x/2} \exp [-(\Delta r)^2/4Dt] \quad (6)$$

x is the "dimensionality" of the problem or the number of degrees of displacement freedom. Let us now assume that the fluctuations are restricted to a curvilinear displacement "path" on which the tip of the internuclear vector migrates. Δr is measured along this path and we put $x = 1$. Formally we can speak of displacement steps "forth" and "back" with reference to the initial state. The probability $W(t)$ is then

$$W(t) = P(o, t) = (2\pi \langle(\Delta r)^2\rangle)^{-1/2} \qquad (7)$$

so that

$$G(t) = C_3 \langle(\Delta r)^2\rangle^{-1/2}. \qquad (8)$$

If the random walk consists of discrete steps, we have the proportionality

$$\langle(\Delta r)^2\rangle \sim \Delta n(t) \qquad (9)$$

where Δn is the total number of steps in a time t. We now let Δn be dependent on another diffusion mechanism the nature of which will be discussed in connection with the subsequent models. If the number Δn is the result of the root mean square displacement of such a primary diffusive motion, we expect

$$\Delta n \sim t^{1/2}. \qquad (10)$$

(Compare the Monte Carlo simulation of reptative displacements in reference [15]).

From equations (8)–(10) we thus obtain

$$G(t) \sim t^{-1/4} \qquad (11)$$

which leads to

$$T_1^{-1} \sim I(\omega) \sim \omega^{-3/4}. \qquad (12)$$

If a further independent process such as, for instance, the global tumbling of the whole molecule is superimposed, the correlation function becomes a product of the functions of the two processes. Assuming an exponential decay [6] of the rotational contribution with a correlation time τ_c, we have

$$G(t) \sim t^{-1/4} \exp (-t/\tau_c) \qquad (13)$$

and

$$I(\omega) \sim \cos [3/4 \arctan (\omega\tau_c)]\, \tau_c^{3/4} (1 + \omega^2\tau_c^2)^{-3/8}. \quad (14)$$

In the limit $\omega\tau_c \ll 1$, relation 14 becomes

$$I(\omega) = \text{const}, \qquad (15)$$

for $\omega\tau_c \gg 1$ we have again

$$I(\omega) \sim \omega^{-3/4}. \qquad (16)$$

In the following two sections we will discuss two models as interpretations of the above formalism.

4. Diffusion in the conformation space (model I)

The dihedral angles ϕ_i and ψ_i of the peptide groups in a protein or polypeptide form a 2N-dimensional conformation space where N is the number of amino acid residues per molecule [16]. We now assume that there are numerous and almost equivalent minima of the conformational energy [17] so that the molecules can hop from equilibrium conformation to equilibrium conformation. Thus the molecules are assumed to experience a sequence of conformations while "diffusing" in the space formed by the angles ϕ_i and ψ_i. The jumplike transitions between the minima of the potential energy occur across the deepest saddle points of the hypersurface. All conformation changes are assumed to contribute to the *average* structure derived from X-ray analysis, of course.

Conformational jumps are connected with changes of the internuclear vectors. As translations are considered to be unlikely in a globular protein we restrict ourselves to the discussion of rotations. Every conformational jump is assumed to cause a rotational displacement of the internuclear vectors either connected with an *increase* of the rotational displacement or with a *decrease* so that there is only one degree of rotational freedom. Hence we can put $x = 1$ in equation (6). Actually we assume that the sequence of conformational states causes a *random* sequence of "forward" and "backward" rotations, i. e. a secondary diffusive process.

The crucial point is now that the conformational jumps and the rotational displacements are assumed to be connected so to say by a *stochastic* link: the mean square rotational displacement $\langle (\Delta r)^2 \rangle$ of a reference spin pair in general is not proportional to the mean square displacement $\langle (\Delta \zeta)^2 \rangle$ in the conformation space as one might expect at the first sight. Rather the rotational displacement increases with a time dependence reduced by a square root operation.

$\langle (\Delta r)^2 \rangle$ should thus be proportional to the *root* mean square "displacement" $\langle (\Delta \zeta)^2 \rangle^{1/2}$ reached in the conformation space in a time t. Since $\langle (\Delta \zeta)^2 \rangle^{1/2} \sim t^{1/2}$, we have the proportionality $\langle (\Delta r)^2 \rangle \sim t^{1/2}$ assumed in section 3, finally leading to the $v^{3/4}$-law 12.

5. Rotational reptation (model II)

There is another interpretation of the formalism of section 2. We have already excluded the relevance of *translational* reptation as suggested by de Gennes [11]. However, there could be a *rotational* analogue.

We assume torsional "defects" independently migrating along the peptide chain. Such torsions could even be identical with the "soliton"-solutions of nonlinear equations of motions occasionally suggested for biopolymers [18–20]. Torsions can be imagined to turn segments to the right or to the left depending on the direction from which they pass the segment. There is again only one degree of rotational freedom. The number Δn of torsions reaching the reference position in a time t is proportional to their root mean square displacement. Relation 10 is thus again valid. The net rotational displacement Δr of a segment then obeys equation (6) with $x = 1$. Thus the $v^{3/4}$-law also is compatible with this quite different notion.

6. Discussion

It has been shown that the $v^{3/4}$-law can be traced back to a double-diffusive fluctuation mechanism. Two tentative interpretations have been suggested. According to these models a certain conformational flexibility is required. There are other sources of information which lead to similar conclusions.

Hydrogen exchange in proteins has been studied extensively [21, 22]. The *solvent penetration* model (in contrast to the *local unfolding* model) is of special interest in the context of smaller amplitude motions. The main arguments for this model are the low apparent activation energies, the insensitivity to denaturing cosolvents and the observation of hydrogen exchange even in crystalline materials such as myoglobin. Models I and II are compatible with these findings. Low apparent activation energies also follow from the temperature dependence of our data [5].

On the other hand, though "diffusion in conformation space" may be the model of choice for globular proteins, it is hard to imagine it for purely α-helical poly-L-alanine in spite of the fluctuations reported in reference [23]. The "rotational reptation" model then appears to be more favourable.

The $v^{3/4}$-law in any case can be valid only over a certain frequency scale. Our observations concern the range $10^4 \leq v \leq 10^8$ Hz. This corresponds to a time scale $10^{-9} \leq t \leq 10^{-5}$ s. Thus the limitations must lie outside of these "windows". The high frequency limitation will be given by the elementary step of the mechanism discussed above. Hence one can conclude that the elementary steps occur in a scale faster than 10^{-9} s.

The opposite limitation will be given by the finite extension of the system. Relation 10 is valid as long as only a part of the states has been reached. After "equilibration" no new states can occur in model I. Rotational reptation of segments (model II), on the other hand, then corresponds to the diffusion of a Brownian particle. We conclude that the equilibration time obeys T_d $> 10^{-5}$ s.

In the frequency range $2\pi\nu < T_d^{-1}$, one would expect a plateau $T_1 = $ const for model I. (Compare the influence of a limitation reported in context with defect diffusion models [7]). The same behaviour is expected for a superimposed rotational tumbling (eq. (15)). Model II would lead to the proportionality $T_1 \sim \nu^{1/2}$ as has been found with polymer melts [10, 12].

Detailed investigations using isotopically labeled proteins are in progress in order to provide further arguments for this discussion. We hope that decisive conclusions concerning the general models will become possible in the near future.

Acknowledgements

The financial support of the Deutsche Forschungsgemeinschaft is gratefully acknowledged. Dr. Förster, Bruker Analytische Messtechnik, has kindly contributed the 300 MHz data. We thank Dr. K.-H. Spohn for providing the purple membrane sample.

References

1. Winter F, Kimmich R (1982) Mol Phys 45:33
2. Kimmich R (1980) Bull Magnetic Resonance 1:195
3. Kimmich R, Nusser W, Winter F (1984) Phys Med Biol 29:593
4. Winter F, Kimmich R (1985) Biophys J 48:331
5. Winter F (1985) Thesis, University of Ulm
6. Abragam A (1961) The Principles of Nuclear Magnetism, Oxford University Press, Oxford
7. Kimmich R, Voigt G (1978) Z Naturforsch 33a:1294
8. Held G, Noack F (1975) (eds) Allen PS, Andrew ER, Bates CA, Proc of the 18th Ampere Congress, North Holland, Amsterdam
9. Pincus P (1969) Solid State Comm 7:415
10. Kimmich R, Bachus R (1982) Coll & Polym Sci 260:911
11. de Gennes PG (1971) J Chem Phys 55:572
12. Kimmich R (1975) Polymer 16:851
13. Spohn KH, Kimmich R (1983) Biochem Biophys Res Comm 114:713
14. Chandrasekhar S (1943) Rev Mod Physics 15:1
15. Kimmich R, Doster W (1976) J Polym Sci, Polym Phys Ed 14:1671
16. Walton AG (1981) Polypeptides and Protein Structure, Elsevier, New York
17. Richards FM (1977) Ann Rev Biophys Bioeng 6:151
18. Davydov AS (1979) Int J Quant Chem 16:5
19. Englander SW, Kallenbach NR, Heeger AJ, Krumhansl JA, Litwin S (1980) Proc Natl Acad Sci USA 77:7222
20. Jardetzky O, King R (1983) Mobility and function in proteins and nucleic acids, Ciba Foundation Symposium 93, Pitman, London
21. Woodward CK, Hilton BD (1979) Ann Rev Biophys Bioeng 8:99
22. Englander SW, Kallenbach NR (1984) Quart Rev Biophys 16:521
23. Go M, Go N (1976) Biopolymers 15:1119

Received May 3, 1985;
accepted July 29, 1985

Authors' address:

R. Kimmich, F. Winter
Universität Ulm
Sektion Kernresonanzspektroskopie
Oberer Eselsberg
D-7900 Ulm, F.R.G.

Progress in Colloid & Polymer Science   Progr Colloid & Polymer Sci 71:71–76 (1985)

Distribution of correlation times in glassy polymers from pulsed deuteron NMR*)

C. Schmidt, K. J. Kuhn, and H. W. Spiess

Institut für Physikalische Chemie, Universität Mainz, and Max-Planck-Institut für Polymerforschung, Mainz, F.R.G.

Abstract: Pulsed deuteron NMR offers new possibilities for the determination of the distribution of correlation times for local motions in glassy polymers. The NMR line shape is a superposition of spectra corresponding to the different values of the correlation times. The weighting factors of these single spectra depend on the distribution function, which therefore can be characterized by a line shape analysis. Moreover, by combination with spin-lattice relaxation experiments the motional behaviour can be probed on a much longer time scale than by a line shape analysis alone. In this way homogeneous and heterogeneous distributions can clearly be distinguished. The method is explained in detail and is demonstrated by a simple example involving the methyl group rotation in glassy polycarbonate. In both polycarbonate itself and in mixtures with low molecular mass additives the results can be described by a log-Gaussian distribution of correlation times. This distribution is heterogeneous in nature, probably resulting from a distribution of the activation energies at spatially different sites between 15 and 21 kJ/mole.

Key words: Solid polymers, molecular motion, solid state NMR, polycarbonate.

1. Introduction

The correlation functions of dynamic processes in glassy polymers generally exhibit non-exponential decays. This is known from dielectric and mechanical relaxation studies as well as from photon correlation spectroscopy and NMR relaxation measurements [1–7]. It is customary to analyze experimental data in terms of a distribution of correlation times. Little is known, however, about the nature of this distribution. In particular, most techniques employed in this area do not allow to distinguish between a *heterogeneous* distribution, where spatially separated groups move with different time constants, and a *homogeneous* distribution, where each monomer unit shows essentially the same non-exponential relaxation. Even worse, relaxation processes resulting from different motional mechanisms often cannot be separated. Thus, if different motions have adjacent or even overlapping distributions of correlation times unselective experiments may easily be misinterpreted to indicate extremely broad distributions.

Pulsed ^{2}H-NMR offers new possibilities in this area. Its value in being able to discriminate different motional mechanisms and to monitor molecular motions via the reorientation of C–H bond directions over an extraordinary wide range of frequencies is well established by now [8,9]. Being a local method it is well suited for the detection of motional heterogeneities in semicrystalline [10] and amorphous polymers [11]. Here we present a first example for the direct determination of a distribution of correlation times by a ^{2}H-NMR line shape analysis. In the next section the NMR line shape calculation is considered theoretically, in section 3 a simple experimental example involving methyl group rotation in polycarbonate (PC) is presented. In section 4 we show how homogeneous and heterogeneous distributions can be distinguished by combination with spin-lattice relaxation experiments.

2. Calculation of NMR line shapes

Deuterons with a spin quantum number $I = 1$ are almost ideal labels for the detection of molecular reorientations in solids [8]. The ^{2}H-NMR spectra are dominated by the anisotropic quadrupole coupling

---

*) Dedicated to Prof. Dr. H.-G. Kilian on the occasion of his 60th birthday.

which is completely intramolecular in nature for covalently bonded molecules. The NMR frequency is given by

$$\omega = \omega_0 \pm \delta (3 \cos^2 \theta - 1 - \eta \sin^2\theta \cos2\phi)$$

$$= \omega_0 \pm \omega_Q \qquad (1)$$

where δ is 3/8 times the quadrupole coupling constant $e^2qQ/\hbar$, η is the asymmetry parameter ($0 \leq \eta \leq 1$), and the polar angles θ and ϕ specify the orientation of the magnetic field in the principal axes system of the electric field gradient tensor (FGT) [12]. The interpretation of spectra is simplified in the case of C-^{2}H bonds, in which the FGT is essentially axially symmetric about the bond direction ($\eta = 0$). Thus, deuteron NMR spectroscopy is particularly suited for the detection of both static distributions and dynamic reorientations of C-^{2}H bond directions in space.

For rigid solids ^{2}H powder spectra have a total width of about 250 kHz. Anisotropic motion leads to partially narrowed spectra with line shapes depending on the type and the time scale of the motion. The most prominent line shape changes occur when the correlation time τ_c of the motion is in the "intermediate exchange region", i. e. of the order of the inverse spectral width ($10^{-6} s \leq \tau_c \leq 10^{-5} s$ for ^{2}H). This region is extended to somewhat faster and somewhat slower motions by the solid echo technique [13]. The true absorption spectrum $g(\omega)$ is the Fourier transform of the free induction decay (FID) $\tilde{g}(t)$ following a single pulse. Due to the dead time of the receiver a direct detection of the full FID of solids is not possible. The solid echo spectrum $j(\omega, \tau_1)$, however, can be obtained by Fourier transformation of the solid echo $\tilde{j}(t, \tau_1)$, which is generated by two $\pi/2$ pulses separated by a time interval τ_1 and shifted in phase by $\pi/2$ [14]. For correlation times $\tau_c < 10^{-8} s$ ("rapid exchange limit") or $\tau_c > 10^{-3} s$ ("rigid solid limit") the line shape becomes independent of τ_c and τ_1, i. e. solid echo spectra and absorption spectra are essentially identical. In the range $T_2^* \leq \tau_c \leq (\delta^2 T_2^*)^{-1}$, where T_2^* is the transverse relaxation time, the intensity of the solid echo spectrum is largely reduced and its line shape differs in general from the true absorption spectrum [13]. This is illustrated in figure 1b and c for the methyl group performing jumps about its threefold axis with single correlation times τ_c.

If the motion cannot be characterized by a single time constant but has to be described by a distribution of correlation times, $P(\ln \tau_c)$, the total solid echo spec-

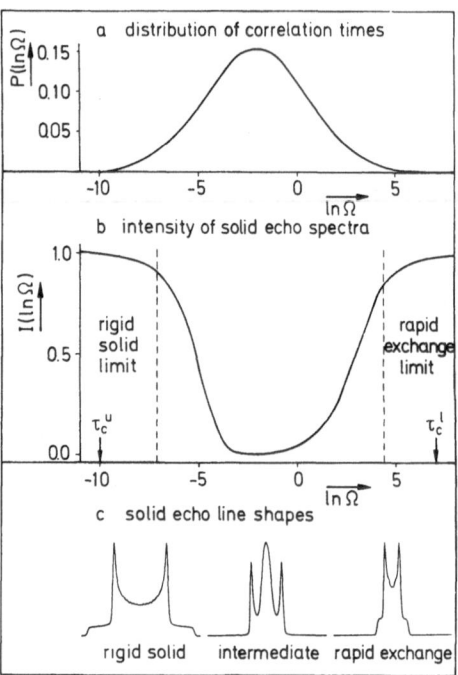

Fig. 1. Scheme for the calculation of spectra for a distribution of correlation times according to equation (2). a) As an example a log-Gaussian distribution with a width of 2.3 decades is plotted. b) Intensity of the solid echo spectra for single correlation times relative to the true absorption spectra, calculated for the three-site-jump of methyl groups; $\Omega = (3 \tau_c \Delta)^{-1}$ is the normalized jump frequency, with $\Delta = 4 \times 10^5 s^{-1}$ for ^{2}H. The curve for $\tau_1 = 80$ µs is shown. The correlation times τ_c^u and τ_c^l are the upper and lower limit of the region where a τ_c-dependence of the intensities and line shapes has to be taken into account (cf. eq. (3)). c) Some typical solid echo line shapes, calculated for the three-site-jump of methyl groups, $\tau_1 = 80$ µs, $\ln \Omega = -5, 1,$ and 6

trum $J(\omega, \tau_1)$ is a weighted superposition of the solid echo spectra $j(\omega, \tau_1, \tau_c)$ for well defined correlation times

$$j(\omega, \tau_1) = \int_{-\infty}^{\infty} j(\omega, \tau_1, \tau_c) P(\ln \tau_c) \, d(\ln \tau_c) \qquad (2)$$

For the numeric evaluation of equation (2) it is necessary to convert the integral into a sum of a finite number of spectra. This is eased by the fact that in the rigid solid limit and in the rapid exchange limit the line shape no longer depends on τ_c. Thus, for correlation times longer than an upper limit τ_c^u ($\tau_c > \tau_c^u$) or shorter than a lower limit τ_c^l ($\tau_c < \tau_c^l$) the line shape function $j(\omega, \tau_1, \tau_c)$ can be replaced by $j(\omega, \tau_1, \infty)$ and $j(\omega, \tau_1, 0)$, respectively, and can be taken out of the integral. As mentioned above, for deuterons typical values are $\tau_c^u =$

10^{-3} s and $\tau_c^l = 10^{-8}$ s. We are then left with the expression

$$J(\omega, \tau_1) \cong j(\omega, \tau_1, 0) \int_{-\infty}^{\ln\tau_c^l} P(\ln\tau_c)\, d(\ln\tau_c)$$

$$+ \sum_{k=1}^{n} j(\omega, \tau_1, \tau_c^{(k)}) \int_{\ln\tau_c^{(k)}}^{\ln\tau_c^{(k+1)}} P(\ln\tau_c)\, d(\ln\tau_c)$$

$$+ j(\omega, \tau_1, \infty) \int_{\ln\tau_c^u}^{\infty} P(\ln\tau_c)\, d(\ln\tau_c), \quad (3)$$

where the time interval $\tau_c^l \leq \tau_c \leq \tau_c^u$ is divided into n parts.

The weighting factors of the single spectra in equation (3) are evaluated by analytic or numeric integration. In general, the number n of the single spectra needs not be larger than 15.

Having calculated the spectra $j(\omega, \tau_1, \tau_c^{(k)})$ once [13, 15] the line shapes for arbitrary distributions of correlation times may be obtained easily. As an example figure 2a displays the simulated spectra of methyl groups for log-Gaussian distributions of correlation times covering 1, 2, 3, 4 and 5 orders of magnitude, respectively. Figure 2b illustrates how the width of the distribution affects the relative intensities corresponding to the spectra in figure 2a. As expected the line shape is most sensitive to the width of the distribution function if the mean correlation frequency Ω_0 is in the intermediate exchange region, i. e. if Ω_0 is comparable with the width of the NMR spectrum in the absence of motion. In addition, the *intensity* of the solid echo spectrum as a function of the pulse spacing τ_1 yields valuable information about the width of the distribution function, cf. figure 2b. The broader the distribution of correlation times the higher the intensity in the intermediate exchange region since the weighting factors for the single spectra of the high intensity regions increase with the width of the distribution (cf. fig. 1 and eq. (2)). On the other hand the relative intensities in the rigid solid limit and in the rapid exchange limit decrease with increasing width of the distribution. Indeed, these intensity effects were first observed in ^{2}H-NMR spectra of the non-crystalline regions of polyethylene [10] and were explained by a distribution of correlation times.

By analyzing line shapes and intensities symmetric distributions, such as the log-Gaussian distribution, and strongly asymmetric ones can clearly be distinguished. For not too broad distributions the width can be determined with high accuracy. Extremely broad distributions with widths exceeding 5 decades, however, may render an exact analysis difficult.

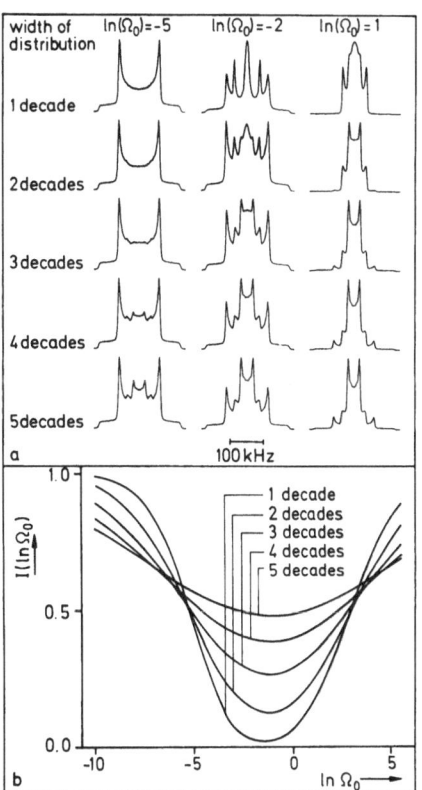

Fig. 2. Spectra (a) and intensities (b) for log-Gaussian distributions of correlation times with different widths, calculated for the three-site-jump of methyl groups, $\tau_1 = 80\ \mu$s. $\ln\Omega_0$ ($\Omega_0 = (3\tau_{c0}\Delta)^{-1}, \Delta = 4 \times 10^5\ \text{s}^{-1}$) is the centre of the distribution

3. Methyl motion in polycarbonate

The power of ^{2}H-NMR for analyzing the distribution of correlation times is demonstrated by the study of the motion of the methyl groups in PC of bisphenol-A.

Sample preparation

4,4'-Dihydroxy-diphenyl-2,2-(propane-d_6) was synthesized from phenol-d_4 and acetone-d_6 by the modified method of references 16 and 17 via 4,4'-dihydroxy-di-(2,6-dideuteron)-phenyl-2,2-(propane-d_6) to avoid exchange of the methyl deuterons during the reaction [18]. The phenyl deuterons were reexchanged in diethylether/HCl. By interfacial polycondensation with phosgene in the system NaOH (aq.)/dichloromethane and triethylamine as a catalyst [19, 20] we obtained a polycarbonate with a molecular weight $M_W = 20\,000$ as determined by GPC. The polymer was dissolved in dichloromethane and precipitated by the addition of petroleum or methanol. A

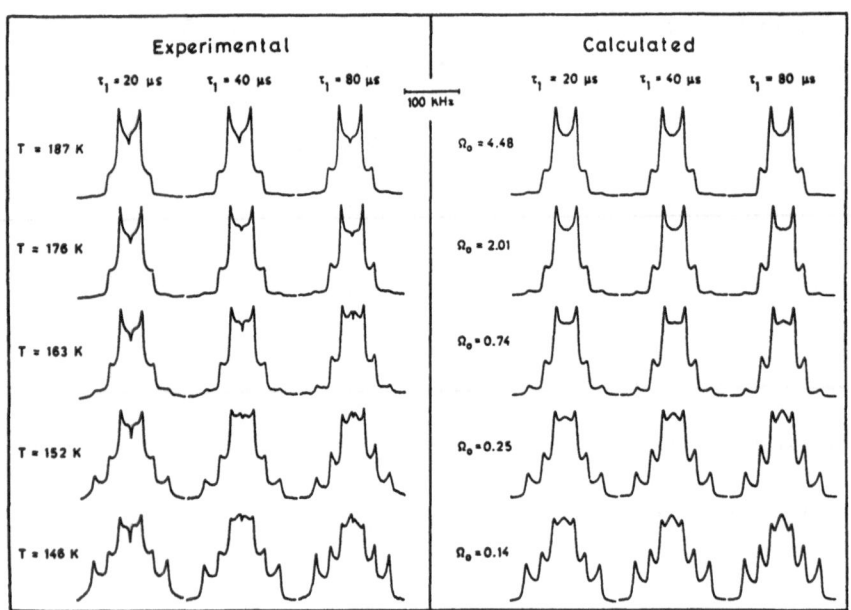

Fig. 3. Experimental and calculated methyl deuteron spectra of PC for different temperatures and different evolution times τ_1. For the definition of Ω_0 cf. figure 2. The width of the distribution of correlation times is 2.3 decades. The repetition time for data acquisition was 1 s so that the spectra at lower temperatures are only partially relaxed

dilute solution in CH_2Cl_2 was filtrated, concentrated to 30%, and cast to an amorphous film of 100 μm thickness. After solvent evaporation at room temperature for 10 h, the film was dried at 100 °C for 10 h. In order to remove paramagnetic impurities such as oxygen the sample was held at 10^{-5} mbar at room temperature for one day and sealed under 500 mbar N_2 in a NMR sample tube.

NMR experiments

Solid echo spectra of the polymer film were studied at temperatures down to 110 K. First changes in the line shape appear at about 190 K indicating a beginning of the freezing in of the methyl motion. At 110 K most of the methyl groups in the sample belong to the rigid solid part. The reason for the relatively high freezing in temperature of the methyl motion presumably is its intramolecular steric hindrance.

In figure 3 experimental and calculated spectra in the transition region are presented for three different values of the evolution time τ_1. First of all, our analysis proves that the methyl motion consists of threefold jumps about the C_3-axis, cf. reference [21]. Other motional mechanisms such as sixfold jumps or rotational diffusion about the C_3-axis can be ruled out, because they do not yield the observed increase in the central region of the spectra at lower temperatures and longer τ_1. The distribution of correlation times was determined from a quantitative line shape analysis by assuming different distribution functions and fitting

the calculated NMR line shapes to the experimental spectra. Symmetric and asymmetric distributions of various widths were considered. The best fit, presented in figure 3, was obtained for a log-Gaussian distribution 2.3 decades in width. All essential features of the experimental line shapes, which occur both with change of temperatures and τ_1, are reproduced remarkably well by this distribution. The relative intensities of the experimental spectra are in accordance with those calculated, too. In this particular case, a highly asymmetric distribution can be excluded. Thus, line shape analysis allows the determination of the distribution of correlation times for the methyl motion in PC in a rather direct way.

4. Motional heterogeneity

So far we have not explicitly distinguished homogeneous and heterogeneous distributions of correlation times because both would lead to the same NMR line shapes obtained via the solid echo, which acts like a snap-shot. In order to distinguish the different types of distributions, we have to monitor the motional behaviour over much longer times. This can be achieved by combining line shape studies with spin-lattice relaxation experiments, because the distribution of correlation times manifests itself not only in the line shape and intensity, but also in a non-exponential spin-lattice relaxation. Different mobility at different sites may lead to spin-lattice relaxation times T_1 differing by at least an order of magnitude, because spin diffusion [12]

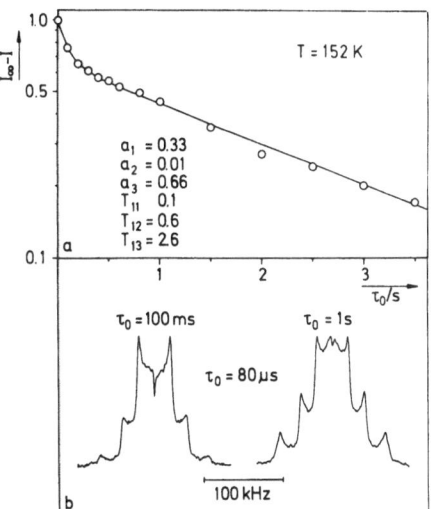

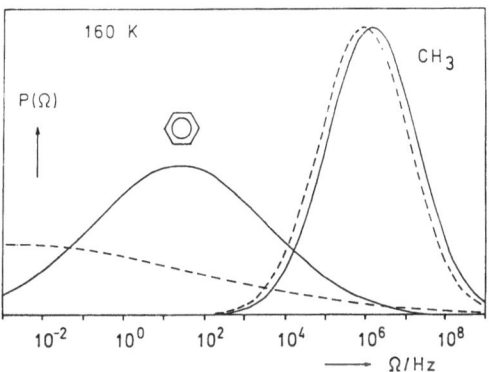

Fig. 4. a) Non-exponential T_1-relaxation. The intensity I is obtained as the height of the solid echo generated at time τ_0 after the end of a 5 × 90° saturation pulse sequence (10). I_∞ is the height of the fully relaxed echo ($\tau_0 \approx 5 \times T_1$). The line is a fit of the function $\sum_{i=1}^{3} a_i \exp(-\tau_0/T_{1i})$ to the experimental values (circles). b) Spectra for saturation pulse sequences with different waiting times τ_0. The τ_0-dependence of the line shape proves the heterogeneous nature of the distribution of correlation times

Fig. 5. Distributions of correlation frequencies $\Omega = 1/(3\tau_c)$ for the methyl and $\Omega = 1/(2\tau_c)$ for the phenylene groups in pure PC (solid lines) and in mixtures of PC with 25% polychlorinated biphenyls (dashed lines). For extremely small correlation frequencies $\Omega < 10^{-2}$ Hz details of the distribution function cannot be obtained from the experiments described here. The line shape analysis, however, yields the integral over the corresponding part of the distribution function, cf. equation (3). It was found to be equal to that resulting from the extrapolation of the log-Gaussian distributions shown to lower frequencies

is relatively inefficient for deuterons [22–24]. The fact that the line shapes of partially and fully relaxed spectra as shown in figure 4 differ markedly proves that the methyl groups retain their differences in mobility with correlation times between 10^{-7} and 10^{-5} s on a time scale of at least several seconds. In fact, the spectra of figure 3 used to determine the distribution of correlation times are partially relaxed and therefore, by suppression of the rigid solid Pake spectrum, yield more characteristic features of the line shapes.

These experiments clearly prove that the distribution of correlation times for the methyl rotation in glassy PC is *heterogeneous* in nature. This heterogeneity reflects the fact that the barrier for rotational motions must be different for spatially separated groups. The NMR experiments described here, however, do not yield any information about the distance between groups with different correlation times. We therefore propose as the most obvious explanation that the physical origin of the distribution of correlation times is due to differences in packing in the amorphous phase. Such a structural heterogeneity will lead to a distribution of the activation energies E_a, which results in a distribution of correlation times. Our line shape analysis suggests a distribution of E_a-values between 15 and 21 kJ/mole, with a mean value of $E_a^0 = 18 \pm 1$ kJ/mole.

Our current investigations of the interrelation between the macroscopic behaviour of polymers and their molecular dynamics suggest that motional heterogeneity strongly influences the mechanical properties [8]. It is of vital importance, however, to work selectively in this area, since not all motions will couple to mechanical deformations. Likewise, in mixtures of the polymer and low molecular mass additives which change the mechanical behaviour [25] typically not all of the molecular motions will be affected. This is demonstrated here for PC and mixtures of PC and polychlorinated biphenyls. In figure 5 we plot the distributions of correlation times for both methyl and phenylene [26] motions of PC and mixtures of PC and 25 weight % polychlorinated biphenyls. Clearly the additive, which strongly suppresses the mechanical relaxation, leaves the methyl mobility essentially unchanged but greatly hinders the phenylene mobility, increasing drastically the width of the distribution of correlation times. Our selective experiments thus clearly suggest a relation between the phenylene mobility and the mechanical properties in this system [25, 26].

Acknowledgements

We wish to thank Dr. G. P. Hellmann for helpful discussions on the interrelation between molecular motions and mechanical properties.

Financial support of the Deutsche Forschungsgemeinschaft (SFB 41 and SFB 213) is gratefully acknowledged.

References

1. Fuoss R (1941) J Am Chem Soc 63:185
2. Davidson DW, Cole RH (1951) J Chem Phys 19:1484
3. Connor TM (1964) Trans Faraday Soc 60:1572
4. Williams G, Watts DC (1970) Trans Faraday Soc 66:80
5. Lindsey CP, Patterson GD (1980) J Chem Phys 73:3348
6. Patterson GD (1983) Advan Polym Sci 48:124, Springer-Verlag Berlin, Heidelberg, New York
7. McBrierty VC, Douglas DC (1980) Phys Rep 63:61 and references therein
8. Spiess HW (1985) Advan Polym Sci 66:23
9. Jelinski LW (1985) (ed) Komoroski RA, "High Resolution NMR of Synthetic Polymers in Bulk", Verlag Chemie, Weinheim
10. Hentschel D, Sillescu H, Spiess HW (1984) Polymer 25:1078
11. Spiess HW (1983) Colloid Polym Sci 261:193
12. Abragam A (1961) The Principles of Nuclear Magnetism, Oxford University Press, Oxford
13. Spiess HW, Sillescu H (1980) J Magn Resonance 42:381
14. Powles JG, Strange JH (1963) Proc Phys Soc 82:6
15. Ebelhäuser R (1985) Dissertation, University of Mainz
16. Meyer KH, Schnell H (1958) D.B.P Nr. 1027205
17. Rauth (1978) diploma thesis, University of Mainz
18. Kuhn KJ (1981) diploma thesis, University of Mainz
19. Schnell H (1964) Chemistry of Polycarbonates, Polym Rev 9, Interscience Publishers, New York, p 31–40
20. Margotte (1980) Bayer Uerdingen, private communication
21. Torchia DA, Szabo A (1982) J Magn Resonance 49:107
22. Alla M, Eckmann R, Pines A (1980) Chem Phys Lett 71:148
23. Schajor W, Pislewski N, Zimmermann H, Haeberlen U (1980) Chem Phys Lett 76:409
24. Suter D, Ernst RR (1982) Phys Rev B 25:6038
25. Fischer EW, Hellmann GP, Spiess HW, Hörth FJ, Ecarius U, Wehrle M (1985) Makromol Chem Suppl 12:189
26. Wehrle M (1985) Dissertation, University of Mainz

Received May 2, 1985;
accepted June 22, 1985

Authors' address:

H. W. Spiess
Max-Planck-Institut für Polymerforschung
Postfach 3148
D-6500 Mainz, F.R.G.

Progress in Colloid & Polymer Science
Progr Colloid & Polymer Sci 71:77–85 (1985)

A reinvestigation of the hypersonic properties and the specific heat of PMMA around the quasi-static glass transition*)

I. High performance Brillouin investigations

J. K. Krüger, R. Roberts, H.-G. Unruh, K.-P. Frühauf, J. Helwig, and H. E. Müser

Fachbereich Physik, Universität des Saarlandes, Saarbrücken, F.R.G.

Abstract: The hypersonic behaviour of the longitudinal and transverse acoustic phonon frequency has been studied in the glass transition region of PMMA. An ultra slow acoustic relaxation process is reported which seems to be intimately connected to the main glass transition. The correlation with volume relaxation processes is discussed. The hypersonic frequency recovery process 20 K below T_g, reported in literature, could not be confirmed.

Key words: Polymethylmethacrylate, glass transition, Brillouin spectroscopy.

1. Introduction

The hypersonic properties of polymethylmethacrylate (PMMA) around its quasi-static glass transition temperature, T_g, have been widely discussed in literature [1–5]. The most essential findings are summarized as follows:

1. It is generally accepted that the glass transition temperature T_g is reflected by a kink in the longitudinal and transverse hypersonic velocity curves $v_l(T)$ and $v_t(T)$ respectively [2, 6–8].

2. For PMMA, however, the transverse phonons have so far only been reported at room temperature [9, 10].

3. The formation of the kink in the sound velocity curves is often attributed to the change in the volume expansion coefficient α at T_g [2, 11]. Brody et al. [12] have discussed the possibility of changed anharmonic interactions at T_g. The anharmonicity around T_g is described by a mode-Grüneisen parameter $\gamma_{p,q}$ (q: wave vector of the sound wave, $p = l, t$: polarization of the sound wave, l: longitudinal, t: transverse) which interconnects within a linear approximation the thermal volume expansion coefficient with the relative sound velocity gradient $\delta_{p,q}$:

$$\gamma_{p,q} = \delta_{p,q}/\alpha . \tag{1}$$

Brody et al. connect their own Brillouin data with expansion coefficients from literature and conclude that a decision about the behaviour of $\gamma_{p,q}$ at T_g can finally not be drawn because of the uncertainty of the volumetric and hypersonic data.

4. Patterson [13, 14] and Coakley et al. [18] have studied the influence of the sample history on the location of T_g. Patterson concluded from his Brillouin measurements on PMMA that he realized a T_g-depression of about 25 K compared with a DSC experiment (Perkin-Elmer DSC-II at 10 K/min, $T_g^{DSC} \approx 378$ K [13]) on the same material. Patterson said nothing about the molecular weight of his sample. We believe that he used a high molecular weight sample, as concluded from the following facts i) the glass temperature T_g^{DSC} he found is relatively high, ii) the author used a cast and polished block material which could be annealed at $T \approx T_g + 50$ K without losing mechanical stability. Patterson reported that a period of one month was the longest time he needed for reaching thermal equilibrium of his sample after a temperature step. He was able to detect the extrapolated equilibrium sound frequency of the liquid state after annealing the sample for one month at $T \approx T_g^{DSC} - 25$ K. In the same paper

---

*) Dedicated to Prof. Dr. H.-G. Kilian on the occasion of his 60th birthday.

he reported a similar depression of T_g for polystyrene (PS) of two different molecular weights.

Coakley et al. [18] have also studied the long-time sound velocity behaviour of PS around T_g but they could not confirm the results of Patterson. They mentioned the possibility that the differences between their own findings and those of Patterson could be due to the fact that samples of different molecular weights have been used.

In this paper (part I) we present high performance Brillouin investigations within the glass transition region of PMMA[1]. The aim of this paper is 1. to elucidate the question to what extent the quasi-static glass transition can be shifted to lower temperatures by annealing the sample in the vicinity of T_g, 2. to report the isothermal relaxation behaviour of the adiabatic elastic stiffness constant c_{11} using extremely long measuring times (adequate to the effective relaxation process, 3. to compare the sound velocity response as a function of time with the volume recovery, 4. to discuss extrinsic and intrinsic influences on the width of the glass transition interval, 5. to present the sound frequency behaviour of the transverse phonon around the glass transition.

In Part II of this paper [19] we report long-time adiabatic calorimetric investigations $\left(\dot{T} = \dfrac{dT}{dt} \approx 0.01 \text{ K/min} \right)$ on the same sample with the same history using a recently developed automatic adiabatic calorimeter [28]. The results will be discussed in the context of those from Brillouin spectroscopy.

We have also investigated the behaviour of the mode-Grüneisen parameters near T_g. Using the 90 A and 180 N scattering geometries [15] simultaneously, we have found strong evidence for a discontinuity in $\gamma_l(T)$ at T_g: $\gamma_l (T < T_g) = 4$, $\gamma_l (T > T_g) = 2.4$. These results will be discussed in more detail in a subsequent paper [16].

2. Experimental

The Brillouin investigations were performed with a completely automatized high performance 5-pass Brillouin spectrometer [17]. The free spectral range (FSR) of the Fabry-Pérot interferometer was calibrated using a new calibration technique leading to an increased accuracy (error of the mirror distance: $\Delta d = \pm 1 \, \mu\text{m}$) avoiding any sound frequency standard [16]. The investigations of the sound frequency responses as a function of time (figs. 5–12) were performed at a FSR of 8 GHz. Any hypersonic frequency was determined using a non-linear least squares fit procedure to fit a Gauss function

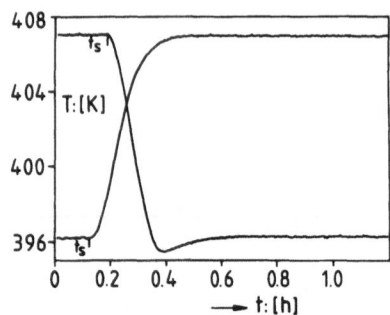

Fig. 1. Temperature recovery of the reference sample on cooling and heating after a temperature step of 10 K. t_s is the set-time of a new temperature set

to the measured 'phonon line'. To avoid laser damage of the sample the power of the laser beam passing trough the sample did not exceed 40 mW. A Pockels cell was used to decrease the laser light on the sample during the passage of the interferometer through non interesting parts of the spectrum. In between two measurements a shutter was used to keep the laser light away from the sample. The secondary beam of the Pockels cell was used in connection with an additional scatterer to maintain the necessary Rayleigh light which is needed for the long-time stabilization of the interferometer. The long-time measurements were performed in the 90 N scattering geometry [15]. The sample was a cast and polished block of $12 \times 5 \times 4 \text{ mm}^3$. The thermal contact between the sample and the sample holder was made by Apiezon H[2] to avoid any mechanical influence of the sample holder on the sample during the experiment. It should be mentioned that the glass forming process is strongly influenced by mechanical constraints. Glass slides which were glued above T_g through pure adhesion on the surface of the sample changed the glass transformation behaviour significantly [8, 16]. During the experiment the sample was kept in dry nitrogen. The temperature of the sample was measured with a Chromel-Alumel thermocouple placed in the center of a reference sample of the same material and size. Figure 1 shows the temperature response in the sample with time after a change of temperature setting of $\Delta T \approx \pm 10$ K. Recovery of temperature equilibrium is obtained after a thermal stabilization time τ_T of ≈ 25 min. The intrinsic temperature recovery time of our sample after a temperature jump of 10 K has been calculated to be $\tau_i \approx 100$ s [23] for temperatures around the glass transition. The thermal stabilization time τ_T of ≈ 25 min is therefore predominantly determined by the thermal slowness of the thermostat. Before starting any measurement the samples were annealed and dried under vacuum at 400 K for at least 4 weeks.

The Brillouin measurements on PMMA/PVDF[3] (80/20) were performed in the 90 A-scattering geometry [15] relating the hypersonic frequencies to a fixed sound wavelength $\Lambda^{90\text{A}} = 363.81$ nm.

A spectrum of a transverse acoustic phonon of PMMA at room temperature (295 K) is shown in figure 2. Although the measurement was performed realizing the VH-scattering situation residual 'longitudinal phonon lines' were observed in the spectrum. The phonon frequency f_t is in fairly good agreement with that reported by Lindsay et al. [10]. The small difference of $\approx 5\%$ between the sound frequencies is probably due to different sample materials, but the Poisson ration ($\nu = 0.328$) is in excellent agreement.

---

[1] PMMA-233, kindly provided by Röhm GmbH, $M_w \approx 4 \cdot 10^6$ g/mol, residual monomers less than 0.8 %.

[2] Apiezon Products Limited, London.
[3] PVDF: Polyvinylidene fluoride.

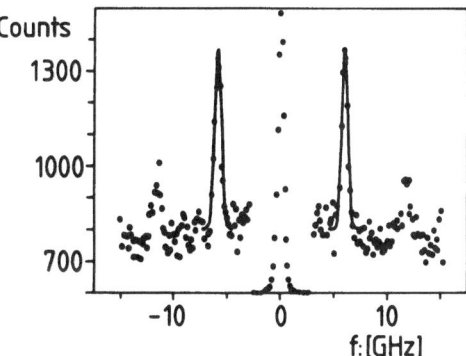

Fig. 2. VH-Brillouin spectrum of PMMA at room temperature showing the transverse phonon lines at $\pm$ 5.9 GHz and the residual longitudinal phonon lines at $\pm$ 11.6 GHz

3. Results and discussion

Near T_g, Brillouin spectroscopy measures the quasi-static elastic properties and their changes with temperature in contrast to investigations performed at higher temperatures in the hypersonic relaxation region [14]. The extrapolation of a quasi-static glass transition temperature T_g from hypersonic measurements needs comment: There is experimental evidence that, far from the glass transition region, the sound frequency curves can in general be described as linear functions of temperature within temperature regions of several 10 K [2–4, 8, 16]. Linear extrapolations of these linear functions to their intersecting point result in a temperature denoted T_g. The so defined quantity T_g is first of all an operational quantity whereby the physical meaning of this quantity is still a matter of discussion. Within the glass transition region the sound frequency curves are more or less bent and the temperature interval in which the transformation takes place is more or less extended. The width of the transformation interval is influenced, for example, by the cooling or heating rate during the experiment and by mechanical constraints imposed on the sample. An infinite long PMMA rod with a diameter of 10 mm which is cooled down with a constant rate of $T = 5$ K/ min shows around T_g a temperature difference of more than 5 K between center and surface. Therefore, experimental methods using very small volumes as information volumes might be advantageous compared with methods using the complete sample volume to obtain the desired information. Depending on the sample size, experiments with constant cooling rates can produce very complicated situations around $T_g(\dot{T})$: the case may occur that the sample surface is

already in the glassy state imposing mechanical constraints on its fluid center.

Apart from these extrinsic influences the width of the transformation interval depends on the material itself. As will be discussed below PMMA shows a rather large transformation interval of $\approx$ 15 K even though possible extrinsic influences were minimized. Melt mixing of 80 wt% PMMA with 20 wt% PVDF (polyvinylidene fluoride) leads to a PMMA/PVDF blend with a strongly reduced glass transition interval; figure 3 a shows the hypersonic frequencies of the longitudinal acoustic phonon as a function of temperature. The measured data are excellently described by two intersecting straight lines

$$f(T) = a - bT - c \, |T - T_g| \, . \tag{2}$$

The fit parameters are given in the figure caption of figure 3 a.

The standard deviation σ of the data is only 0.004 GHz. From the small standard deviations of the data

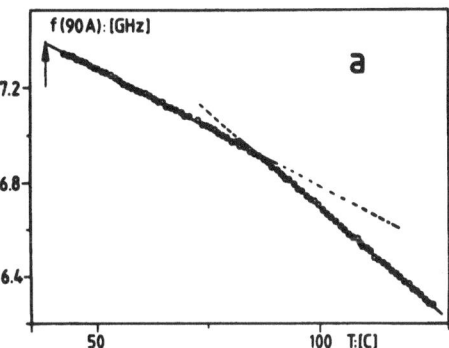

Fig. 3. a) Longitudinal sound frequency f of a PVDF/PMMA (20 wt%/ 80 wt%) in the glass transition region. The fit function is given by equation (2). The fit parameters are: $a = (8.037 \pm 0.003)$ GHz, $b = (1308 \pm 3.5) \, 10^{-5}$ GHz/K, $c = (324 \pm 3.5) \, 10^{-5}$ GHz/K, $T_g = (84.7 \pm 0.3)$ °C

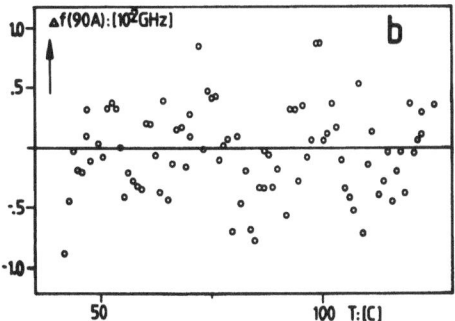

Fig. 3. b) The residuals corresponding to figure 3 a

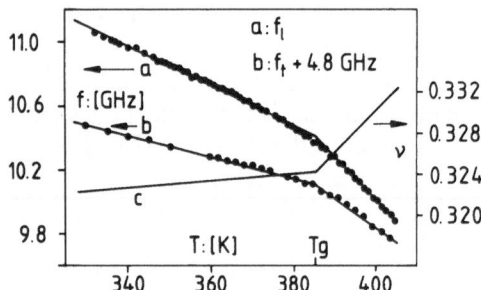

Fig. 4. Longitudinal (f_l)- and transverse (f_t) sound frequency of PMMA as a function of temperature T. ν Poisson ratio, f_t has been shifted as indicated

and the parameters and from the distribution of the residuals (fig. 3 b) we conclude a transformation interval of less than 1 K for this material.

In figure 4 the longitudinal and transverse hypersonic frequencies of PMMA are shown as functions of temperature for the phonon wavelength $\Lambda^{90\,\mathrm{N}} = (363.81\, n(T))$ nm ($n(T)$: temperature dependent refractive index of the sample). Using the linear approximation we have calculated the Poisson ratio $\nu(T)$ from both frequencies (fig. 4). The measurements were performed on cooling from ≈ 405 K to ≈ 330 K. Periods of 15 hours and 68 hours were spent measuring the longitudinal and transverse sound frequency curves respectively. The corresponding sound frequency gradients are listed in table 1 together with values from literature. Our transverse sound velocity data slightly exceed those of Kono [26], measured at 2.25 MHz on a sample material of unknown molecular weight.

A description of the longitudinal and transverse sound frequency data of figure 4 through two straight intersecting lines (eq. (2)) is only a rough approximation. Around T_g the sound frequency curves are bent, whereby the longitudinal sound frequency data deviate stronger from the fit function (eq. (2)) than the transverse ones. This different behaviour of the longitudinal and transverse polarized acoustic phonons in the glass transition region could be due to either a different coupling of the mechanical deformations involved with the glass transition process or with the different times involved in the measurements of the longitudinal and transverse sound frequency curves.

A convenient way to study structural relaxations in the glass transition region is to use discrete step changes in temperature and to analyze the responses of the physical properties of interest. This technique has been applied particularly to study the volume recovery

of glasses by dilatometric experiments [21, 22]. However, it has not yet been applied to study the hypersonic properties. We have performed such measurements in the following way (fig. 5): Starting well above the quasi-static glass transition temperature T_g we have changed the temperature of the sample by steps ΔT. Simultaneously we measured the longitudinal sound frequency Δf in response to the temperature disturbance. To ensure that the curve $f(T)$ constructed from these step measurements is an equilibrium curve at least for $T > T_g$ and reflects an internal equilibrium state of the sample for $T \lesssim T_g$, we have performed the steps ΔT by heating and cooling, as is shown schematically in figure 5, and we have prolonged any measurement until the finally measured sound frequencies $f(t)$ (t: time) deviated only insignificantly from the extrapolated asymptotic value $f_T^\infty = \lim\limits_{t \to \infty} f_T(t)$ (see also below). The sound frequencies indicated through the data points (●) in figure 5 are the extrapolated values f_T^∞, the two straight lines were fitted as mentioned above.

Figure 6 shows schematically the temperature disturbance and the sound frequency response as a

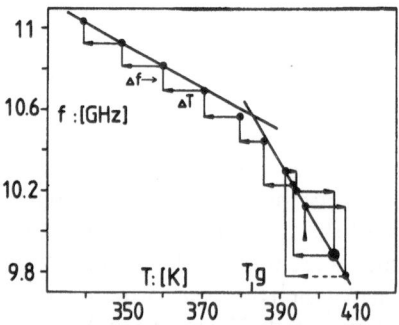

Fig. 5. Longitudinal sound frequency as a function of temperature T (explanation see text). ⊙ indicates that the further measurements were performed with decreasing temperature

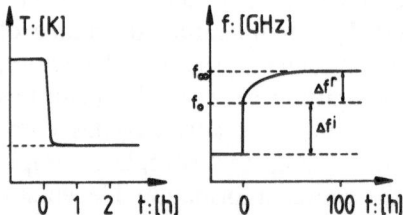

Fig. 6. Schematic representation of a temperature step (left figure) imposed on the PMMA sample and the consecutive hypersonic frequency response (right figure)

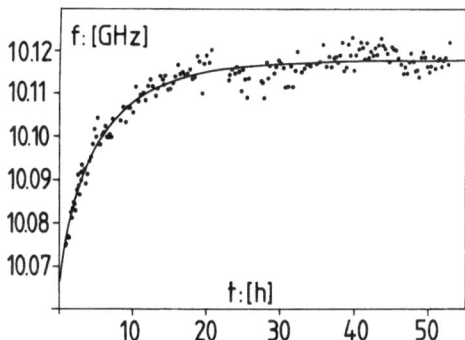

Fig. 7. Relaxing part of the sound frequency response as a function of time after a temperature step from 407 K to 396.3 K

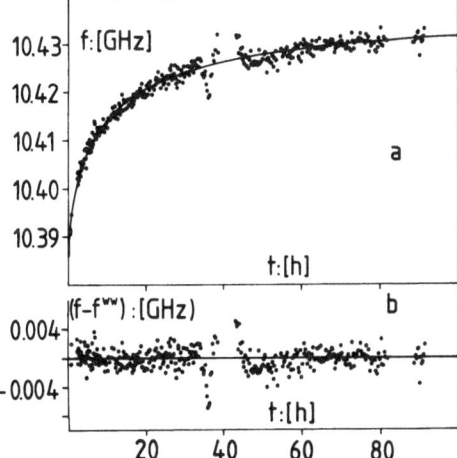

Fig. 8. a) Relaxing part of the sound frequency response as a function of time after a temperature step from 393.3 K to 385.8 K fitted by a Williams-Watts function (explanation see text). b) The residuals corresponding to figure 8 a

function of time. The sound frequency response Δf for temperature steps above T_g seems to consist of two parts: the quasi-instantaneous part Δf^i is realized as quickly as the sample temperature is settled (sec. II), the development of the relaxing part $\Delta f^r(t)$ with time depends strongly on the temperature distance from T_g as well as on the amount of ΔT. The relaxing parts of the sound frequency for two temperature steps are shown in figure 7 and figure 8 a. Each relaxing part $\Delta f^r(t)$ could be fitted using only one empirical Williams-Watts function [20]

$$\phi_f(t) = \frac{f^\infty - f(t)}{f^\infty - f^o} = \exp\left(-\left(\frac{t}{\tau}\right)^\beta\right), \qquad (3\,a)$$

leading to the modified fit function

$$f(t) = f^\infty - \Delta f^r \cdot \phi_f(t) \qquad (3\,b)$$

where τ is a characteristic relaxation time, β a parameter specifying the distribution of relaxation times (the smaller β, the broader the distribution of relaxation times), f^∞ is the equilibrium value for long times and $f^\infty - f^o = \Delta f^r$ is the amplitude of the relaxing part (fig. 6). From the relaxation time τ and the distribution parameter β the mean relaxation time $\langle \tau \rangle$ is derived:

$$\langle \tau \rangle = (\tau/\beta)\, \Gamma(1/\beta), \qquad (4)$$

where $\Gamma(1/\beta)$ is the gamma function. The applicability of equation (3 b) to the measured data is demonstrated by the residuals shown in figure 8 b. To ascertain that the Williams-Watts function also fits those data points collected immediately after the temperature stabilization of the sample we have chosen the representation of $\log\left[-\ln\left(\phi_f(t)\right)\right]$ versus $\log(t)$. Figure 9 shows the result for the sound frequency response curve measured after a temperature step from 403.8 K to 393.3 K. In this figure only the data of the first quarter of the total time of measuring (Δt) are shown.

Because the extremely slow relaxation process producing $\Delta f^r(t)$ has not yet been reported in literature, we denote this the α_o-process.

A quasi-instantaneous sound frequency response Δf^i (fig. 6) was calculated from the relation

$$\Delta f_j^i = f_j^o - f_{j-i}^\infty,\, j = 1, 2, 3, \ldots \qquad (5)$$

where j designates successive temperature steps. A pseudo sound frequency gradient $G_i = \Delta f_j^i / \Delta T_j$ can be

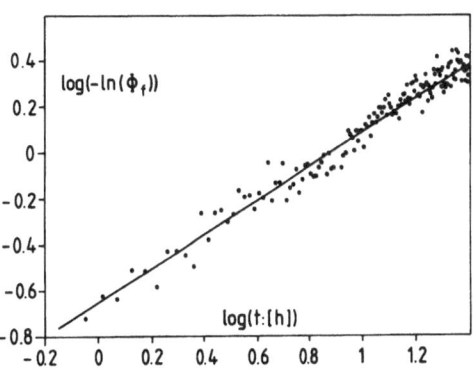

Fig. 9. Explanation see text

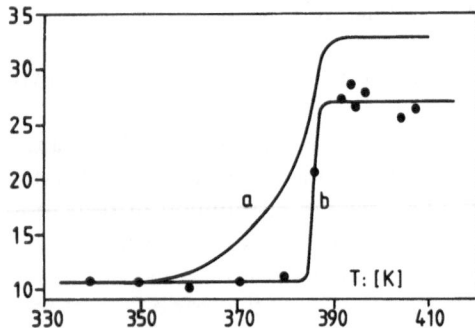

Fig. 10. Sound frequency gradient $G_r = -\dfrac{\Delta f}{\Delta T}$ (curve a) of the fully relaxed sound frequency curve (fig. 5) and of the quasi-instantaneous response (explanation see text) $G_i = -\dfrac{\Delta f^i}{\Delta T}$ (curve b) as a function of temperature

Table 1. Sound frequency gradients of PMMA. $f_y^x = -\left(\dfrac{df_y}{dT}\right): [10^{-3}$ GHz/K], $x = a: T > T_g; x = b: T < T_g; y = l, t$

f_l^a	f_l^b	Reference
24.3	10.6	2
26.2	10	3
23.9	–	13
27.2	12.4	figure 4
32.8	10.7	figure 5
27.1	–	equation (4)

f_t^a	f_t^b	
18.6	6.7	figure 4

derived (fig. 10, curve b) using only the quasi-instantaneous sound frequency responses Δf_j^i from equation (5) in connection with the related temperature steps $\Delta T_j := T_j - T_{j-1}$. The sound frequency gradient $G_i(T)$ at temperatures above T_g agrees well with those gradients reported for PMMA in literature and our own, derived from figure 4 (see table 1). Near T_g, $G_i(T)$ decreases rapidly with decreasing temperature and becomes constant again within the remaining low temperature interval.

We also evaluated the sound frequency gradient $G_r(T)$ from the "relaxed" sound frequency data of figure 5 as a function of temperature (fig. 10, curve a). Above T_g, $G_r(T)$ becomes constant and exceeds values reported in literature and our own derived from figure 4 by about 20% (table 1). Around T_g, $G_r(T)$ decreases with decreasing temperature and converges against $G_i(T)$ around $T_o = 350$ K. Furthermore, $G_r(T)$

behaves very asymmetrically around T_g. The relaxing amplitude Δf^r decreases continuously between 386 K and 350 K approaching zero around 350 K.

The possibility to resolve the relaxing sound frequency response $\Delta f^r(t)$ of the α_o-process by Brillouin spectroscopy depends on the relation between the mean relaxation time $\langle\tau\rangle$, the thermal stabilization time τ_T and the necessary data accumulation time for a Brillouin spectrum τ_a (table 2). For the investigations of longitudinal acoustic phonons in PMMA $\tau_T > \tau_a$ holds in the temperature interval of interest. $\Delta f^r(t)$ can in principle be resolved only if $\langle\tau\rangle \gg \tau_T$ holds. Another limit of resolution, not principle in nature, is approached if the time Δt spent in the measurement of a frequency response is small compared with $\langle\tau\rangle$. For our measurements above T_g the relation $\tau_T \ll \langle\tau\rangle < \Delta t$ was always fulfilled (table 3). Well below T_g the calculated relaxation times $\langle\tau\rangle$ largely exceeded the time of measurements. We therefore renounce a detailed discussion of the time behaviour in the low temperature region.

From the course of $G_i(T)$ and $G_r(T)$ in the glass transition region it is clear that besides the α_o-process at least one other relaxation process is operative in $f(T)$. This additional process is at least partially responsible for the step like behaviour of $G_i(T)$, (fig. 10). It is obvious to identify the additonal process with

Table 2. Characteristic time constants for 'temperature step-sound frequency response' measurements

		Typical value
$\langle\tau\rangle$	Mean relaxation time (eq. (4))	e. g. table 3
τ_T	Thermal stabilization time of the sample and the thermostat	25 min
τ_a	Accumulation time of a Brillouin spectrum	4 min
Δt	Time for measuring a sound frequency response $f(t)$ after imposing a temperature step ΔT on the sample	e. g. table 3

Table 3. Characteristic time constants for two 'temperature step-sound frequency response' measurements above T_g

$\Delta T_j := T_{j-1} \to T_j$	407 K → 396.3 K	393.3 K → 385.8 K
$\langle\tau\rangle$	6.1 h	23.5 h
Δt	55 h	90 h

the classical α-process which is intimately correlated with the elastic and volume properties near the glass transition [24, 29–31]. Near T_g the relaxation time of the α-process, τ_α increases strongly and becomes of the order of τ_T [31]. A qualitative similar hypersonic behaviour has recently been found for polystyrene in the glass transition region [16].

Our investigation of the sound frequency response at $T = 359.9$ K $(T_g - T \approx 23$ K) corresponds to a situation similar to that discussed by Patterson [13] for his hypersonic investigations at the lowest temperature he measured: figure 11 shows the remaining relaxing hypersonic response due to a temperature step from $T = 370.4$ K to $T = 359.9$ K. A Williams-Watts fit (eq. (3)) of our data leads to an upper limit of 10.9 GHz for f^∞ and a mean relaxation time $\langle\tau\rangle \approx 7500$ h $(\beta = 0.2)$. The standard deviation of the data points in figure 11 is about 10^{-3} GHz. The calculated frequency f^∞ may be compared with the fictive equilibrium value $f^{ex} = (11.32 \pm 0.04)$ GHz extrapolated linearly from the equilibrium region to $T = 359.9$ K. The difference $f^{ex} - f^\infty = (0.52 \pm 0.14)$ GHz is far outside the margin of error of f^∞. To illustrate the situation in figure 12 we have drawn, besides the measured data of figure 11, the value f^{ex}, the value f^∞ and the Williams-Watts function up to 80 h. This representation supports the conclusion that the sound frequency at this temperature will never reach the value f^{ex}. Taking into account that the Williams-Watts function might not deliver the correct description of our data below T_g, because of the long relaxation times and the small distribution factor involved, we have tried to estimate an ultimate shortest time necessary to realize f^{ex}. For this purpose we have fitted linearly the data points of the last 30 hours of figure 11 and then calculated the time τ^* after which f^{ex}

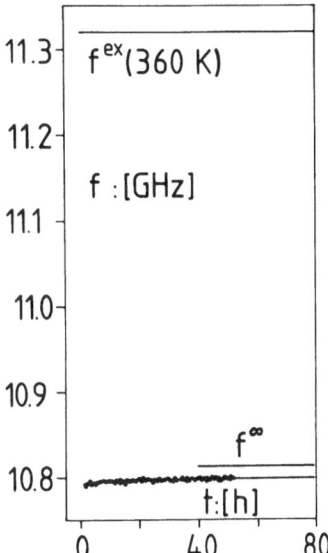

Fig. 12. Relaxing part of the sound frequency response as a function of time at $(T_g - T) \approx 23$ K. The meaning of f^{ex} and of f^∞ is explained in the text

(359.9 K) is obtained. Even this very rough estimate, which in any case delivers essentially too small time constants, leads to $\tau^* = 7500$ h, a time interval which exceeds largely that discussed by Patterson [13].

The question now appears, as to whether other physical properties which are in general sensitive to the glass transition, like the specific volume v, the volume expansion coefficient α and the specific heat at constant pressure C_p, reflect the peculiarly long-time behaviour found for the longitudinal hypersonic frequency. Knowledge about the temperature dependence of v and α is here of special interest because they influence the sound frequency behaviour in a more or less direct way. The longitudinal sound frequency f_l for a given wavelength depends directly on the elastic stiffness coefficient c_{11} and the specific volume (for the following discussion we exclude acoustic dispersion effects):

$$f_l = (c_{11}\, v)^{0.5}/\Lambda. \qquad (6)$$

In a linear approximation the temperature dependent elastic stiffness coefficient is given by

$$c_{11}(T) = c_{11}(T_o)\,(1 - (2\delta_l + \alpha)\,(T - T_o)), \qquad (7)$$

T_o reference temperature.

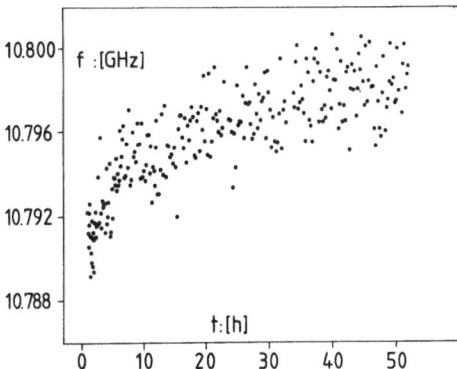

Fig. 11. Relaxing part of the sound frequency response as a function of time at $(T_g - T) \approx 23$ K

The sound velocity gradient δ_l has to be determined at fixed wavelength (e. g. with the 90 A scattering geometry). With equation (6), equation (7), and the Lorenz-Lorentz relation the temperature dependence of the sound frequency can be derived for any scattering geometry even if the sound wavelength depends on temperature through the refractive index [16]. For the temperature dependent frequency, measured in the 90 N-scattering geometry, the following relations hold:

$$f(T) = f(T_o) \left(1 - (\delta_l + \alpha K) (T - T_o)\right) \qquad (8)$$

with

$$K = (n^2(T_o) - 1) (n^2(T_o) + 2)/6n^2(T_o).$$

Thus it is obvious that a volume recovery process affects the behaviour of the sound frequency. As has been mentioned above, we have strong evidence for a discontinuity of the mode-Grüneisen constant $\gamma_l(T)$ around T_g. It seems possible that harmonic and anharmonic interactions can relax in the course of structural rearrangements within the glass transition region independently of the volume recovery process. To separate the contributions to the relaxing sound frequency, systematic longtime measurements of the specific volume are desirable for our sample material, but not yet available.

Schwarzl et al. [24, 25] have recently extensively studied the influence of the glass transition process on the volumetric properties of a PMMA similar to ours. The PMMA 240 (Röhm GmbH) they used has a molecular weight of about $5 \cdot 10^6$ g/mol and contains practically no additives. Thus it should be possible to discuss our results in the context of their data. The measurements of Schwarzl et al. were performed, however, with a different cooling treatment from ours. They measured the specific volume v at constant cooling rates $\dot{T}$ with $\dot{T}$ between 0.01 K/min and 2 K/min in a temperature range from at least 338 K to 405 K. For $\dot{T}$ = 2 K/min they report data between 98 K and 448 K. The specific volume reaches its thermodynamic equilibrium value below 405 K even at the fastest cooling rate of 2 K/min indicating a much faster volume response at temperatures below 405 K compared with the sound frequency response. In this case factors other than volume changes are responsible for the slow relaxing part of the sound frequency. Changes of the elastic constants of second or higher order seem to be operative even at constant volume, and changes of the mode-Grüneisen parameters are then expected. However, some discrepancies should be noted:

i: Schwarzl et al. have stated that their samples contain water. This is supported by the fact that they found significantly different T_g's for the same cooling rate and different expansion coefficients α_l in the liquid [24, 25]. Our dry sample shows a T_g of 383 K which is supported by vacuum adiabatic caloric measurements [19] performed on our sample. *ii*: We have measured the temperature dependence of the refractive index around T_g by Brillouin spectroscopy. Using the Lorenz-Lorentz relation we have calculated the volume expansion coefficient α_l above T_g, and α_g below T_g: $\alpha_l = 9.5\,10^{-4}\,\mathrm{K}^{-1}$, $\alpha_g = 2.5\,10^{-4}\,\mathrm{K}^{-1}$. α_g agrees well with the value given by Schwarzl et al. Our α_l exceeds that of Schwarzl by $3.3\,10^{-4}\,\mathrm{K}^{-1}$. On the other hand Schwarzl et al. report a further increase of α_l above ≈ 420 K [25]. *iii*: Schwarzl et al. report even for their slowest cooling rate of $\dot{T} = 10^{-2}$ K/min a T_g-shift of 3.3 K per decade of $\dot{T}$. Caloric investigations on our samples at the cooling rates of 0.47 K/min and $8.3\,10^{-3}$ K/min showed no T_g shift at all [19] in agreement with our statement that the relaxed sound velocity gradient curve $G_r(T)$ reflects quasi-equilibrium properties of the sample even near T_g. A final comparison of the hypersonic with the volume expansion behaviour around T_g should therefore be postponed till volume expansion data on dry samples of our material are available. Such measurements are in progress [27].

The question arises whether an interrelation between the α_o-process and the liquid-liquid transition, introduced by Boyer (e. g. [33]), exists. It is obvious that in the vicinity of T_{ll} the mean relaxation time of the α_o-process of PMMA becomes comparable with usual times of measurements. Therefore, in the case of PMMA, one could try to interpret the transition at T_{ll} as a cross-over phenomenon. However, such a conclusion is not supported by our Brillouin investigations on atactic polystyrene [16].

An interpretation of the α_o-process in terms of molecular mechanism would be of interest. This process does not seem to follow an Arrhenius law near T_g, indicating that the underlying molecular mechanism becomes increasingly cooperative [30] for T approaching T_g from above. Because of the large relaxation times of the α_o-process it is reasonable to assume that the molecular units which rearrange by this process are larger than those involved in the classical α-process. However, at this time we have neither sufficient precise information about the temperature dependence of the relaxation times nor any knowledge about the relation between the α_o-process and the molecular weight of the material. A further discussion of the α_o-process will be reserved until more detailed measurements on PMMA and other glass forming polymers of different molecular weights are finished.

4. Conclusion

High performance Brillouin spectroscopy has been used to investigate the properties of the longitudinal and transverse acoustic phonon frequencies in the glass transition region of PMMA. We report an ultra slow relaxation process (α_o-process) for the longitudinal sound frequency which is possible not due to volume relaxation processes. The reason for this long-time relaxation process is not yet clear. However, a qualitative similar hypersonic behaviour has been found in atactic PS. From our measurements it is evident that already 20 K above T_g the mean relaxation time of the longitudinal hypersonic frequency of PMMA is of the order of hours and it is not sufficient to wait for the temperature stabilization to realize the thermal equilibrium as is often done. Furthermore it follows from our results (table 1) that for PMMA the hypersonic data published so far do not reflect the equilibrium properties above T_g. Moreover, the sound frequency gradients given in literature coincide with ours, deduced from the fast frequency response $(\Delta f^i)_j$, and T_g is reflected by the slow as well as by the fast process observed. Tentatively we attribute $(\Delta f^r)_j$ to the changes of the conformational state and $(\Delta f^i)_j$ to the solid like properties of the liquid, an idea which is supported by the discontinuous behaviour of the Grüneisen parameter γ_l at T_g. The hypersonic frequency recovery process 20 K below T_g reported by Patterson could not be confirmed.

Acknowledgements

This work was kindly supported by the Deutsche Forschungsgemeinschaft and the Sonderforschungsbereich 130. The authors are grateful to Prof. R. Siems for fruitfull discussion.

References

1. Peticolas WL, Stegeman GIA, Stoicheff BP (1967) Phys Rev Letters 18:1130
2. Friedman EA, Ritger AJ, Andrews RD (1969) J Appl Phys 40:4243
3. Romberger AB, Eastman DP, Hunt JL (1969) J Chem Phys 51:3723
4. Jackson DA, Pentecost HTA, Powles JG (1972) Mol Phys 23:425
5. Mitchell RS, Guillet JE (1974) J Polym Sci, Polym Phys Ed 12:713
6. Patterson GD (1976) J Polym Sci, Polym Phys Ed 14:1909
7. Durvasala LN, Gammon RW (1979) J Appl Phys 50:4339
8. Krüger JK, Marx A, Roberts R, Unruh HG, Bitar MB, Trong HN, Seliger H (1984) Makromol Chemie 185:1469
9. Hunt JL, Huang YY, Mitchel RS, Stevens JR (1973) J Opt Soc America 63:1308
10. Lindsay SM, Halawith B, Patterson GD (1982) J Polym Sci, Polym Letters Ed 20:583
11. Work RN (1956) J Appl Phys 27:69
12. Brody EM, Lubell CJ, Beatty CL (1975) J Polym Sci, Polym Phys Ed 13:295
13. Patterson GD (1975) J Polym Sci, Polym Letters Ed 13:415
14. Patterson GD (1980) Methods of Experimental Phys, Vol 16, Academic Press, New York, p 170 ff
15. Krüger JK, Peetz L, Pietralla M (1978) Polymer 19:1397
16. Krüger JK, Roberts R, Unruh HG, to be published
17. Krüger JK, Kimmich R, Sandercock J, Unruh HG (1981) Polymer Bulletin 5:615
18. Coakley RW, Mitchell RS, Stevens JR, Hunt JL (1976) J Appl Phys 47:4271
19. Frühauf KP, Helwig J, Müser HE, Roberts R, Krüger JK, Part II of this paper, to be published
20. Williams W, Watts DC (1970) Trans Faraday Soc 66:80
21. Kovacs AJ (1963) Fortschr Hochpolym Forsch 3:394
22. Kovacs AJ, Aklonis JJ, Hutchinson JM, Ramos AR (1979) J Polym Sci, Polym Phys Ed 17:1097
23. Carslaw HS, Jaeger JC (1959) Conduction of heat in solids, 2nd Edition, Oxford Press
24. Schwarzl FR, Zahradnik F (1980) Rheol Acta 19:137
25. Greiner R, Schwarzl FR (1984) Rheol Acta 23:378
26. Kono R (1960) J Phys Soc Japan 14:718
27. Arndt H, to be published
28. Sauerland E, Helwig J, Müser HE (1983) Thermochimica Acta 69:253
29. McCrum NG, Read GE, Williams G (1967) Anelastic and Dielectric Effects in Polymeric Solids, John Wiley, New York
30. Donth EJ (1981) Glasübergang, Akademie-Verlag, Berlin (Ost)
31. Patterson GD, Carroll PJ, Stevens JR (1983) J Polym Sci, Polym Phys Ed 21:605
32. Roberts R (1985) Dissertation, Saarbrücken
33. Boyer RF (1980) J Macromol Sci Phys B18:461

Received February 6, 1985;
accepted April 11, 1985

Authors' address:

J. K. Krüger
Fachrichtung 11.2-Experimentalphysik
Universität des Saarlandes
D-6600 Saarbrücken, F.R.G.

Progress in Colloid & Polymer Science   Progr Colloid & Polymer Sci 71:86–95 (1985)

Atomistic calculation of chain conformations and crystal structures of polyoxymethylene*)

R. Aich and P. C. Hägele

Abteilung Angewandte Physik, Universität Ulm, Ulm, F.R.G.

Abstract: A set of potential functions for hydrocarbons has been extended to oxygen interactions by fitting the parameters to a large number of experimental data of small ether molecules. This set is applied to the conformational and packing analysis of polyoxymethylene (POM).

The energy minimization yields several helix conformations of the unperturbed chain. The experimentally well-known 2 * 9/5 helix (or 2 * 29/16) helix is not obtained in single chain calculations; the problem, however, is solved by means of crystal calculations. It is shown that there is an influence of crystalline packing on chain conformation: the 2 * 2/1 helix is changed to a 2 * 9/5 helix in a hexagonal packing. This is a case where Natta's second postulate is not strictly satisfied.

In contrast to this result the orthorhombic modification of POM is built up of chains (2 * 2/1) in the minimum of intramolecular energy.

The lattice calculations give some further hexagonal and triclinic modifications of POM, which have not yet been observed experimentally.

Key words: Polyoxymethylene, atomistic calculations, conformational analysis, crystal structure.

1. Introduction

Atomistic calculations with semiempirial potential functions on polymers are a useful tool to determine chain conformations, conformational energies and other molecular parameters, and also crystallographic data, such as lattice constants and cohesion energies.

In order to apply such calculations to polyethers, a previously used set of potentials for hydrocarbons [1, 2] has been extended by including oxygen interactions [3].

This paper deals with systematic investigations of chain conformations and crystal lattices of polyoxymethylene (POM). In the course of the calculations it turned out that the influence of crystalline packing on chain conformation should be taken into account.

2. Method of semiempirical atomistic calculations

In semiempirical atomistic calculations the total energy of molecules and crystals is composed of several potential terms, which describe the contributions of intra- and intermolecular interactions:

$$E_{\text{tot}} = E_{\text{nb}}^{\text{intra, inter}} + E_{\text{val}} + E_{\text{rot}}$$

We use the following potential functions:

$$E_{\text{nb}}^{\text{intra, inter}} = \sum_{j < k} - A_{jk}\, r_{jk}^{-6} + B_{jk}\, e^{-c_{jk}\, r_{jk}} + \frac{q_j \cdot q_k}{r_{jk}}$$

$$E_{\text{val}} = \sum_j \frac{k_l}{2}\, (l_j - l_{j0})^2 + \sum_j \frac{k_{\vartheta}}{2}\, (\vartheta_j - \vartheta_{j0})^2$$

$$E_{\text{rot}} = \sum_j \frac{U_{03}}{2}\, (1 - \cos 3\, \varphi_j)$$

where

A, B, C potential constants (Buckingham potential)
q partial charges
k_l, k_{ϑ} force constants for bond length and bond angle deformation

---

*) Dedicated to Prof. Dr. H.-G. Kilian on the occasion of his 60th birthday.

U_{03} rotational force constant
r_{jk} distance between nonbonded atoms j and k
l_j bond length
ϑ bond angle
φ dihedral angle

The complete set of potential parameters is given in table 1. The potential parameters for hydrocarbon interactions have been successfully used in earlier investigations [1, 2] and remain unchanged.

Table 1. Potential parameter set

Bond length deformation

	k_l/kJ/mol nm	l_o/nm
C − H	24.28	0.0959
C − C	13.49	0.1219
C − O	27.21	0.1300
O − LP	rigid	0.0481

Valence angle deformation

	k_ϑ/kJ/mol rad	$\vartheta_o/°$
C − C − C	418.6	
C − C − H	217.7	
H − C − H	314.0	
C = C − C	385.1	
C = C − H	226.1	
$=C\begin{smallmatrix}C\\\\C\end{smallmatrix}$	418.6	109.47
$=C\begin{smallmatrix}C\\\\H\end{smallmatrix}$	217.7	
$=C\begin{smallmatrix}H\\\\H\end{smallmatrix}$	314.0	
H − C − O	293.0	
C − C − O	397.7	
O − C − O	460.5	
C − O − C	460.5	93.50
C − O − LP	418.6	103.55
LP − O − LP	rigid	140.0

Bond rotation

	U_o/kJ/mol
C − C	10.88
C = C	164.51
$C^{sp2} − C^{sp3}$	9.42
C − O	8.62

Table 1 (continued)

Steric interaction

− Buckingham potential

	$A/10^6$ kJ/mol nm^6	B/kJ/mol	C/nm^{-1}
C … C	2093.0	221858	0.360
C … H	535.8	58604	0.367
H … H	138.1	12558	0.374
C … O	1277.6	177535	0.381
H … O	364.2	47737	0.394
O … O	1068.0	413917	0.427

− Coulomb potential

		q (in multiples of the elementary charge e_o)
$C^{sp2} − C^{sp3}$	q_C	0.036
$C^{sp2} − H$	q_H	0.080
C − O	q_C	0.164
	q_O	− 0.126
	q_{LP}	− 0.019

LP: electron lone pair

These potential parameters are fitted to a large number of experimental data of small molecules being representative for the polymers of interest (polyethers). These molecules are listed in table 2.

Table 2. Molecules used for the fit of the potential parameters

Dimethylether	CH_3OCH_3
Ethylmethylether	$CH_3CH_2OCH_3$
Diethylether	$CH_3CH_2OCH_2CH_3$
2,4-dioxapentane	$CH_3OCH_2OCH_3$
2,5-dioxahexane	$CH_3OCH_2CH_2OCH_3$

The details of the fit procedure as well as the complete results of the calculations on these small ether molecules are given elsewhere [4].

The conformational analysis of single chains was done in the usual way by minimizing the conformational energy with respect to the dihedral and bond angles. The summation of energy is performed with a fast matrix algorithm [5]. It extends over a representative chain segment (one or a few repeating units) and − with a factor 0.5 − over adjacent repeating units, the number of which is chosen large enough to avoid cutoff errors.

Similarly the crystalline packing is calculated. The intermolecular energy is minimized with respect to the lattice constants a and b, as well as the angles describing the relative orientation of the (rigid or flexible) chains.

Table 3. Calculated conformations of the POM chain

Type of helix	2 ✳ 2/1	2 ✳ 29/10	2 ✳ 1/1	4 ✳ 1/1
Conformation (appr.)	gg	tg	tt	tgtḡ
Dihedral angles/°				
C−O−C−O	115.5	1.6	0.0	− 3.8
	115.4	123.5	0.0	127.3
				3.8
				−127.3
Valence angle/°				
C−O−C	113.2	113.2	113.9	113.3
O−C−O	112.9	112.9	111.2	112.3
Identity period/nm	0.353	5.682	0.233 (bent)	0.408
Radius of helix/nm	0.077	0.073	−	0.062
Step angle/°	180.5	124.1	360.0	361.2
Conformational energy in kJ/mol CH$_2$O	0.0	2.22	14.57	4.10

A basis of one chain is surrounded with one, two, or three „shells" of 8, 24, or 48 chains. If there are two chains in the basis, the shells consist of 16, 48, or 96 chains [6]. In most cases two shells are sufficient for the calculation of the lattice parameters and the cohesion energy.

3. Results and discussion

3.1 Conformations of the undisturbed POM chain

Polyoxymethylene (POM) has the chemical unit

CH$_2$−O−

and can therefore be characterized as an *AB*-helix, i. e. a repeating unit of two atoms in the backbone. The calculations have been performed with this repeating unit and multiples of it.

Three stable chain conformations are obtained with the smallest repeating unit (table 3). These conformations change only slightly ($\Delta\varphi$, $\Delta\vartheta \leq 0.1°$) if longer units are taken.

On the other hand the longer repeating units lead to a variety of additional other helical conformations, e. g. *ABCD*-helices. Most of these helices have low conformational energies but large helical diameters, which prevent a compact crystalline packing. An exception is the *tgtḡ*-helix (type *ABCD*) of POM having a small helical diameter, but a slightly bent helix axis. The amount of energy, however, necessary to straighten the helix axis is quite small (+ 0.3 kJ/mol CH$_2$O). The data of this straightened helix are given in table 3.

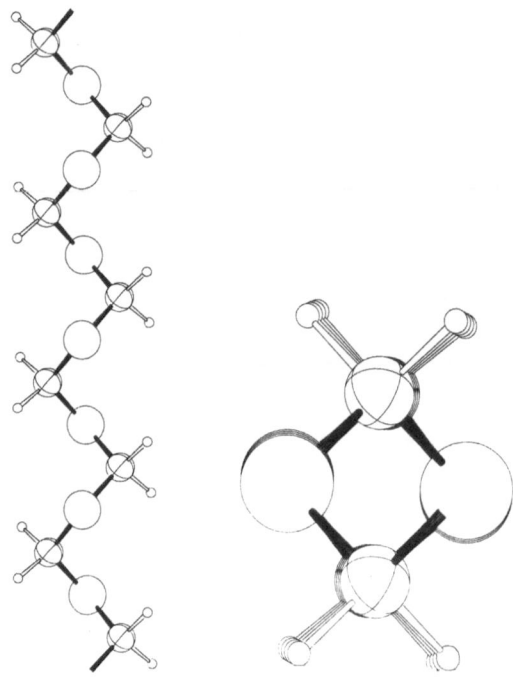

Fig. 1. Polyoxymethylene, *gg*-helix (2 ✳ 2/1)

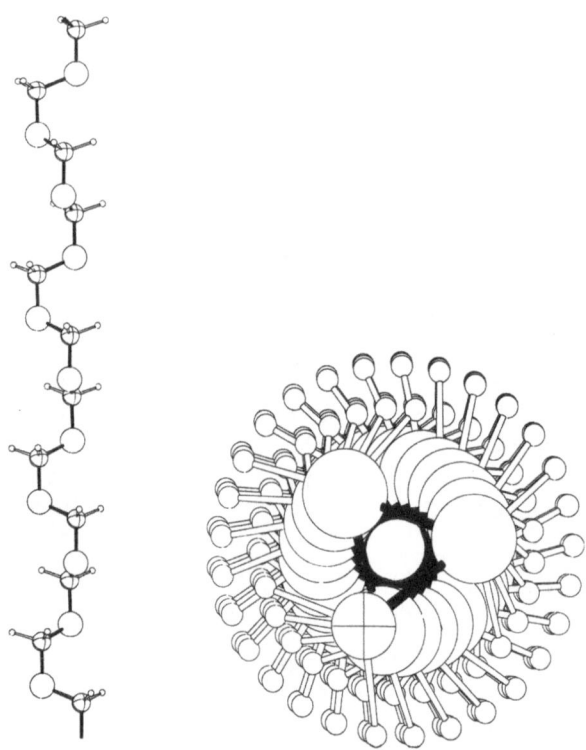

Fig. 2. Polyoxymethylene, *tg*-helix (2 ✳ 29/10)

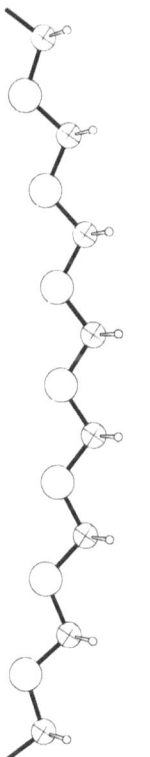

Fig. 3. Polyoxymethylene, *tt*-helix (2 ∗ 1/1)

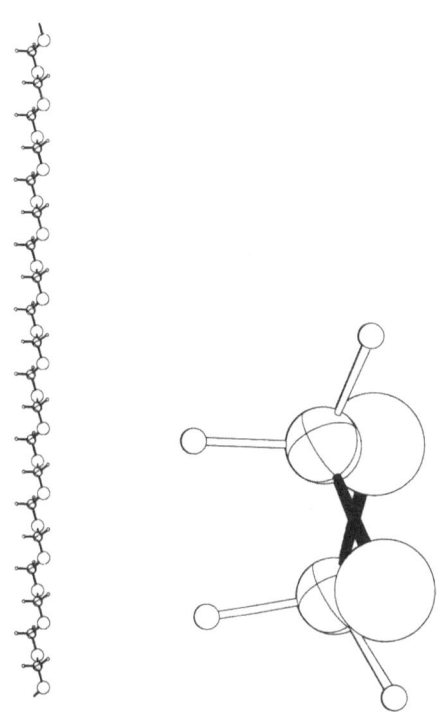

Fig. 4. Polyoxymethylene, *tgtḡ*-helix (4 ∗ 1/1)

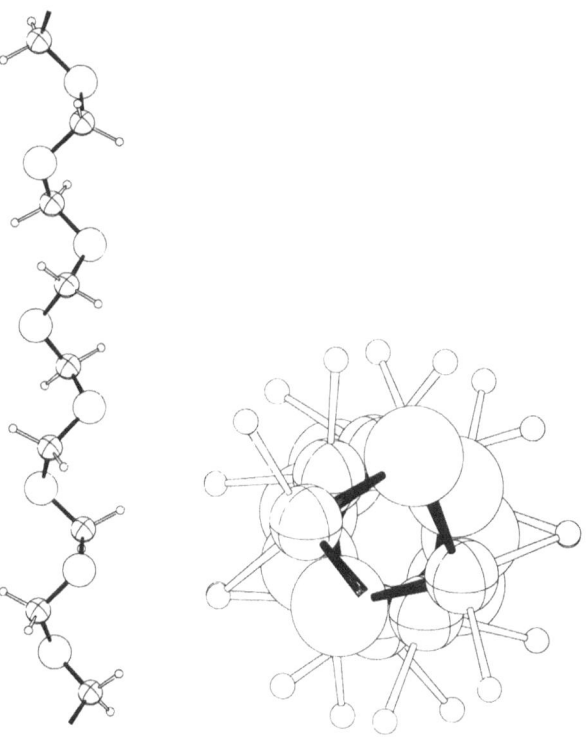

Fig. 5. Polyoxymethylene, *gg*-helix (2 ↗ 9/5; 2 ∗ 29/16)

For some of these single chain conformations computer plots (ORTEP graphics, [7]) are given in figures 1 to 4.

Investigations by X-ray diffraction analysis suggest the existence of two different crystalline modifications of POM having (quasi)hexagonal and orthorhombic unit cells [8, 9, 10]. The two corresponding helical chain conformations are summarized in table 4 (see also figs. 1 and 5).

The characteristic differences between these two conformations are the values of the dihedral angles. The differ by about 13°.

By comparison of these experimental data (table 4) with the calculated results (table 3) it is concluded:

1. The calculated 2 ∗ 2/1 helix is in excellent agreement with the experimentally established conformation.

2. The calculations give no evidence for a 2 ∗ 9/5 helix or a (nearly equivalent) 2 ∗ 29/16 helix [11, 12].

There are several possible reasons for this apparent failure of the conformational analysis:

1. Although it is not surprising that the rather simple potential has no additional minima as close as 13° to the gauche minimum, we might have overlooked a second energy minimum.

Table 4. Conformations of the POM chain proposed by X-ray analysis

Type of helix	Crystalline modification	
	Orthorhombic 2 ✳ 2/1	(Quasi)hexagonal 2 ✳ 9/5 2 ✳ 29/16
Conformation (app.)	gg	gg
Dihedral angles/° C–O–C–O	115.6(5)	102(2)
Valence angles/° C–O–C O–C–O	114.1(1) 112.8(1)	114(1) 112(1)
Identity period/nm	0.356	1.734: 2 ✳ 9/5 5.602: 2 ✳ 29/16
Radius of helix/nm	0.079	0.068

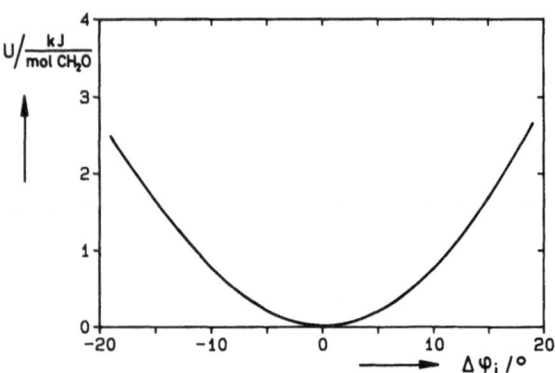

Fig. 6. Rotational potential of the POM single chain vs. all dihedral C–O bond angles

2. There might be other helix types which could explain the X-ray diffraction pattern as well as the 2 ✳ 9/5 helix.

3. The interactions between a chain and its crystalline surroundings might lead to a modified conformation. Evidently such a conformation can not be found in single chain calculations.

These explanations have been checked carefully. Figure 6 shows the rotational potential of the POM single chain vs. all dihedral C–O bond angles (simultaneous rotation). Energy minimization is done with respect to all other variables. It is clearly seen that there is no second minimum next to the gauche minimum (115.6°).

This result is confirmed by the investigation of other cuts through energy hypersurface: For example, the

Table 5. Summary of crystalline structures of POM obtained by X-ray diffraction

Inter/intramolecular parameters	Carazzolo [8,9]		Uchida [11]	Andrews [18]	Takahashi [12]	Gramlich [10]	
	Hexagonal	Orthorhombic	Hexagonal	Hexagonal	Hexagonal	Orthorhombic	Orthohexagonal
Lattice constants/nm							
a	0.447	0.477	0.447	0.448	0.447	0.477	0.447
b	0.447	0.765	0.447	0.448	0.447	0.765	0.774
c	1.73; 5.559(1)	0.356	1.739; 5.602	0.173	1.739; 5.602	0.356	7.34
Proposed helix type	2 ✳ 9/5; 2 ✳ 29/16	2 ✳ 2/1	2 ✳ 9/5; 2 ✳ 29/16		2 ✳ 9/5; 2 ✳ 29/16	2 ✳ 2/1	2 ✳ 38/21
Helical pitch	0.349	0.35	–	0.346	–	0.356	0.350
Radius of helix nm	–	0.079	0.068	0.072	–	–	–
Bond length/nm							
C–O	0.143	–	0.1421; 0.1418	0.1430	0.1433; 0.1429	0.1410	0.1405
C–H	–	–	–	–	0.106	0.098	0.105
Valence angle/°							
∢ COC	–	112.68	112.40; 113.00	106.88	112.3; 112.9	1114.1	114
∢ OCO	–	112.68	110.82; 111.37	–	109.9; 110.4	112.8	112
Dihedral angle/°							
CH_2–O	102.62; 103.65		101.78; 102.55	106.3	102.4; 103.0	115.6	102

rotational potential vs. every second dihedral $C-O$ bond angle, also does not reveal any indication of a double minimum [3]. A variation of the potential parameters does not alter the above result qualitatively.

The two remaining possible explanations of the $2 \ast 9/5$ helix are discussed in the following sections.

3.2 Crystalline structures of POM

Table 5 summarizes the results of X-ray investigations of crystalline polyoxymethylene.

We now report on the calculations of POM crystals. In all cases (but see Section 3.2.2.1) the chains are rigid, i. e. their (intramolecular) conformations have been taken from single chain conformational analysis and have not been changed during minimization of the intermolecular energy. Translational periodicity is taken into account („ideal crystal mode" with a basis of two chains), but no additional symmetries have been assumed. The summation of interactions extends up to 50 chains, and typically includes 50 chemical units in chain direction.

The lattice parameters and the cohesion energy result from the minimization of the intermolecular energy.

3.2.1 Orthorhombic modification

The $2 \ast 2/1$ helix from the intramolecular conformational calculations is taken to determine the lattice data. The results are compared with experimental data in table 6. These data are taken from [8, 10] (see table 5). A computer plot of this lattice is shown in figure 7.

The lattice parameters have been extrapolated to 0 K with the help of the expansion coefficient [13].

The calculated results agree well with the experimental ones. It is again emphasized that the lattice angles are obtained by calculation and have not been kept fixed. The 20% discrepancy in the value of the cohesion energy might be due to the fact that the experimental value is gained by extrapolation from low-molecular data [14].

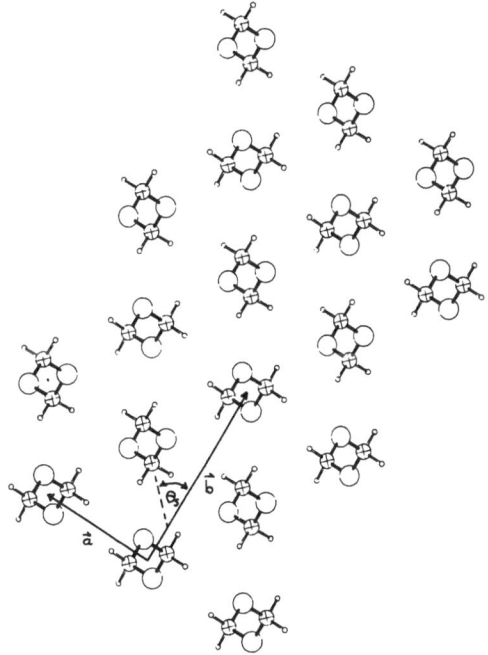

Fig. 7. Orthorhombic polyoxymethylene, gg-helix ($2 \ast 2/1$)

3.2.2 Hexagonal modification

3.2.2.1 The $2 \ast 29/10$ helix conformation

The results of the intramolecular calculations are applied to the determination of the lattice parameters, as shown in table 7 and figure 8.

The results are in good agreement with the experimental data except for the helix type. In order to see the influence of the helix type, the same lattice calculation is performed with the $2 \ast 29/16$ helix proposed by X-ray analysis (see fig. 9). As has been pointed out in Section 3.1, this helix exhibits no minimum of the intramolecular energy.

Inspection of table 7 shows that the resulting lattice parameters are nearly equal for both helix types. Hence the X-ray pattern could be explained by the $2 \ast 29/10$ helix as well as by the $2 \ast 29/16$ helix.

This assumption has been tested by structure factor calculations [3]. The results reveal significant differ-

Table 6. Orthorhombic modification of POM ($2 \otimes 2/1$ helix) (θ_s - setting angle with respect to the b-axis)

| | Lattice parameters (in nm and degrees) | | | | | | | Cohesion energy in kJ/mol CH_2O |
	a	b	c	α	β	γ	θ_s	
Experiment	0.469	0.753	0.356	90.0	90.0	90.0		$-15.9(1.7)$
Calculation	0.472	0.758	0.353	90.0	90.0	90.0	46.1	-13.0

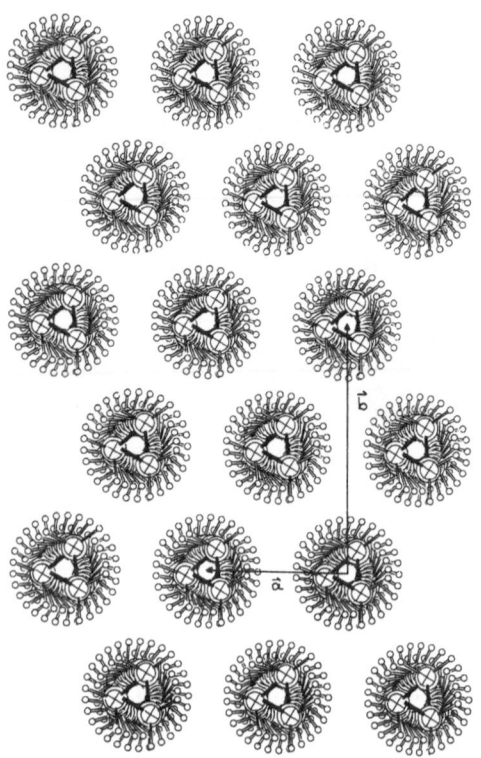

Fig. 8. Hexagonal polyoxymethylene, *tg*-helix (2 ⋇ 29/10)

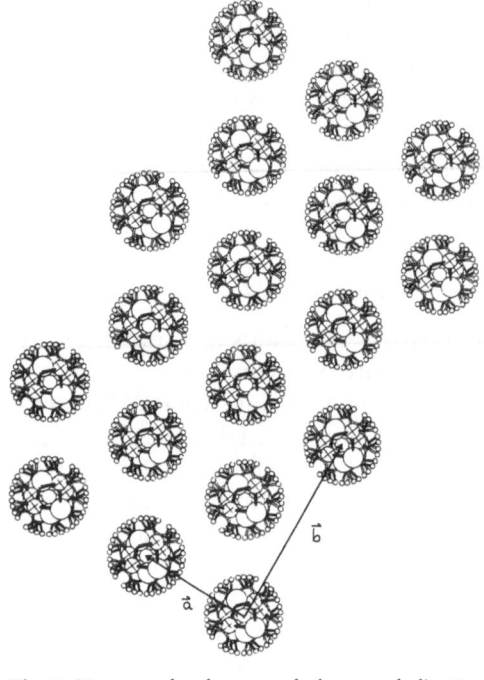

Fig. 9. Hexagonal polyoxymethylene, *gg*-helix (2 ⋇ 29/16)

ences in the region of $2\theta = 20° \ldots 40°$ (θ - diffraction angle). Therefore the 2 ⋋ 29/10 helix has to be ruled out as the chain conformation of the hexagonal modification. Consequently the attention is focussed on the remaining possibility of an influence of crystalline packing on the chain conformation.

3.2.2.2 Influence of crystalline packing on chain conformation

Because of the striking similarity of both helix types, 2 ⋇ 2/1 ($\equiv$ 2 ⋇ 32/16) and 2 ⋇ 29/16, it has been examined whether a hexagonal chain packing could turn

the 2 ⋇ 2/1 helix into a 2 ⋇ 29/16 helix. Hence the concept of rigid chains in crystal calculations had to be left.

According to the experimental lattice parameters of table 5 the POM helices of the type 2 ⋇ 2/1 are arranged in orthorhombic and in hexagonal lattices. Subsequently the total intra- and intermolecular energy U was calculated as a function of the internal dihedral C−O angles φ_i. The resulting "rotational potentials" are shown in figure 10.

In the case of orthorhombic surroundings the energy minimum is located at the dihedral angle of the 2 ⋇ 2/1 helix known from single chain calculations (see table 3). So there is no conformational change, which justifies the reported calculations with rigid 2 ⋇ 2/1 helices.

Table 7. Hexagonal modification of POM ($b \approx a \sqrt{3}$) with two helix types

| Helix conformation | Lattice parameters in nm and degrees | | | | | | Cohesion energy in kJ/mol CH$_2$O |
	a	b	c	α	β	γ	
Experiment 2 ⋇ 29/16	0.447	0.774	5.599	90	90	90	−
Calculation 2 ⋇ 29/16	0.453	0.775	5.602	90.2	89.9	89.9	− 10.86
2 ⋇ 29/10	0.453	0.770	5.682	89.4	90.8	89.4	− 11.23

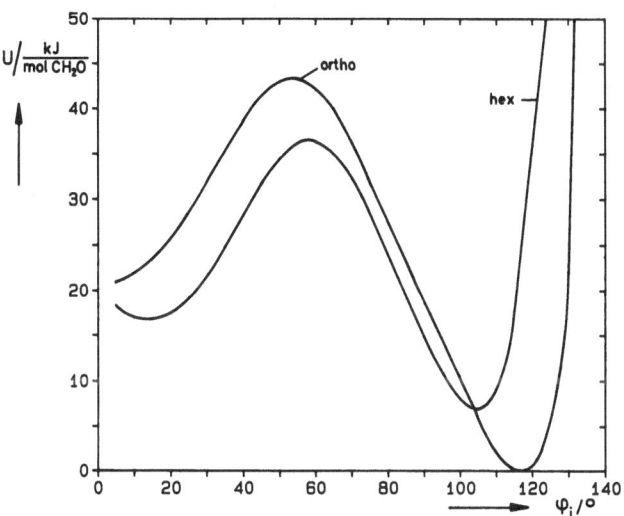

Fig. 10. Rotational potential of POM chains in crystalline surroundings

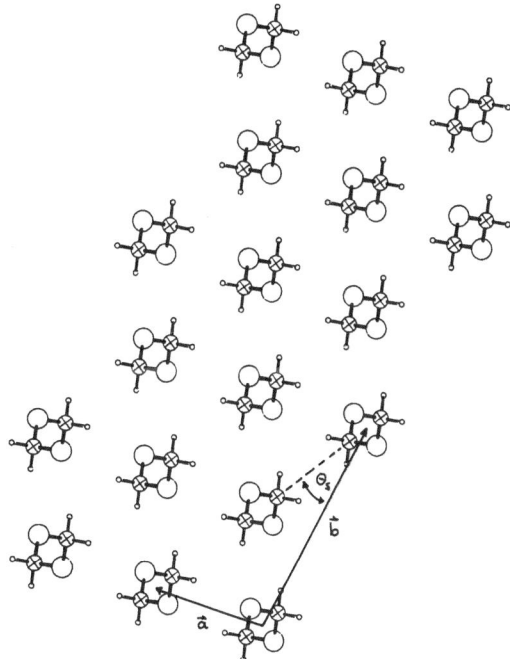

Fig. 11. Triclinic polyoxymethylene, gg-helix (2 ∗ 2/1)

On the other hand the minimum of the total energy U is shifted to a value of the dihedral angle $\varphi_i \approx 104°$, which is very near to the experimental value 103° (see table 5) of the 2 -29/16 helix.

This means that under the influence of the hexagonal crystalline packing the 2 ∗ 2/1 helix is changed to a 2 ∗ 29/16 helix.

It is concluded that unlike the orthorhombic modification the hexagonal modification of POM is formed by chains, the conformations of which are intramolecularly unstable, but which are stabilized by the cooperation of intra- and intermolecular interactions. So this is a case where Natta's second postulate [15] is not strictly satisfied.

The result is sensitive to the quantitative relation between these interactions and indicates the reliablility of the potentials used. As can be seen from figure 10, the orthorhombic modification corresponds to the lower energy. The total energy difference between the two modifications is about 6.9 kJ/mol CH_2O. This is in qualitative agreement with an endothermic transition orthorhombic → hexagonal (monotropic, 75...80 °C) stated by Zamboni and Zerbi [16].

So far the calculations do not explain the observed higher stability of the hexagonal modification of POM. Entropy considerations and further experiments, e. g. calorimetric and lattice vibrational mode measurements, are required to answer that question.

3.2.3 Other calculated modifications

In the course of the investigations other crystal structures besides the experimentally known orthorhombic and hexagonal modifications have been found. These calculated structures are energetically (intra- and intermolecular energy) less favourable, however. Corresponding experimental data have not yet been reported. So the following additional results might stimulate or support further experimental investigations.

3.2.3.1 Lattices with the 2 ∗ 2/1 helix (gg and $\bar{g}\bar{g}$)

Besides the orthorhombic lattice (cf. Section 3.2.1) there are a triclinic lattice (see table 8 and fig. 11) and a

Table 8. Other calculated POM lattices with 2 ∗ 2/1 helices (θ_s-setting angle with respect to the *b*-axis). Approximation: 18 chains

Helix conformation	Lattice parameters in nm and degrees						Cohesion energy in kJ/mol CH_2O
	a	b	α	β	γ	θ_s	
gg, gg	0.458	0.864	86.1	93.8	100.9	25.0	− 10.63
gg, $\bar{g}\bar{g}$	0.494	0.756	90.0	90.0	90.0	37.5	− 11.97

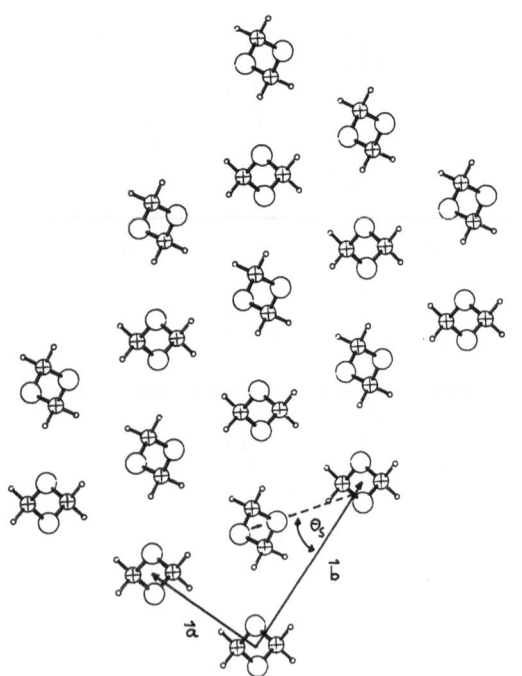

Fig. 12. Orthorhombic polyoxymethylene, *gg*- and *ḡḡ*-helices (2 ✳ 2/1)

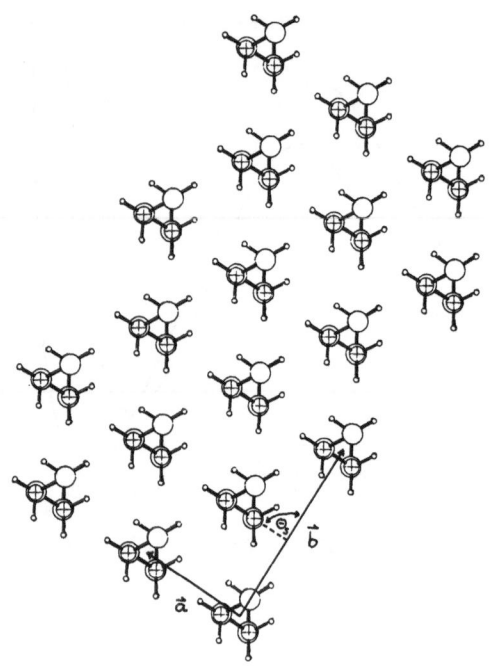

Fig. 13. Hexagonal polyoxymethylene, *tg*-helix (2 ✳ 3/1)

second orthorhombic lattice with a basis of the *gg*- and the enantiomorphic *ḡḡ*-helix (see table 8 and fig. 12).

3.2.3.2 Lattices with the 2 ✳ 29/10 helix (tg)

A hexagonal lattice with the 2 ✳ 29/10 helix of POM has been discussed in Section 3.2.2 (cf. table 7).

In this case also the influence of crystalline packing on the chain conformation has been investigated. The 2 ✳ 29/10 helix turns out to an almost exact 2 ✳ 3/1 helix. The lattice is nearly hexagonal (see table 9 and fig. 13).

3.2.3.3 Lattices with the 4 ✳ 1/1 helix (tgtḡ)

Finally, results of lattice calculations with the 4 ✳ 1/1 helix are reported: The lattice cells (see table 10 and fig.

14) are triclinic (no. 5), monoclinic (nos. 2 and 4) and approximately orthorhombic (nos. 1 and 3).
Lattices of *tgtḡ*-helices have become important in the interpretation of the high temperature phase of polyethylene [17].

Acknowledgement

The financial support of the Deutsche Forschungsgemeinschaft (DFG) is gratefully acknowledged.

For stimulating discussions we thank Dr. T. Debaerdemaeker, Dr. H. P. Grossmann, Prof. Dr. W. Pechhold, Drs. J. and C. Schmieg, and Prof. Dr. W. Wilke.

References

1. Hägele PC, Pechhold W (1970) Kollid-Z u Z Polymere 241:977
2. Bautz G, Leute U, Dollhopf W, Hägele PC (1981) Coll & Polym Sci 259:714

Table 9. Another calculated POM lattice with the 2 ✳ 3/1 helix. Approximation: 50 chains

| Helix conformation | Lattice parameters in nm and degrees | | | | | | Cohesion energy in kJ/mol CH$_2$O |
	a	b	α	β	γ	θ_s	
tg	0.436	0.753	90.0	90.1	90.0	90.0	-12.2

Table 10. Other calculated POM lattices with 4 ✳ 1/1 helices. Approximation: 18 chains

Helix conformation	No.	Lattice parameters in nm and degrees						Cohesion energy in kJ/mol CH$_2$O
		a	b	α	β	γ	θ_s	
$tgt\bar{g}$	1	0.432	0.768	90.8	89.6	92.0	27.3	− 10.5
	2	0.408	0.793	89.9	78.2	90.1	17.1	− 11.3
	3	0.445	0.732	90.0	88.2	90.0	43.0	− 12.7
	4	0.405	0.800	90.0	74.2	90.0	20.0	− 14.8
	5	0.450	0.720	79.1	84.2	96.0	19.4	− 12.7

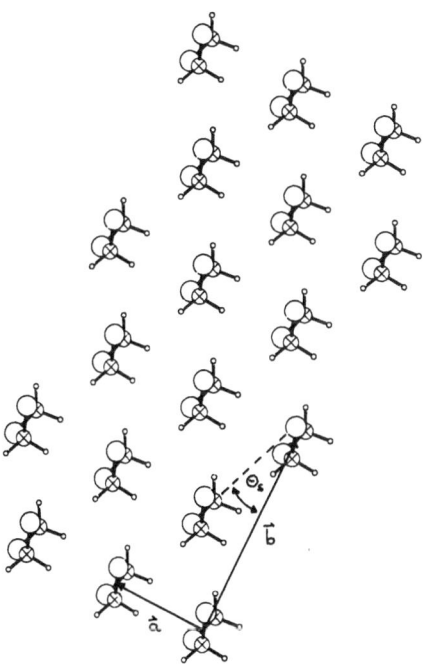

Fig. 14. Monoclinic polyoxymethylene, lattice 2, $tgt\bar{g}$-helix

3. Aich R (1985) thesis, Ulm
4. Aich R, Hägele PC (to be published)
5. Schmieg C, Hägele PC, Beck LM (1982) J Comp Phys 48(1):45
6. Schmieg J (1984) thesis, Ulm
7. Johnson, CK (1965) Acta Cryst, 18:1004
8. Carazzolo G, Mammi M (1963) J Polym Sci, A, Vol 1, 965
9. Carazzolo G (1963) J Polym Sci, A, Vol 1, 1573
10. Gramlich V (1977) ECM-4, 4th Europ Cryst Meeting, Oxford, Abstract Book, 531
11. Uchida T, Tadokoro H (1967) J Polym Sci, A-2, Vol 5, 63
12. Takahashi Y, Tadokoro H (1979) J Polym Sci Polym Phys Ed, 17:123
13. White GK, Smith TF, Birch JA (1976) J Chem Phys, 65(2):554
14. Boyd RH (1961) J Polym Sci 50:133
15. Natta G, Corradini P (1960) Nuovo Cimento, 15, Suppl No 1, 9
16. Zamboni V, Zerbi G (1964) J Polym Sci, C No 7, 153
17. Grossmann HP (1977) thesis, Ulm
18. Andrews EH, Martin GE (1973) J Mat Sci, 8:1315

Received July 8, 1985;
accepted July 25, 1985

Authors' address:

P. C. Hägele
Universität Ulm
Abteilung Angewandte Physik
Oberer Eselsberg
D-7900 Ulm, F.R.G.

Progress in Colloid & Polymer Science   Progr Colloid & Polymer Sci 71:96–102 (1985)

Polyethylene-polystyrene gradient polymers*)

I. Structure modification of polyethylene matrix

P. Milczarek and M. Kryszewski

Centre of Molecular and Macromolecular Studies, Polish Academy of Sciences, Lódz, Poland

Abstract: In this paper the influence of styrene diffusion and its polymerization during the preparation of gradient polymers from LDPE is discussed. It is demonstrated that the monomer showing a partial dissolving ability modifies the structure of the host polymer. These structural changes are determined by the diffusion temperature T_D and consist in the reorganization of amorphous and disordered crystalline phases which are accessible to the monomer.

Key words: Gradient polymers, low density polyethylene, crystallization.

Introduction

Mixing of polymers is a widely used method in the production of different polymeric materials. The mixtures are usually obtained by hot extrusion of molten polymeric components or by precipitation from a common solvent. An interesting preparation method of new polymeric materials has been recently reported [1–4]. Its technology makes use of the diffusion process of a chosen monomer (liquid or gaseous) into the host polymer matrix. The diffusing monomer can be polymerized after a controlled time in order to retain the concentration gradient produced.

Such multicomponent systems have a structure and composition which are not macroscopically homogeneous and their physical properties vary as a function of position in the sample. The monomer should be polymerized at a rate which is rapid in comparison with the rate of diffusion.

As opposed to known gradient polymers consisting of two amorphous components a new system has been obtained using partially crystalline polyethylene (LDPE) as a host polymer [5]. The diffusion of styrene (ST) into LDPE was carried out at room temperature.

It has been shown that the diffusion gradient profile can be modified by the diffusion time, method of the polymerization initiation and changes of the matrix structure. It is obvious that the diffusion temperature should be a parameter which strongly influences the gradient profiles and properties.

In this paper it will be shown that the variation of this parameter makes it possible to obtain gradient polymers with different characteristics because monomeric ST modifies the structure of the matrix.

Experimental

Samples

Since the sturcture of LDPE films strongly depends on their thermal pretreatment it is neccessary to describe the method of sample preparation. Branched LDPE, Lupolene 1800 S (BASF) was used (molecular weight (M_V) 28000, the number of CH_3 groups/1000 C is 35, melting point 110 °C). Molten LDPE was pressed between two metallic plates at 130° and cooled at 0 °C.

The spherulitic samples, 0.5 mm thick, were kept at room temperature for a few days. A small thickness of these foils makes it possible to control ST penetration. Figure 1 schematically illustrates the thermal history of the samples. Two samples were placed in liquid styrene. The diffusion process occurred at eleven different temperatures from the range $T_D = 25$ to 70 °C, for 3 h. One of these samples (N) was taken out of the styrene bath (non-polymerized) and the second sample (P) was polymerized by UV light (two mercury lamps, 1.6 kW power). The temperature of polymerization was equal to the diffusion temperature T_D and its duration was chosen

---

*) Dedicated to Prof. Dr. H.-G. Kilian on the occasion of his 60th birthday.

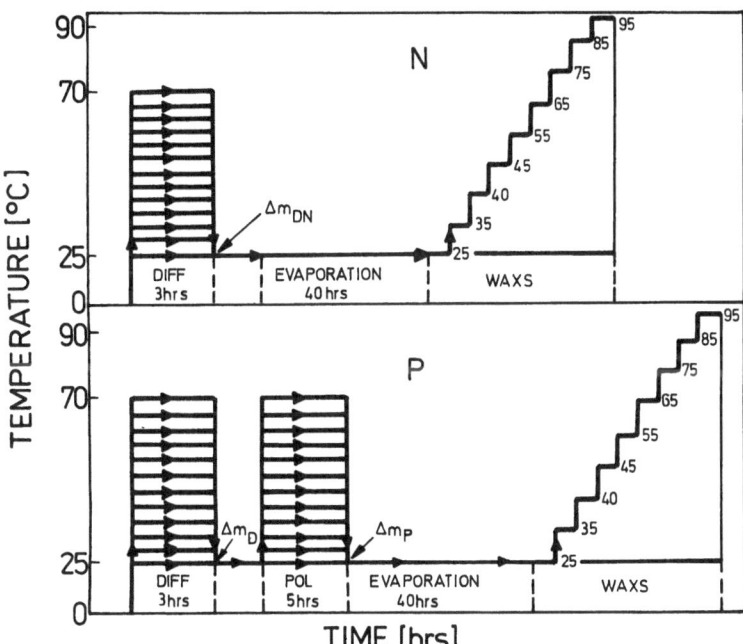

Fig. 1. Scheme of thermal history of ST treated LDPE samples polymerized (P) and non-polymerized (N)

to be 5 h. After the polymerization the weight increase (Δm_p) of the sample was measured. In order to avoid monomer evaporation during the polymerization, the samples were placed between quartz plates. Subsequently, both (N) and (P) films were dried in vacuum for 40 h.

Poylstyrene content in polymerized samples

The content of polystyrene in PE matrix for P samples was measured using the Specord-Unicam spectrophotometer. Thin 10

µm layers of these films were placed in the apparatus and the changes in the absorbance peak of 269 nm, characteristic for Ps, were measured (figs. 2a and 2b). Using the spectra presented in the figures one can calculate the content of polystyrene, taking into consideration the extinction coefficient and sample thickness.

The relative content polystyrene $\left(1 + \dfrac{C_P}{C_{PS}}\right)$, where C_P is the absorbance of the obtained sample and C_{PS} is the absorbance of polystyrene standard film, was plotted as a function of T_D (fig. 3, curve I). A maximum at $T_D = 50$-$57\,°$C and minimum at $T_D = 60$-

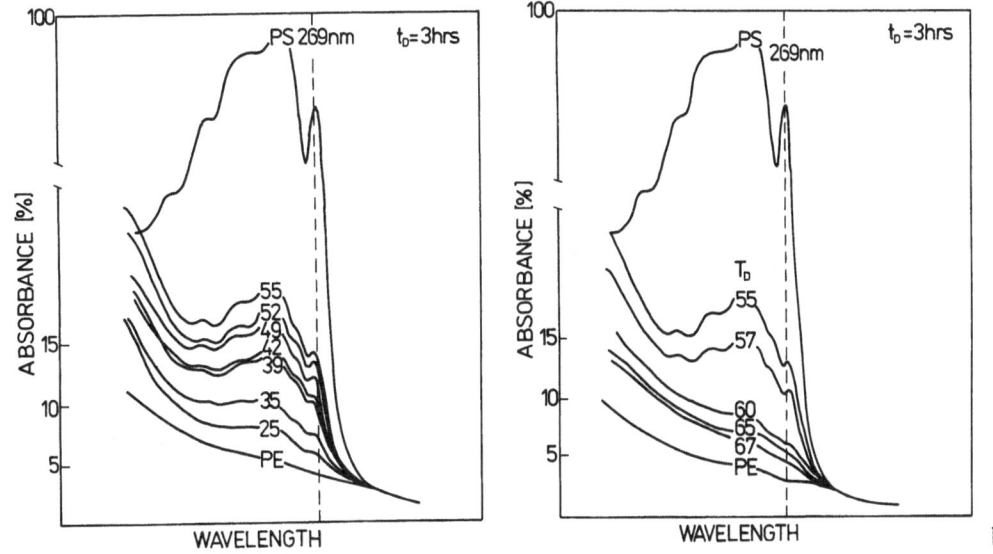

Fig. 2. Changes of the 269 nm absorbance peak for P samples prepared at different diffusion temperatures T_D; a = temperature range from 25° to 55 °C; b = temperature range from 55° to 67 °C

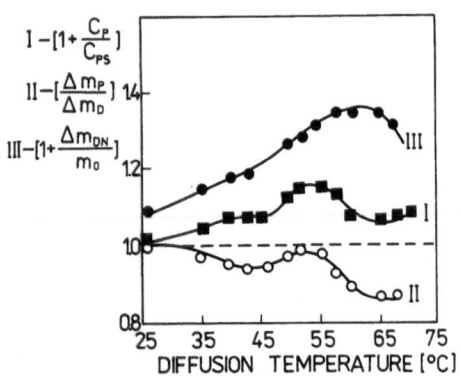

Fig. 3. Diffusion temperature dependence of $\left(1 + \dfrac{C_P}{C_{PS}}\right)$ (I), of $\left(\dfrac{\Delta m_P}{\Delta m_D}\right)$ (II), and of $\left(1 + \dfrac{\Delta m_{DN}}{m_o}\right)$ (III)

65 °C clearly appear. Curve II represents the changes of the ratio of Δm_P (mass increase of the sample after polymerization) to Δm_D (mass increase of the sample after ST diffusion in the matrix) determined gravimetrically.

For N samples the content of styrene Δm_{DN} was determined gravimetrically immediately after its diffusion into the film.

Curve III in figure 3 depicts the dependence of $\left(1 + \dfrac{\Delta m_{DN}}{m_o}\right)$ as function of T_D.

Temperature dependence of crystallinity

The degree of crystallinity (W_R) was measured by X-ray diffraction from 110 and 200 reflections using DRON-2 diffractometer (U.S.S.R.), equipped with a controlled temperature chamber. The samples were heated from 25 °C to melting temperature. The results for N samples are shown in figures 4 a and 4 b, and for P type samples in figures 5 a and 5 b. It can be seen that the $W_R = f(T_D)$ functions in the case of P samples are similar to those for N type

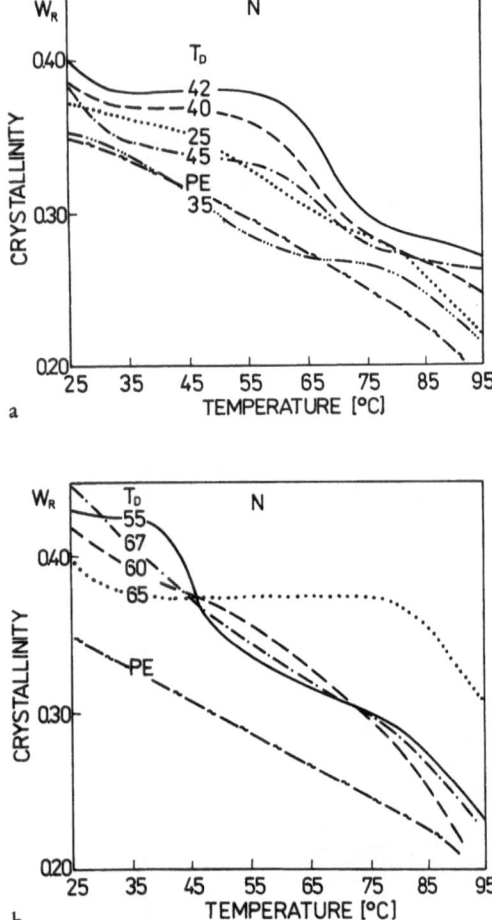

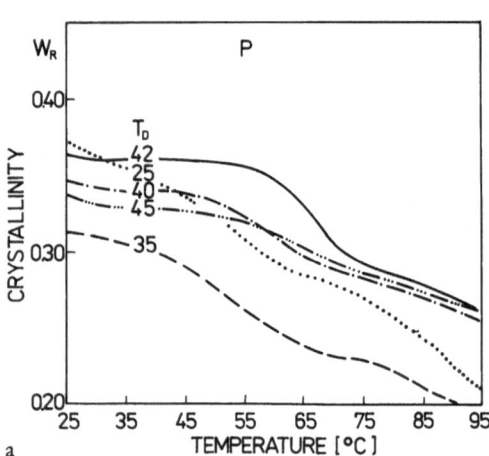

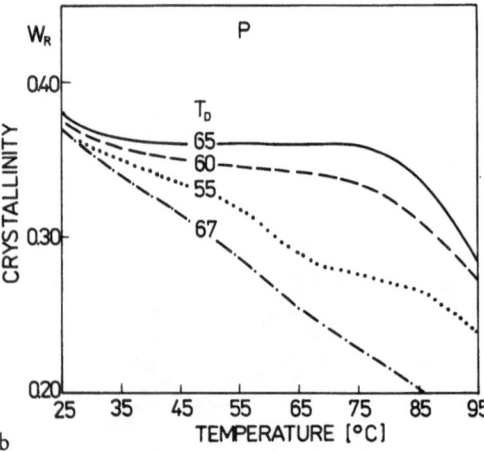

Fig. 4. Temperature dependence of crystallinity of non-polymerized samples (N) saturated with ST at different temperatures T_D; a = temperature range from 25° to 45 °C; b = temperature range from 55° to 67 °C

Fig. 5. Temperature dependence of crystallinity of non-polymerized samples (P) saturated with ST at different temperatures T_D; a = temperature range from 25° to 45 °C; b = temperature range from 50° to 67 °C

samples. It should also be noticed that the crystallinity of *P* films decreases in comparison with *N* samples measured at the same temperature.

Discussion

The specific properties of macromolecules containing a small amount of noncrystallizable units are responsible for the controversy concerning the nature of the crystalline state of such polymers. In the case of branched polyethylene various points of view have been presented. According to the first, the crystalline phase consists of lamellae of varying perfection and stability [6, 7]. The contrary view presented by Fischer et al. [8] proposes an increase of the thickness of the interlamellar amorphous layers which appears in the long spacing growth during annealing. Theoretical and experimental data given by Strobl et al. [9] lead to the other model of LDPE structure. A model of the interlamellar amorphous phase is proposed and the influence of that phase on the secondary crystallization effect, which can be observed even at room temperature, is discussed. According to this work, during the crystallization of lamellae all noncrystallizable chain elements are transported to their surface and the concentration of these disordered unit rises. Electron density measurements have shown that there is a thin paracrystalline core (fig. 6, region I) on the lamella surface, the density of which is similar to the lamellae. The concentration of the extruding elements decreases with the distance from the lamella surface (fig. 6, region II) and approaches a plateau (fig. 6, region III). In the plateau region, the secondary crystallization is possible. Lamellae formed at a lower temperature are characterized by a lower content of long, unperturbed chain sequences than in the case of primary lamellae. The perfection and stability of the primary lamellae are higher. However, this model cannot fully explain the melting processes and the structural changes induced

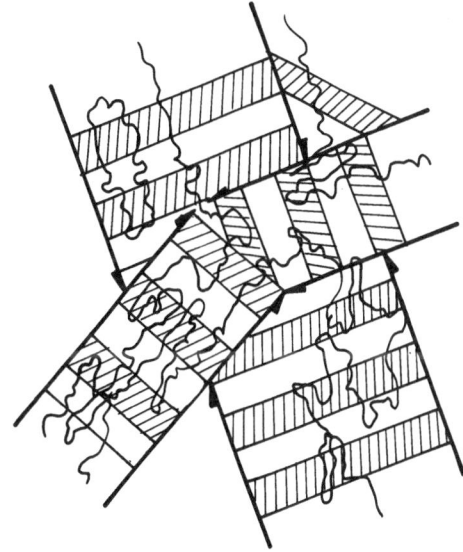

Fig. 7. Model of the cluster network [10]

by swelling and subsequent removal of the solvent, e. g. by evaporation.

Interesting new possibilities of analysis of such systems are offered by the colloidal multicomponent model proposed by Kilian [10, 11]. Low density polyethylene can be regarded as a pseudo-eutectoid copolymer with a large concentration of the noncrystallizing co-units. The superstructure of such a polymer is assumed to be built up by clusters consisting of a certain number of stacked lamellae, as depicted in figure 7. The lamellae in the cluster are arranged parallel. The structure of such systems consists of a preponderance of polymer chains between the crystals which form a cluster network. In the unswollen state the amorphous parts of the chains occur in a disordered state. The swelling of these systems can result in the rearrangement of their configuration into more extended anisotropic ones. The description of the observed structural changes in the swollen state can be mode using the multicomponent colloidal approach proposed above, which offers consistent thermodynamic treatment.

Our structural studies concern the LDPE samples from which ST has been removed by evaporation (completely for *N* samples, partially for *P* samples during the polymerization process), thus they are not really examined in the swollen state. For such films it is possible to take into consideration the existence of an amorphous phase and crystalline phases of different order as in the model proposed by Pakuła [12]. This model can be treated as a limiting case of Kilian's multicomponent colloidal systems.

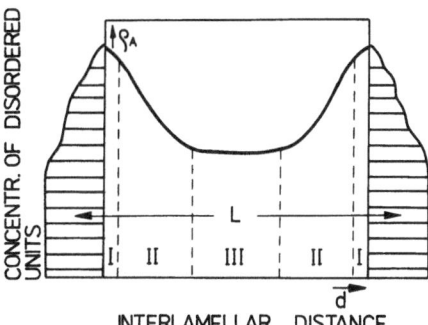

Fig. 6. Model of the interlamellar amorphous phase [9]

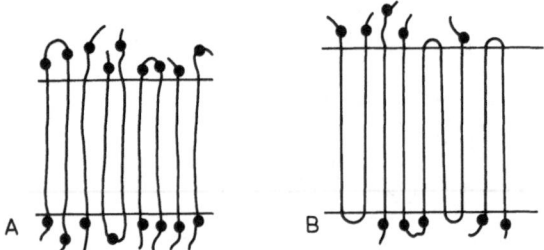

Fig. 8. Model of crystalline phases in LDPE [12]

Discussing temperature dependence of the long spacing, Pakuła has suggested a model of the crystalline phase in LDPE. Lamellae formed at a lower temperature consist of these chain segments which have a length greater than the critical one at this temperature (fig. 8, scheme A). The remaining parts of the segments contribute to the amorphous phase. One can suppose that this kind of crystalline unit should be responsible for the secondary crystallization. Primarily existing lamellae (fig. 8, scheme B) are formed from much longer chain segments than the critical length. In this case the polymer chain can be folded, and the lamellae formed show a higher thermal stability and perfection than lamellae presented in scheme A. The above proposed models seem to be useful for describing the effects presented in this work.

If LDPE includes two different kinds of crystalline elements, styrene as a solvent should reveal these structural differences. At low temperatures this solvent should at first swell the amorphous phase, causing its anisotropic rearrangement. The increase of diffusion temperature causes the deviation of the $W_R = f(T_D)$ function from its linear course, which was observed for nontreated LDPE (fig. 4a). Above $T_D = 35\,°C$ the curves have a two-step character and at $T_D = 42\,°C$ a characteristic trace-like shape is observed. Two plateaux of $W_R = F(T)$ at this temperature T_D can be related to the presence of the different ordered phases. The melting of the less stable one is observed at the 65–75 °C region and the more perfect lamellae melt above 85 °C. Below a temperature range $T_D = 40–45\,°C$ the recrystallization effect is stronger and the volume of the amorphous phase containing styrene gradually decreases. The course of the curves presented in figures 4 and 5 correspond to that predicted by Kilian's model (see fig. 3 in this work) [11].

These facts become clearly visible in the decrease of the content of styrene polymerized in the matrix (fig. 3, I) and in a decrease of polystyrene mass as compared with a mass of absorbed, liquid styrene (fig. 3, II). In the temperature range $T_D = 40–45\,°C$ samples show a higher content of both ordered phases. Maxima observed at temperatures $T_D = 50–55\,°C$ (fig. 3) can be connected with the highly disordered phase, formed from the swollen, para-crystalline elements presented in figure 8, scheme A. Thus the styrene absorption is the highest and attains 38 % before and 18 % after polymerization. Above $T_D = 55\,°$ the extraction of the amorphous phase starts to be observed and the mass of the sample rapidly decreases (fig. 3, III). It can be expected that at higher T_D the diffusion of ST makes it possible to obtain samples containing only one kind of crystalline element, the B type. Such a sample is characterized by a high, almost invariable degree of crystallinity ($T_D = 65\,°$), and a very low styrene absorption capacity. Further heating leads to the dissolution of the films.

The structural changes in LDPE films can be better visualised analysing the dependencies depicted in figures 9a and 9b. These curves were obtained from the data presented in figures 4 and 5. One can discuss, as an example, the curve corresponding to $T = 65\,°$. It shows that at lower temperatures T_D, the crystallinity decreases as in nontreated LDPE. Above $T_D = 35\,°$ a distinct increase of crystallinity can be noticed. Maximum at $T_D = 40–45\,°$ is caused by swelling-induced crystallization and a higher content of both kinds of crystalline lamellae. The decrease of crystallinity, at still higher temperatures, may be connected with the appearance of a more disordered phase originating from the A-type para-crystalline elements. At $T_D =$ from 50° to 60 °C a further increase of crystallinity can be related to the strong effect of secondary crystallization and washing-out of the new-formed discordered phase. The last W_R maximum (at $T_D = 65\,°$) is related to the presence of the B crystalline elements only. Experimental data given by Pakuła [12] allow us to calculate the theoretical temperature dependence of crystallinity for two model polyethylenes containing only one (type A or B) isolated ordered phase in the amorphous matrix (fig. 10, curves W_A and W_B). If $T_D = 67\,°$ appears to be the temperature of melting of B-type lamellae, the curve $W_R = f(T)$ for films obtained at this diffusion temperature should be similar to the W_B plot. Figure 10 confirms this suggestion. The knowlege of the $W_R = f(T)$ for nontreated LDPE, containing both A and B type lamellae (fig. 10, PE), makes it possible to calculate (by subtraction) the curves characteristic of the polymer containing only A-type lamellae (W'_A-curve in fig. 10). The similarity of the courses $W_A = f(T)$ and $W'_A = f(T)$ is easily seen in this figure.

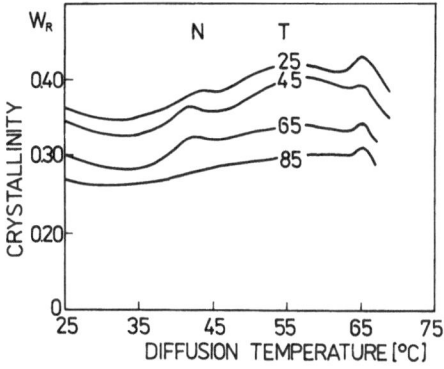

a

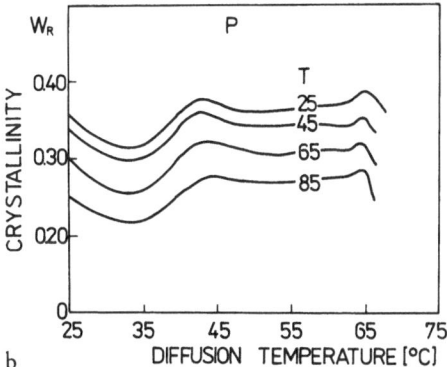

b

Fig. 9. Crystallinity changes of investigated films as function of T_D; a = non-polymerized samples (N); b = polymerized samples (P)

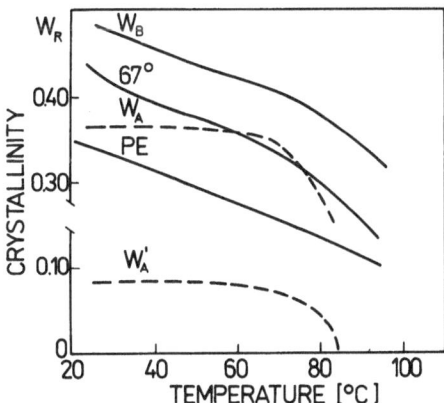

Fig. 10. Comparison of temperature dependence of crystallinity of ST treated at $T_D = 65\,°C$ film and LDPE sample with theoretical curves calculated from the date given in [12]

These observations seem to show that this model, assuming the existence of two different crystalline phases, may be useful in the analysis of the influence of a diffusant, with a limiting solvent power, on the structural changes in this polymer.

Less ordered crystalline phases seem to be similar to the anisotropic amorphous microphase formed during the swelling process as discussed in Kilian's model. This phase can recrystallize and is responsible for structural changes observed at lower temperatures in *P* and *N* samples.

Conclusions

The results described above show that the diffusion temperature of styrene into LDPE determined the structural modification of the host polymer. This treatment, dependent on diffusion temperature T_D, causes:

1. washing-out of the amorphous and disordered phases,

2. modification of the LDPE structure leading to a change of its crystallinity degree,

3. in the range of T_D for 45° to 60° corresponding to intense penetration of ST into LDPE, the swelling of the less ordered crystalline phase enables a secondary crystallization of the matrix during both polymerization and evaporation of ST. This recrystallization leads to the formation of a reorganized, more perfect paracrystalline phase,

4. at temperature T_D above 65° the relative content of more perfect lamellae increases.

These conclusions on the influence of the ST treatment of LDPE are supported by the studies of samples from which ST was eliminated by evaporation. However, it should be noticed that these structural modifications are more pronouced for samples (*N*) than for films containing polymerized ST. The presence of polystyrene partially hinders these structural transformations. These results show that gradient polymers obtained from semi-crystalline polymers exhibit properties which are not only determined by the gradient of the diffusant but also by the modified structure formed during diffusion and interaction with the partially dissolving monomer.

References

1. Shen M, Berger B (1972) J Mater Sci 7:741
2. Jasso CF, Hung SD, Scen M (1978) ACS Polymer Preprints 19:1, 63
3. Akovali G, Biliyar K, Shen M (1976) J Appl Polym Sci 20:2419
4. Czarczyńska H, Trochimczuk W (1974) J Polym Sci Symp 47:111
5. Kryszewski M, Czeremuszkin G (1980) Plaste u Kautschuk 11:605

6. Flory PJ (1955) Trans Farad Soc 51:848
7. Kilian HG (1969) Kolloid Z u Z Polym 231:534
8. Fischer EW, Martin B, Schmidt GF, Strobl GR (1968) IUPAC Toronto, Preprints A-6:17
9. Strobl GR, Schneider MJ, Voigt-Martin JG (1980) J Polym Sci, Polym Phys Ed 18:1361
10. Kilian HG, Wenig W (1974) J Makromol Sci Phys B 9:463
11. Kilian HG, Maier E (1979) J Polym Sci Polym Phys Ed 7:1531
12. Pakuła T (1982) Polymer 23:1300

Received June 11, 1985;
accepted August 9, 1985

Authors' address:

P. Milczarek and M. Kryszewski
Centre of Molecular and Macromolecular Studies
Polish Academy of Sciences
90-362 Łódz, Boczna 5, Poland

Progress in Colloid & Polymer Science   Progr Colloid & Polymer Sci 71:103–112 (1985)

Desmeared, slit-smeared and projected SAXS: a comparison of various methods of evaluation for segmented polyurethanes*)

H. Meyer and R. Bonart

Institut für angewandte Physik, Universität Regensburg, Regensburg, F.R.G.

Abstract: Evaluation methods for X-ray small angle scattering of segmented polyurethane elastomers are discussed using the assumption of isolated, randomly distributed hard segment domains in order to test the reliability of evaluation.

Projection of the intensity distribution onto the meridian of the scattering diagram reduces the analysis to a one-dimensional problem, since the projection is related to a chord through the three-dimensional structure. Furthermore, the chord may be identified with the one-dimensional structure of an ideal lamellar stack, so that the projection of the intensity is comparable with the well-known result of J. J. Hermans for lamellar stacks.

The projection has to be calculated in the three dimensional Fourier space either by desmearing the slit-smeared scattering intensity or by projecting it directly onto the meridian. Practically identical results are obtained by both procedures, so that the desmearing operation may be omitted in the present case. We show that the damping function, which is due to domain boundary diffuseness, can be suitably and directly eliminated from the slit-smeared intensity by applying a method proposed by Koberstein.

A reproducibility of the final structural parameters within 10–15 % can be achieved using the methods described.

Key words: Two-phase structure, "sponge" structure, small angle X-ray scattering, slit-smeared scattering curves, polyurethanes.

Introduction

In a previous paper a procedure for evaluating small-angle X-ray scattering curves of isotropic, "sponge-like", two-phase systems was presented [1]. The procedure, which is essentially based on the validity of Porod's asymptotic law and makes use of a projection of the scattering effect on a straight line in reciprocal space, will be applied to segmented polyurethane elastomers containing glassy or semicrystalline hard segment domains dispersed in an amorphous soft segment matrix. A knowledge of the structural parameters of the two-phase system (volume fractions, chord lengths, inner surface and the density difference between the phases) is an important prerequisite for being able to relate mechanical properties of the substances with their morphology.

The projection on the meridian in reciprocal space is related to the chord through the scattering structure. For isotropic sponge structures this chord can be identified with the chord through an ideal lamellar packet. Accordingly, the intensity projection can be compared with the well-known scattering formula due to J. J. Hermans, the parameters of which must be found by means of suitable fitting to the experimental data. Previously, Poisson distributions were employed as distribution functions for the chord lengths. In this way one obtains not only the Porod constant and the Porod invariant, but also the structural parameters mentioned above without the need for additional information. To what extent other distribution functions fit the experimental data better will not be discussed for the present.

*) Dedicated to Prof. Dr. H.-G. Kilian on the occasion of his 60th birthday.

The starting-point for the evaluation procedure is a slit-smeared small-angle X-ray scattering curve such as is obtainable from measurements from a Kratky camera. Since it is advantageous to consider the structural parameters for idealized two-phase systems, it is necessary to calculate the scattering curve for these idealized conditions. Firstly, the diffuse background due to molecular density fluctuations within the individual phases must be subtracted. In addition, diffuse transition zones between the phases influence the form of the scattering curve via a damping function D. After mathematical elimination of these effects, the scattering curve then obtained is based on a two-phase system with homogenous densities in the phases and with sharp boundaries between the phases. In order to ascertain the volume fractions and size distributions of the individual phases, the above-mentioned projection is made and the problem thereby reduced to one dimension.

There are several different ways of carrying out the analysis. The slit-smeared intensity distribution J can first be desmeared ($J \rightarrow I$) in order to calculate, starting from the desmeared scattering curve I, the two-dimensional intensity projection P on the meridian in reciprocal space ($I \rightarrow P$). So, a smearing must first be eliminated ($J \rightarrow I$) in order to reintroduce it subsequently ($I \rightarrow P$). Hence, it was obvious to ask whether one could circumvent — if applicable — the initial desmearing. Since the desmearing involves a truncation effect, the reliability of evaluation could decrease by this. In a further paper [2], a method was therefore outlined which aims at an evaluation of the SAXS curve by avoidance of a desmearing procedure.

On the other hand, the damping function D due to the diffuse phase boundaries raises problems. This function must be obtained by fitting the scattering intensity to Porod's asymptotic law and must be eliminated by division, which has to be done, strictly speaking, on the desmeared scattering curve.

In order to ascertain if either the "theoretically exact", but numerically problematical evaluation with desmearing, or a suitable approximation which can be handled more easily numerically works more reliably, different ways of evaluation were chosen and are discussed in comparison in the present work. The following scheme gives a survey of possibilities of evaluation, in which in each case the diffuse scattering background is first separated.

$$J \rightarrow J - U = J' \rightarrow I' \rightarrow I'/D = I'' \rightarrow P'' \qquad \text{(A)}$$
$$\rightarrow J'/D_v = J'' \longrightarrow P'' \qquad \text{(B)}$$
$$\rightarrow P' \longrightarrow P'/D_p = P'' \qquad \text{(C)}$$

A) The intensity J' above the background can be desmeared and, after desmearing, divided by the damping function D. Subsequently the quotient $I'' = I'/D$ has to be smeared doubly in order to obtain the two-dimensional intensity projection P'' on the meridian of reciprocal space, which is comparable to Hermans's formula.

B) On the other hand, one can do without the desmearing and subsequently double smearing and can directly calculate the intensity projection P'' from the slit-smeared scattering intensity J' after this has been divided by a damping function D_v.

C) Finally, one can first obtain the projection P' from the slit-smeared scattering J' and then divide by a damping function D_p.

The present work will now explain the evaluation steps especially necessary for methods A and B and will show that in both cases the same results are obtained within limits of error. Method C is described in [2]. The obtained structural parameters are discussed in connection with the chemical composition.

Experimental

Samples

Polyurethane elastomers were prepared by the two-step or prepolymer procedure and comprise as hard component 4,4'-diphenylmethane diisocyanate (MDI) with a low molecular weight diamine as chain extender and as soft component polytetramethylene ether (PTME) with an average molecular weight of either 2000 or 1000. The compositions of the materials can be found in table 1. Polymer films cast from solution were dried in a vacuum oven for several weeks.

Small-angle X-ray scattering

SAXS measurements were carried out by means of monochromatic CuK_α radiation with a Kratky camera equipped with a position sensitive detector. For the determination of absolute intensities the primary beam is recorded simultaneously with the scattering curve.

Table 1. Substances

Sample	Hard segment	Soft segment	Reaction ratio molar	Weight-% hard segment
	MDI + CE[a]		PTME:MDI:CE[a]	
PU 1	MDI+EDA[a]	PTME 2000	1:1.87:0.87	20.6
PU 2	MDI+EDA	PTME 2000	1:2:1	21.9
PU 3	MDI+EDA	PTME 2000	1:2.25:1.25	24.2
PU 4	MDI+EDA	PTME 1000	1:2:1	35.9

[a]) MDI = 4,4'-diphenylmethanediisocyanate; CE = chain extender; PTME = polytetramethylene ether; EDA = ethylenediamine

In order to avoid overloading the counting tube, the primary beam must be attenuated by an absorber in a defined manner. Parasitic scattering is eliminated by subtracting the scattering of the empty camera.

Theoretical foundations

The absolute intensity $i(b)$ of small-angle X-ray scattering is described with usual approximations by

$$i(b) = \frac{1}{a^2} \, P_o f_e \, \text{Int}(b). \tag{1}$$

Here, a is the distance between sample and detector, f_e is the Thomson factor, P_o is the primary intensity and $\text{Int}(b)$ is the interference function with $\vec{b} = (b_1, b_2, b_3)$ and $|b| = 2 \sin \theta / \lambda$ as scattering vector for the scattering angle 2θ and wavelength λ. When the interference function is normalized to the number of electrons participating in the scattering, one thus obtains with the mean electron density ϱ in the illuminated sample volume V

$$i(b) = \frac{1}{a^2} \, P_o f_e \varrho V \, [\text{Int}(b)]. \tag{1a}$$

where $[\text{Int}(b)]$ shall indicate the interference function related to one electron. The known factors and those which have to be determined from the scattering geometry are summarized to a constant C, so that the absolute intensity is given by

$$i(b) = C \, \varrho \, [\text{Int}(b)] = C \cdot I(b). \tag{1b}$$

In what follows we work with the intensity $I(b)$, which is obtained in absolute units (electrons^2 per volume) after normalization of the experimentally found scattering curve.

For isotropic real two-phase systems one can imagine the SAXS curve $I(b)$ to be composed of the scattering effect $I''(b)$ of a corresponding idealized two-phase system, a background $U(b)$ due to molecular density fluctuations within the individual phases and a damping function $D^2(b)$ for diffuse phase boundaries [3, 4].

$$I(b) = I''(b) \cdot D^2(b) + U(b). \tag{2}$$

Here, and in the following, if a distinction is necessary, nonprimed symbols will refer to real conditions. Symbols with one prime refer to singly idealized structures, i. e. molecular density fluctuations in the phases are eliminated. Symbols with two primes refer to doubly

idealized conditions, i. e. where additionally the phase boundaries are assumed to be sharp.

The small-angle scattering of an ideal two-phase system with sharp phase boundaries was treated by Porod [5] and gives for large values of b a fall-off in intensity as $1/b^4$.

$$I''(b) \xrightarrow[b \to \infty]{} \frac{K}{b^4}. \tag{3}$$

Here, K is the so-called Porod constant which is related to structural parameters of the system such as the specific inner surface Ω and the individual electron densities.

$$K = \frac{1}{(2\pi)^3} \, (\varrho_D - \varrho_M)^2 \cdot \Omega \tag{4}$$

where ϱ_D represents the density of the domains and ϱ_M the density of the matrix. If w is the volume fraction of the domains, then for the mean density ϱ of the substance is valid

$$\varrho = w \cdot \varrho_D + (1 - w) \cdot \varrho_M. \tag{5}$$

The specific inner surface Ω is connected with the average chord lengths x_D and x_M through domains and matrix according to Porod and Mittelbach [6] by

$$\Omega = \frac{4}{x_D + x_M}. \tag{6}$$

The relative volume fraction of the domains results once more from the average chord lengths

$$w = \frac{x_D}{x_D + x_M}. \tag{7}$$

The integral of the scattering curve $I(b)$ affords the Porod invariant in the form

$$Q = \int\int\limits_{-\infty}^{\infty}\int I(b) \, d^3 b = \int\limits_0^{\infty} 4\pi \, b^2 \cdot I(b) \, db$$
$$= w(1-w) \, (\varrho_D - \varrho_M)^2. \tag{8}$$

For the description of the transition zones between the phases different models are discussed in the literature [3, 4, 7, 8]. In general, these transition zones can be taken in consideration by convoluting the electron density distribution ϱ'' of the idealized system — repre-

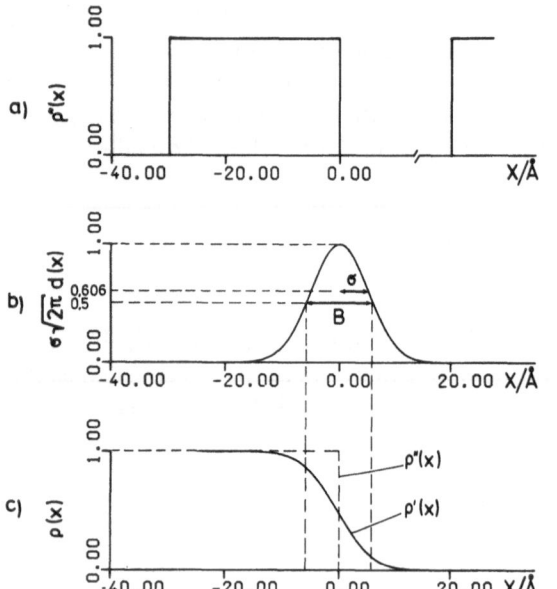

Fig. 1. a) Electron density distribution in the ideal two-phase system; b) Gaussian curve as damping function with $\sigma = 5$ Å; c) Phase transition zone with diffuse phase boundary produced by convolution of the step function a) with the Gaussian curve b)

sented in any chord by step functions (cf. fig. 1a) — with a smoothing function $d(x)$

$$\varrho'(x) = \varrho''(x) * d(x).$$ (9)

In the following, as model for transition zones exclusively a form according to figure 1c will be considered, i.e the smoothing function $d(x)$ is represented by a normalized Gaussian curve (fig. 1b)

$$d(x) = \frac{1}{\sigma\sqrt{2\pi}}\, e^{-x^2/2\sigma^2}.$$ (10)

As width B of the transition zone, the full width at half height of the Gaussian curve is used:

$$B = 2\sqrt{\ln 4}\,\, \sigma.$$ (11)

Since the intensity function is calculated as the Fourier transform of the self-convolution of the electron density, one obtains for the damping function in equation (2) using the convolution theorem of Fourier transformation

$$D^2(b) = e^{-4\pi^2\sigma^2 b^2}.$$ (12)

In Porod's law region one then obtains after subtracting the background $U(b)$ an intensity profile of the form

$$I'(b) = \frac{K}{b^4}\, e^{-4\pi^2\sigma^2 b^2}.$$ (13)

The background $U(b)$ is taken to a first approximation as independent of b [3, 4].

Treatment of slit-smeared scattering curves

Using a line focus for X-ray patterns such as in the Kratky system causes a smearing of the scattering curves. Porod's asymptotic law in the form of equations (3) and (13) refers, however, to three-dimensional intensity functions isotropic in reciprocal space such as they are obtainable by use of an ideal pin-hole collimation. One must therefore desmear the scattering profiles obtained from Kratky slit collimation by means of mathematical methods or adapt the corresponding equations of the preceding paragraph to the conditions of line collimation. The use of an infinite slit produces an intensity distribution of the form

$$J'(b) = 2\int_0^\infty I''(\sqrt{b^2+y^2}) \cdot D^2(\sqrt{b^2+y^2})\, dy.$$ (14)

For the region of validity of Porod's law and by assuming a Gaussian curve equation (12) as damping function it was shown by Ruland [3] that the intensity function of equation (14) can be written as

$$J'(b) = \frac{K'}{b^3}\, D_v^2(b) \text{ with } K' = \frac{\pi K}{2}$$ (15)

where

$$\begin{aligned} D_v^2(b) = {} & (1 - 8\pi^2\sigma^2 b^2) \cdot \text{erfc}(2\pi\sigma b) \\ & + 4\sqrt{\pi}\,\sigma b \exp(-4\pi^2\sigma^2 b^2) \end{aligned}$$ (16)

was given in terms of the complementary error function erfc.

In order to simplify the practical evaluation, Koberstein [7] has fitted the expression of equation (16) by the simple relation

$$D_v^2(b) = e^{-a(\sigma b)^n}$$ (17)

and has found thereby $a = 38$ and $n = 1.81$, i.e. a plot $\ln(J'b^3)$ vs. $b^{1.81}$ should afford a straight line in Porod's law region. The applicability of this empirical relation for determination of phase boundary diffuseness has been discussed by Koberstein [7, 9, 10].

The Porod invariant corresponding to equation (8) can be calculated from slit-smeared scattering curves as

$$Q = \int_0^\infty 2\pi b \cdot J(b) \, db. \tag{18}$$

Projection of scattering curves

In order to be able to completely characterize the two-phase system only from the knowledge of the SAXS curve we calculate the projection of the scattering effect onto a straight line in reciprocal space [1, 2] and thus reduce the problem to one dimension.

After separation of the background and determination of the phase boundary widths the experimental scattering curves are calculated for idealized conditions. The projection curve of the ideal spongelike two-phase system subsequently obtained is related to the density distribution along a chord and will be compared with a lamellar packet (fig. 2) to which the Hermans interference function is applicable [11, 12].

The projection $P(b_3)$ of a three-dimensional scattering curve $I(b)$ is given by

$$P(b_3) = \int\int_{-\infty}^{\infty} I(b) \, db_1 \, db_2. \tag{19}$$

Owing to the spherical symmetry this can be transformed to

$$P(b_3) = \int_{b_3}^\infty 2\pi b \cdot I(b) \, db. \tag{20}$$

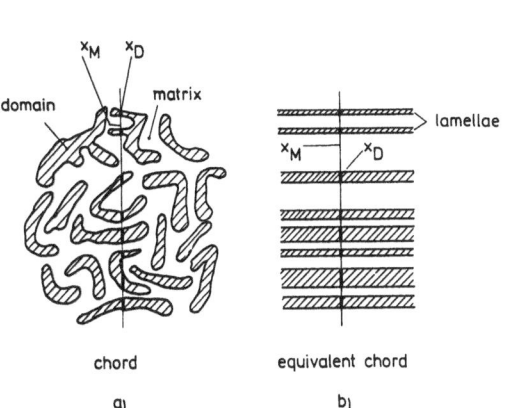

sponge structure   equivalent lamellar packet

Fig. 2. Comparison of the chords of sponge structure and lamellar packet

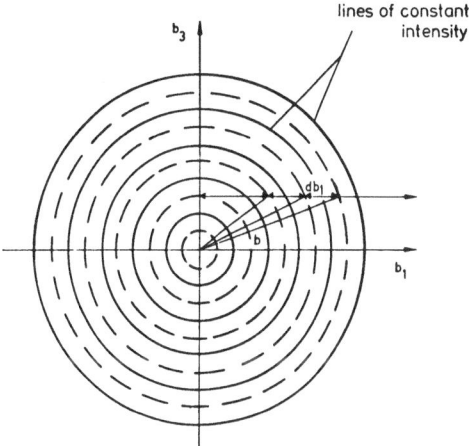

Fig. 3. Projection of slit-smeared scattering curves: the intensity function in reciprocal space radially symmetrical in the b_1-b_3-plane is projected along b_1 on the b_3-axis

For Porod's law region one obtains

$$P(b_3) \xrightarrow[b \to \infty]{} \int_{b_3}^\infty 2\pi b \, \frac{K}{b_4} \, db = \frac{\pi K}{b_3^2}. \tag{21}$$

The projection curve can also be calculated directly from the slit-smeared scattering curve. The scattering effect, radially symmetrical in reciprocal space in the b_1-b_3-plane (slit-smearing in the b_2 direction), is projected onto the b_3 axis by (cf. fig. 3)

$$P(b_3) = 2 \int_0^\infty J(b_1, b_3) \, db_1. \tag{22}$$

With the substitution $b_1 = \sqrt{b^2 - b_3^2}$ it follows that

$$P(b_3) = 2 \int_{b_3}^\infty J(b) \, \frac{b}{\sqrt{b^2 - b_3^2}} \, db. \tag{23}$$

For Porod's law region one obtains again

$$P(b_3) \xrightarrow[b \to \infty]{} 2 \int_{b_3}^\infty \frac{\pi K}{2b^3} \, \frac{b}{\sqrt{b^2 - b_3^2}} \, db = \frac{\pi K}{b_3^2}. \tag{24}$$

The Porod invariant can also be calculated, alternatively to equations (8) and (18), after forming the projection by means of simple integration of $P(b_3)$.

$$Q = 2 \int_0^\infty P(b_3) \, db_3. \tag{25}$$

In order to obtain information about the chord lengths from the projection curve equations (20) or (23), the Hermans formula equation (26) is fitted to it.

$$L(b_3) = \frac{1}{2\pi^2 b_3^2} Re \left\{ \frac{(1 - V_D(b_3))(1 - V_M(b_3))}{1 - V_D(b_3) V_M(b_3)} \right\} \quad (26)$$

where

$$V_D(b) = F\{v_D(x)\} \text{ and } V_M(b) = F\{v_M(x)\} \quad (27)$$

are the Fourier transforms of the distributions $v_D(x)$ and $v_M(x)$ of both domain and matrix chords, which have to be set up suitably for practical evaluation.

For this we have used Poisson distributions because on the one hand they have a suitable form, and on the other hand they can easily be mathematically processed. This distributions for the domain and matrix chords will therefore be described by

$$v_D(x) = \alpha^2 x e^{-\alpha x} \text{ and } v_M(x) = \beta^2 x e^{-\beta x} \quad (28)$$

where the mean values for these distributions and the average chord lengths are given by

$$x_D = \frac{2}{\alpha} \text{ and } x_M = \frac{2}{\beta}. \quad (29)$$

The required Fourier transforms of the distributions are given by

$$V_D(b) = \left(1 + \frac{2\pi i b}{\alpha}\right)^{-2} \text{ and}$$

$$V_M(b) = \left(1 + \frac{2\pi i b}{\beta}\right)^{-2}. \quad (30)$$

In the selection of these distribution functions there is a certain arbitrariness involved. However, the possibility remains of substituting other functions into equation (26) in order to improve the quality of fitting.

Experimental results and discussion

Background separation

The first step to evaluate the scattering curves which were normalized to absolute units and corrected for parasitic scattering effects consists in the separation of a background due to molecular density fluctuations in the individual phases. This background is assumed, to a first approximation, to be constant [3, 4] and can be

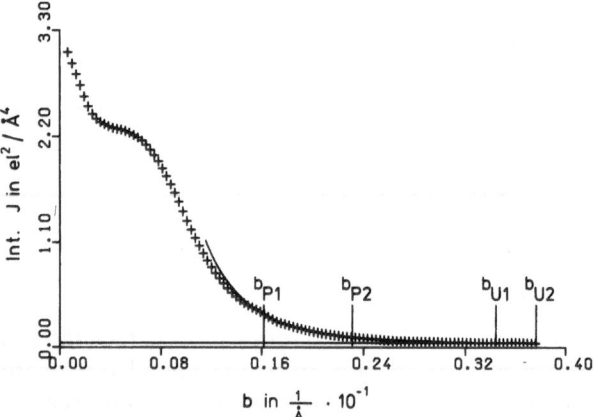

Fig. 4. Slit-smeared SAXS curve with constant background fitted in the interval b_{U1}-b_{U2} and with Porod's asymptotic function fitted in the interval b_{P1}-b_{P2} (solid line)

found from the tail region of the scattering curve at large b. Figure 4 shows a slit-smeared scattering curve with a background fitted in the region b_{U1}-b_{U2}.

Other approaches to separate the background, which assume an increasing of the scattering effect at larger values of b and are discussed in the literature [7, 8, 13], seem to make no improvements for the substances examined here. In addition, the Porod's law region (cf. later) can be found at significantly smaller values of b than those for background separation. Since the intensity increases with decreasing b, the scattering curve lies markedly above the background level in the chosen Porod's law region so that a possibly incorrect background separation is only marginally noticeable in the evaluation. Moreover, the tail region of the scattering curve is substituted for further evaluation by the analytical relation of Porod's law. With this, it is guaranteed that for further considerations the background level is in fact only reached in the limiting case $b \to \infty$.

On the other hand, from the Porod plot one has the possibility of optimizing the background separation. Irrespective of different effects (such as finite slits, detector linearity) which influence particularly the tail region of the scattering curves, but can be minimized, an extension of the linear region in figure 7 a–c can be obtained by variation of the background. The curves of figure 7 allow one in any case to recognize a linear region. From the straight line fitted in an interval b_{P1}-b_{P2} the scattering curve in the tail region deviates on an average upwards if the background was chosen too low and deviates downwards if the background was chosen too high.

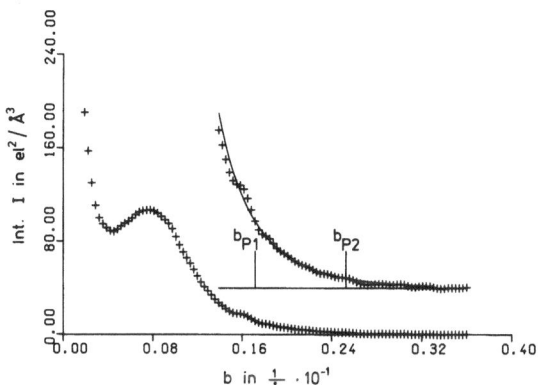

Fig. 5. Desmeared SAXS curve with tail region plotted on a larger scale and with Porod's asymptotic function fitted in the interval b_{P1}-b_{P2} (solid line)

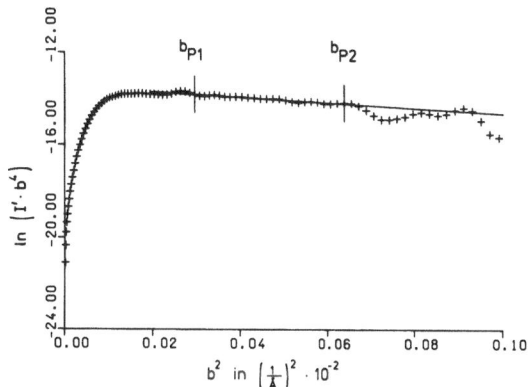

Fig. 6. Plot $\ln (I' b^4)$ vs. b^2 of the scattering curve figure 5 with Porod's asymptotic function fitted in the interval b_{P1}-b_{P2} (solid straight line)

Fitting in the Porod's law region

For further evaluation the scattering curves were either desmeared after background separation according to a procedure of Strobl [14], or they were analyzed without desmearing. The results obtained by both methods will be compared with each other. An example of a desmeared scattering curve is given in figure 5. One should mention that a distinct maximum cannot be observed in all cases.

Corresponding to equation (13) the desmeared scattering curve in a plot $\ln (I' b^4)$ vs. b^2 should yield in the Porod's law region a straight line (fig. 6) from the slope of which one can obtain the width B of the phase boundaries and from the intercept the Porod constant K. The parameters B and K determined by means of linear regression in a b_{P1}-b_{P2} region (fig. 6) are listed in table 2 for some samples.

For direct determination of the phase boundary diffuseness from the slit-smeared scattering curve one chooses a plot of the form $\ln (J' b^3)$ vs. $b^{1.81}$ according to equations (15) and (17). In a b_{P1}-b_{P2} region once again a straight line is fitted permitting the determination of B and K (fig. 7).

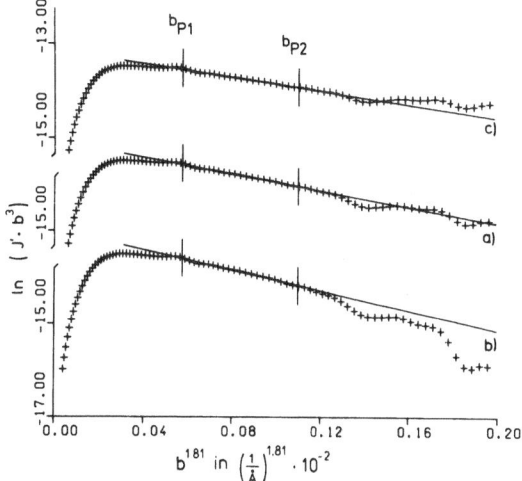

Fig. 7. Plot $\ln (J' b^3)$ vs. $b^{1.81}$ of the scattering curve figure 4 with Porod's asymptotic function fitted in the interval b_{P1}-b_{P2} (solid straight line). a) A background U_1 was separated from the scattering curve J; b) a background $U_2 = U_1 + 20\%$ was separated from the scattering curve J; c) a background $U_3 = U_1 - 20\%$ was separated form the scattering curve J

It is true that equation (15) combined with equation (16) could be fitted directly to the experimental scattering curve by means of least-square methods, but a plot

Table 2. Structural parameters obtained by evaluation with desmearing (symbols and dimensions cf. text)

Sample	U 10^{-2} Å^{-4}	B Å	K 10^{-6} Å^{-7}	Q 10^{-3} Å^{-6}	Q_{id} 10^{-3} Å^{-6}	w %	Ω m^2/cm^3	x_D Å	x_M Å	$\Delta\varrho$ el/Å^3
PU 1	5.05	13.4	1.05	1.20	1.90	21.5	231	37.1	136	0.106
PU 2	4.61	12.9	1.41	1.25	2.14	22.8	288	31.6	107	0.110
PU 3	4.72	12.4	1.35	1.40	2.25	23.7	271	35.1	113	0.111
PU 4	5.67	12.2	1.68	1.51	2.52	23.1	293	31.5	105	0.119

Table 3. Deviations of the structural parameters obtained by evaluation without desmearing compared to that obtained with desmearing (table 2) in per cent (symbols cf. text)

Sample	B	K	Q	Q_{id}	w	Ω	x_D	x_M	$\Delta\varrho$
PU 1	+11	+ 4	−12	− 5	−2	+8	−9	−7	−2
PU 2	+15	+ 3	−12	− 4	−4	+4	−7	−3	0
PU 3	+ 3	−10	−11	−11	+2	+2	−1	−3	−5
PU 4	− 7	−15	− 9	−14	+3	0	+2	−1	−8

aiming at linear relationships has the advantage that a more careful selection of a fitting region is possible. The validity of the approximation equation (17) instead of equation (16) is fulfilled very well in the present cases.

The deviations of the parameters ascertained directly from slit-smeared scattering curves in comparison to those found from the desmeared scattering curve are listed for several substances in table 3.

It has to be emphasized once more that for a reliable determination of the parameters B and K it is necessary in every individual case to carefully select the fitting region for Porod's law. The accuracy of background separation is not so important in comparison. In general we expect an uncertainty of at least ± 10 % in determining the width of the phase boundaries.

For further evaluation, i. e. for determination of the invariant and calculation of the projection, the tail region of the scattering curve is substituted beyond the upper limit b_{P2} of the fitting region by the analytical relation of Porod's law in order to be able to perform the required integrations up to $b \to \infty$.

Moreover, the complete scattering curve must be divided by the determined damping function equation (12) or (17) as a result of the considerations on idealized systems. In the evaluation without desmearing, an approximation is involved in that the damping function equation (17) was derived only in the region of validity of Porod's law for the considered model for phase boundaries, whereas equation (12) can be applied to the total desmeared scattering curve. This has as a result that without desmearing regularly a smaller invariant Q_{id} for the respective idealized two-phase system is calculated than with desmearing, provided that in both cases the same width B of the phase boundaries and the same invariant Q for the singly idealized system were found. But it appears that this effect is negligible in view of the uncertainties in B and Q. It might become more significant if the influence of the phase boundaries is only calculated after projection

formation; this corresponds to method C outlined in the introduction.

If one compares the values of B and K determined in both methods, one then sees from table 3 that statistically deviations of several per cent arise, which are obviously compatible with the general uncertainty in the determination of these values. However, the invariant Q estimated without desmearing seems to be systematically smaller than that with desmearing. The reason for this might lie in the desmearing procedure itself. Indeed the use of a finite slit could also have such an effect which can be easily taken into account by including the primary profile in the calculation in the applied desmearing procedure. One is allowed, in the present arrangement, to proceed from the approximation of infinite slit height, so that a calculation of too small an invariant seems improbable for the sake of incomplete integration caused by slit focus. The form of the primary profile was not taken into account in the evaluation without desmearing.

Projection and fitting of the Hermans formula

After dividing the intensity function by a damping function due to diffuse phase boundaries, the projection is calculated from the desmeared or slit-smeared scattering curve according to equations (20) or (23). The integrations for $b \to \infty$ must be performed by inserting for the tail region of the scattering curves the analytical relation of Porod's law in the corresponding form. For fitting the Hermans formula equation (26) a plot of the form $b^2 \cdot P(b)$ vs. b is chosen (fig. 8). The constant level for large b results from Porod's law according to equations (21) or (24).

To this projection curve one now fits equation (26) in conjunction with equation (30) with the parameters α and β as well as a scaling factor K'' by means of least-square methods. As reported in a previous paper [1] one chooses a fitting region from $b = 0$ up to the begin-

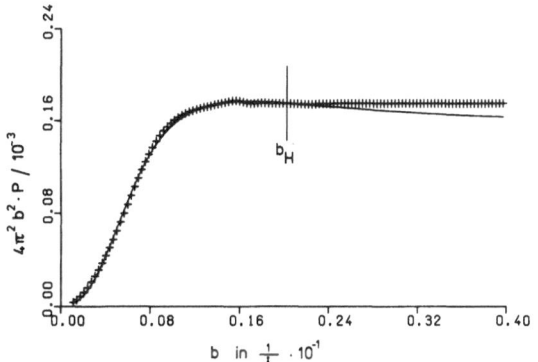

Fig. 8. Projection plot of the scattering curve with Hermans's function fitted in the interval $b = 0$ to $b = b_H$ (solid line)

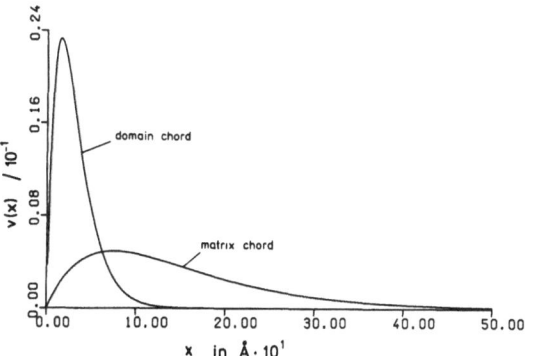

Fig. 9. Poisson distributions for domain and matrix chord with $x_D = 31.6$ Å and $x_M = 150$ Å as parameters

ning of the constant level approximately. Figure 8 shows a fitted curve in the form of a solid line with fitting region from $b = 0$ to $b = b_H$, where in this case the chord lengths were determined to be $x_D = 31.6$ Å and $x_M = 150$ Å. The appropriate Poisson distributions are represented in figure 9.

It appears that the form of the fitting curve changes only slightly, even for marked variations of the larger chord length (in our case therefore of the matrix chord), whereas a variation of the smaller chord length shows a more sensitive reaction to the form of the fitting curve [1, 15]. One must reckon therefore with great inaccuracies in determining the matrix chord from the fitting curve while the domain chord might be obtained more reliably. However this difficulty can be ignored since the system is overdetermined.

From the combination of equations (4), (6), (7) and (8) the domain fraction, with the magnitudes Q, K and x_D now known, is given by

$$w = 1 - \frac{Q}{2\pi^3 x_D K}.\tag{30}$$

The inner surface is then

$$\Omega = \frac{4w}{x_D}.\tag{31}$$

The matrix chord has the form

$$x_M = \frac{4}{\Omega} - x_D.\tag{32}$$

Finally, one obtains for the electron density difference between the phases

$$\Delta \varrho = \sqrt{\frac{Q}{w(1-w)}}.\tag{33}$$

The parameters found for some substances through this scheme complete tables 2 and 3. From table 3, which shows the deviations of the values obtained without desmearing in contrast to those obtained with desmearing, one sees that the projection curves calculated in different ways yield the same results within limits of error. The particularly interesting geometrical parameters of the system, such as volume fractions and chord lengths, seem to be determined in a reproducible way within a few per cent.

Comparison with the chemical composition

The absolute values found are, of course, based upon the model of Poisson distributions, with which a completely satisfying fit of the projection curves cannot be obtained (cf. fig. 8). The application of the described evaluation procedure permits, however, according to the present results, even by avoidance of desmearing, the determination of all parameters in order to characterize sponge-like two-phase systems. In particular, the comparison of measurements on different substances or differently pretreated samples is interesting; this can be easily performed as long as the parameters of the evaluation are not changed.

The values found also lie, absolutely seen, in a "reasonable" order of magnitude, as shown by simple comparison with the chemical composition. Thus, the domain chord of about 30 to 40 Å corresponds to the mean length of a hard segment sequence. The sum of domain and matrix chord is reflected in a long period maximum provided that one can be observed. Similarly the electron density difference can be estimated to be in just the same order of magnitude from the chemical composition and the densities of the single phases on the condition of phase separation in the elastomer.

Only the calculated volume fraction of the domains appears to be too high as absolute value. However from table 2 one can state the tendency for the substances *PU*1 to *PU*3 to be that with increasing reaction ratio the domain content is enhanced. In agreement with this are the results from differential thermo analysis (DTA), where in all three cases the same glass temperature of the matrix is found. It corresponds approximately with the glass temperature of the pure soft segment. This means that no remarkable phase mixing occurs, so that higher hard segment contents also lead to higher domain contents.

On the contrary, the domain volume fraction of *PU*4 is noticeable because it does not differ from the other values correspondingly despite a higher hard segment content. This can only be interpreted as being due to a considerable fraction of the hard segments incorporated into the soft segment matrix. The density fluctuations within the matrix due to such phase mixing lead to an elevated background level. Comparison with the DTA diagram, showing a correspondingly increased glass temperature of the matrix, confirms this hypothesis.

Conclusion

Small angle X-ray scattering curves of polyurethane elastomers, as an example of sponge-like two-phase systems, were evaluated by application of Porod's asymptotic law and the projection of the scattering curve. The width of the transition zone between the phases, the chord lengths, the volume fractions of the phases and the density differences were determined. Subsequent to the works of W. Neumüller and R. Bonart [1, 2] emphasizing the theoretical aspects to the projection formation, suitable ways of evaluation for practical use were demonstrated here. Thereby, it was shown that for slit-smeared small angle scattering curves one can avoid desmearing, since the projection curve obtained directly from the smeared scattering curve leads to the same results as the projection obtained after previous desmearing. A damping function taking into consideration the diffuse transition zones between the phases has to be applied in a suitable form.

The avoidance of a desmearing procedure is profitable in so far as statistical fluctuations, especially in the tail region of scattering curves at small intensities, are amplified additionally on desmearing. This might be important for suitable selection of a fitting region for Porod's asymptotic law.

The structural parameters evaluated on the basis of Poisson distributions for domain and matrix chords are compatible with the chemical composition. In particular, changes of structural parameters and, thus, morphological variations based on thermal treatments of the samples, can be oberserved with the described procedure; this will be discussed in a following work.

Acknowledgement

Part of this work has been supported by the Deutsche Forschungsgemeinschaft.

References

1. Neumüller W, Bonart R (1982) J Macromol Sci Phys B21(2):203–217
2. Neumüller W, Bonart R (1984) J Macromol Sci Phys B23(1):1–16
3. Ruland W (1971) J Appl Cryst 4:70
4. Bonart R, Müller EH (1974) J Macromol Sci Phys B10(1):177–189
5. Porod G (1951) Koll Z 124:83
6. Mittelbach P, Porod G (1965) Kolloid Z Z Polym 202:40
7. Koberstein JT, Morra B, Stein RS (1980) J Appl Cryst 13:34–45
8. Vonk CG (1973) J Appl Cryst 6:81
9. Koberstein JT, Stein RS (1983) J Polym Sci, Polym Phys Ed 21:1439–1472
10. Koberstein JT, Stein RS (1983) J Polym Sci, Polym Phys Ed 21:2181–2200
11. Hermans JJ (1944) Rec Trav Chim Pays-Bas 63:5
12. Hosemann R (1949) Z Phys 127:16
13. Ruland W (1977) Coll & Polym Sci 255:417–427
14. Strobl GR (1970) Acta Cryst A26:367
15. Neumüller W (1982) Röntgenkleinwinkeluntersuchungen an segmentierten Polyurethanen, Dissertation, Regensburg

Received May 28, 1985;
accepted July 26, 1985

Authors' address:

Professor R. Bonart
Institut für Physik III
Angewandte Physik
Universitätsstraße 31
D-8400 Regensburg, F.R.G.

Progress in Colloid & Polymer Science   Progr Colloid & Polymer Sci 71:113–118 (1985)

Phase separation in incompatible polymer blends: polypropylene-polyethylene system*)

W. Wenig and Th. Schöller

Universität-GH-Duisburg, Laboratorium für Angewandte Physik, Duisburg, F.R.G.

Abstract: The morphology of polypropylene-polyethylene blends has been investigated in the as-prepared state by employing interface distribution functions to evaluate their small angle X-ray scattering. While the morphology of the PP (within the investigated composition range) is not altered by the presence of finely dispersed PE, the morphology of the PE component is strongly dependent on composition. Below a PE weight fraction of 0.3 the PE crystal thickness increases with increasing PE content. Concurrently the lattice distortion decreases.

Key words: Polyproylene-polyethylene blends, small-angle X-ray scattering, morphology, interface distribution functions.

1. Introduction

It is commonly known that polypropylene (PP) and polyethylene (PE) are incompatible polymers. The system phase segregates rapidly when the samples are kept at temperatures above the crystallization temperatures of the components.

When polypropylene is melt-blended with polyethylene, the PE acts as impact modifier [1–9]. The responsible parameters for the impact improvement of the material are the dimensions and the morphology of the dispersed phase. The finer the modifier is dispersed, the higher is the resulting impact resistance [9,10]. Also, the polypropylene-modifier interface plays an important role [11,12].

The superstructure of the system can be well investigated by analysing the small angle X-ray scattering using interface distribution functions [13]. This method has been successfully applied to evaluate morphological parameters of various homopolymers [14, 15] and compatible polymer blends [16].

In this paper we report results of X-ray investigations on PP/PE-blends in the "as-prepared" state. The usefulness of the interface distribution function analysis applied to incompatible polymer blends is demonstrated.

2. Experimental

Polypropylene with a molecular weight $\bar{M}_w = 395,000$ was melt-blended with polyethylene ($\bar{M}_w = 127,000$) in a brabender at a temperature of 205 °C for 7 min. The conditions chosen were such that a fine dispersion of PE into PP was ensured. After the blending the samples were not further heat treated except that they were allowed to cool down after the brabender was opened. Pieces of appropriate size were cut for X-ray and microscopic investigations.

Small angle X-ray scattering curves were recorded using a Kratky compact camera. The setting of the camera was such that sufficiently high resolution was ensured (entrance slit width = 30 μm, detector slit width = 75 μm, distance sample-detector slit: 22 cm).

The measurements were controlled by a computer and the cooling water was kept constant through a constant-temperature unit. CuK_α radiation was used and monochromatization was achieved by using a Ni-filter in conjunction with pulse-height analysis.

Since for the evaluation of the interface distribution functions the "tail" of the scattering curve plays an important role, each curve was measured in an angular range corresponding to $0.7 \cdot 10^{-3} \text{ Å}^{-1} \leq s \leq 25 \cdot 10^{-3} \text{ Å}^{-1}$ (s = modulus of the scattering vector in reciprocal space. $s = 2 \sin \theta/\lambda$, 2θ being the scattering angle and λ being the X-ray wavelength). Each curve was measured several times, and no deviations beyond the usual experimental error occurred.

Microscopic observations were carried out using a Mettler hot stage.

---

*) Dedicated to Prof. Dr. H.-G. Kilian on the occasion of his 60th birthday.

3. Evaluation of the interface distribution function

The interface distribution function is defined as follows [13]:

$$g_1(r) = \frac{16\pi^3 t}{V} \int_0^\infty (\lim_{s \to \infty} s^4 I(s) - s^4 I(s))$$
$$\cos 2\pi \, rs \, ds \qquad (1)$$

where $I(s)$ is the corrected intensity and

$$\lim_{s \to \infty} s^4 I(s) - s^4 I(s) \sim G_1(s) \qquad (2)$$

is the interference function.

$G_1(r)$ is proportional to the second derivation of the correlation function, $P_1(r)$:

$$g_1(r) - \frac{2k_1}{d_p} \delta(r) = \frac{\partial^2}{\partial r^2} P_1(r) \qquad (3)$$

where k_1 and d_p are the one-dimesional equivalents of Porod's invariant and the average chord length, respectively.

The correlation function in turn is connected with the one-dimensional intensity, $I_1(s)$, which is related to the measured intensity, $I(s)$, of an isotropic system by $s^2 I(s) \sim I_1(s)$:

$$F_1 \left(\frac{\partial^2}{\partial r^2} P_1(r) \right) = -4\pi^2 s^2 I_1(s). \qquad (4)$$

The intensity $I_1(s)$ is thus proportional to the Fourier transform of the self-convolution of the electron density distribution within a lamellar stack.

Several corrections of the function $I(s)$ have to be made to determine the interference function.

1. The intensity due to density fluctuations (diffuse scattering), I_{Fl}, is subtracted from the measured (background corrected) curve.

2. From a plot $s^4(I - I_{Fl})$ as a function of s^2 the thickness of the domain boundary, d_z, is determined.

3. Using d_z, the intensity is corrected. The curve $s^4 I(s)$ then yields into a straight line for large s values.

From the $s^4 I(s)$ curve $G_1(s)$ is calculated, from which the interface distribution function, $g_1(r)$, is obtained (eq. (1)).

4. Results and discussion

As we have already mentioned above, PE melt-blended with PP improves the impact resistance of PP.

The mechanism is well understood: a crack propagates on a path which is predominantly determined by the degree of dispersity of the PE inclusions, the average particle size and the size distribution being the controlling parameters.

There are generally two approaches to explain the structure-porperties relationship [9]: 1. There are analogies between the behaviour of amorphous impact resistant blends like high impact polystyrene or high impact PVC and semi-crystalline polymer blends, and 2. there are changes in the crystalline phase caused by the melt-blending.

Here the interface between the two components plays as important role. Some authors claim to have found a diffusion of PE into the PP matrix up to a depth of 300 µm [11]. Recent investigations indicate a more complete phase separation with sharp phase boundaries between PE and PP [12]. It can be expected that a broad phase boundary influences the growth rate of the PP spherulites. Figure 1 shows the growth rate as a function of PE weight fraction measured during isothermal crystallization at 130 °C. As one can see, the growth rate shows no dependency on the PE weight fraction. The same holds for samples which had been annealed in the melt at different temperatures for times between 10 min and 50 min. As figure 2 shows, the

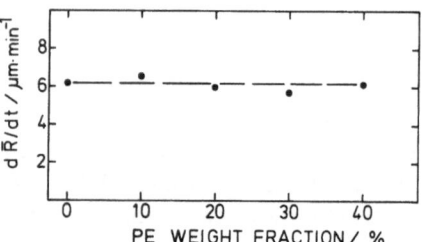

Fig. 1. PP spherulitic growth rate as a function of PE weight fraction

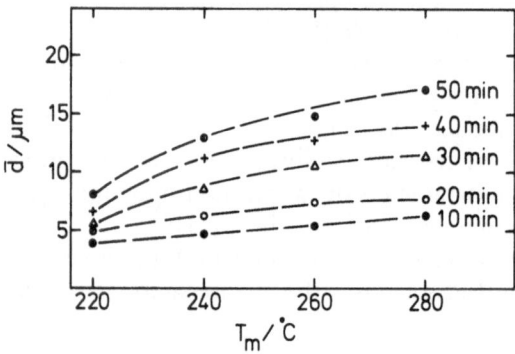

Fig. 2. Mean size of PE particles as a function of melt temperature for different holding times

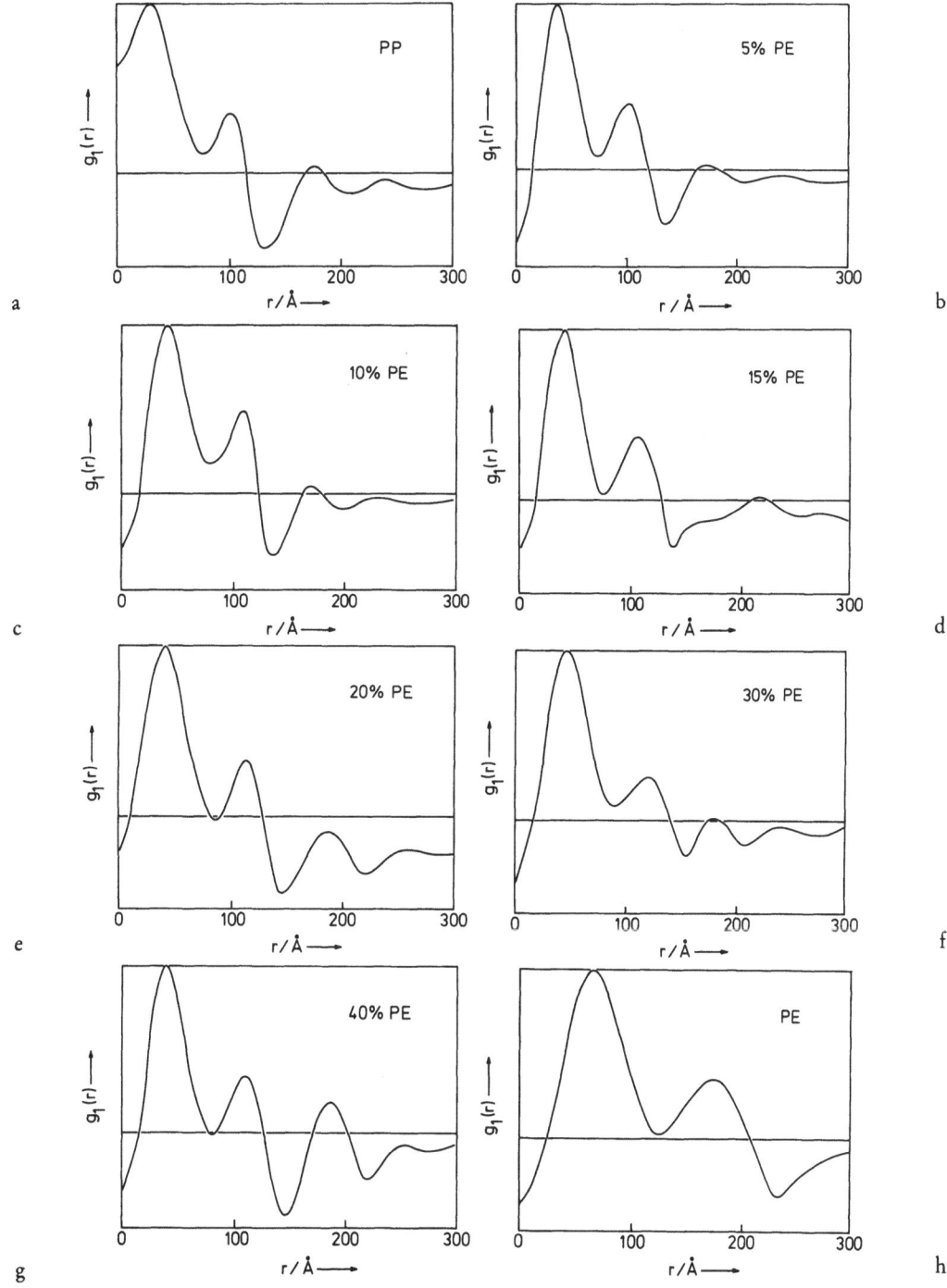

Fig. 3. Interface distribution function of the investigated blends

phases separate during annealing in the melt. In figure 2 the average sizes of the PE inclusions are plotted as they become visible in the polarization microscope after isothermal crystallization at 130 °C for 25 minutes (see table 1). This temperature is above the crystallization temperature of polyethylene; the spherulitic PP structure, however, is allowed to form out. The spherulitic radii and the growth rates calculated from these values are identical with those plotted in figure 1. Considering the wide range of PE mean particle sizes and

Table 1. Mean size of PE particles as a function of melt temperature for different holding times. The particle diameters were measured after isothermal crystallization at 130 °C for 25 min

| Melt temperature T_m/°C | Size of PE particles $\bar{d}$/μm | | | | |
	10 min	20 min	30 min	40 min	50 min
220	3.9	4.9	5.6	6.6	8.0
240	4.7	6.2	8.5	11.1	12.9
260	5.4	7.4	10.5	12.7	14.8
280	6.2	7.7	11.4	14.0	17.1

Table 2. Structure parameters derived from interface distribution functions of the investigated PP/PE blend

PP/PE weight fraction/%	Mean interlamellar distance $\bar{d}_a$/Å	Mean lamellar thickness $\bar{d}_c$/Å	Long period $\bar{L}$/Å	$\bar{d}_3$/Å	$\bar{d}_4$/Å
100/ 0	29	97	131	172	240
95/ 5	35	101	133	172	236
90/ 10	39	102	135	172	235
85/ 15	40	104	140	171	217
80/ 20	41	109	145		258
		186	221		
70/ 30	45	118	147		238
		176	204		
60/ 40	44	108	147		252
		186	223		
0/100	68	170	235		

the drastic change of the inner surface which is connected with it, this is a strong indication that the influence of the interface between the two phases is rather small. Indeed one is allowed to conclude that there is no sizable interpenetration of PE molecules into the PP matrix.

We expect to achieve further insight into the crystallization behaviour by discussing the small angle X-ray results. Figure 3 shows the results of the interface distribution function calculations for all investigated blends and for PE used for the blending. The curves $g_1(r)$ show a series of maxima and minima. The first peak can be attributed to the lamellae with the lower volume fraction, the second to the lamellae with the higher volume fraction, while the position of the first minimum is due to the long period. For incompatible polymer blends, where both components form independent morphologies, the interface distribution functions should be a superposition of two functions. As figure 3 reveals, this is the case for PP/PE blends. Comparing, however, figure 3 a and figure 3 h, the interface distribution function for pure PP and pure PE, the first peaks ar not separable. The position of the peak yields the thickness of the interface layer between two crystals. The respective second peaks yield the mean crystal (lamellar) thicknesses since for both PP and PE the crystallinities are above 50 %. In this case the maximum following the minimum of the long period are of great help.

The third maximum denotes the distance between the "upper" edge of a crystal and the "lower" edge of the overnext crystal, it yields thus the sum of two interlamellar and one lamellar thicknesses. The maxima of higher order are built accordingly. As a result, we derived the values listed in table 2 which represent characteristic parameters of the superstructure of the blends. Some of these parameters are plotted in figures 4 to 6. They allow a discussion on possible changes in the crystalline phase of PP caused by the melt blending.

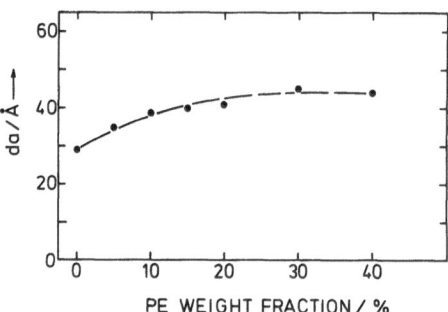

Fig. 4. Mean interlamellar (amorphous) distances of the investigated samples as derived from interface distribution functions

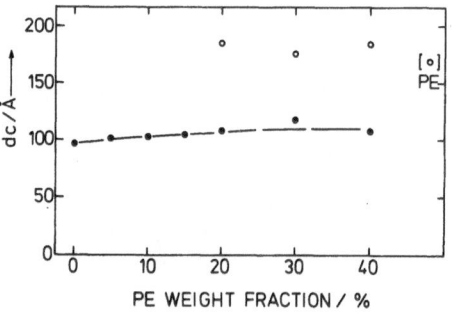

Fig. 5. Mean crystal (lamellar) thicknesses of the investigated samples as derived from interface distribution functions

Figure 4 reveals that the interlamellar thickness increases with increasing PE weight fraction up to a PE content of 30 %. The mean crystal thickness (fig. 5) also increases with PE weight fraction, but this increase is much less pronounced. It is obvious that the long period shows an increase too (fig. 6). In these figures

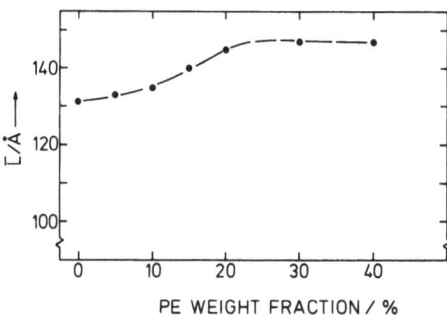

Fig. 6. Long periods of the investigated samples as derived from interface distribution functions

Table 3. Half widths of the (110) and (220) PP and PE peaks corrected for insturmental broadening

PP/PE weight fraction/%	Polypropylene $\delta s_o/\text{Å}^{-1}\,10^{-2}$		Polyethylene $\delta s_o/\text{Å}^{-1}\,10^{-2}$	
	(110)	(220)	(110)	(220)
100/ 0	1.09	2.68		
95/ 5	1.16	2.72	(1.80)	(1.96)
90/10	1.11	3.11	1.90	2.35
85/15	1.06	2.57	2.29	2.35
80/20	1.16	2.38	1.70	2.00
70/30	1.06	2.03	1.60	1.71
60/40	1.01	2.72	1.35	1.51

one can clearly see that above a PE weight fraction of 30 % no further increase occurs. This behaviour is due to the phase separation of the two components and is in agreement with earlier measurements [17] which showed that the coalescence of the dispersed PE is favoured if the PE content is increased.

From these results, a change in crystalline morphology cannot be concluded. To affirm this finding, we measured the wide angle X-ray scattering of the blends. In figure 7 some of the scattering curves are displayed. The figure shows that the (110) and (220) PP peaks and the (110) and (220) PE peaks in the blends with a PE concentration of 10 % or higher can be clearly resolved. The corrected half widths are listed in table 3. The values have been used to separate the mean crystal thickness $\bar{e}_{110}$ from the lattice distortion g_{II} by applying the method of Hosemann and Wilke [18] (for

details of the method and the experimental procedure we refer to reference [19]):

$$(\delta s)_o^2 = (\delta s)_c^2 + (\delta s)_{II}^2 = \frac{1}{\bar{l}_{hkl}^2} + \frac{(\pi g_{II})^4 m^4}{d_{hkl}^2} \qquad (5)$$

where

$\bar{l}_{hkl}$ = crystal thickness in *hkl* direction
m = order of the reflection
d = "d" spacing
g_{II} = lattice distortion parameter (distortion of the *II* kind).

The results are listed in table 4 and plotted in figures 8 to 10. For polypropylene, neither crystal thickness nor lattide distortion show any dependency on sample composition (figs. 8 and 10).

The polyethylene crystal thickness increases with increasing PE weight fraction by about 15 % (fig. 9), while concurrently the lattice distortion decreases (fig. 11). The values, however, are smaller than those of the PP, probably due to the lower crystallization tempera-

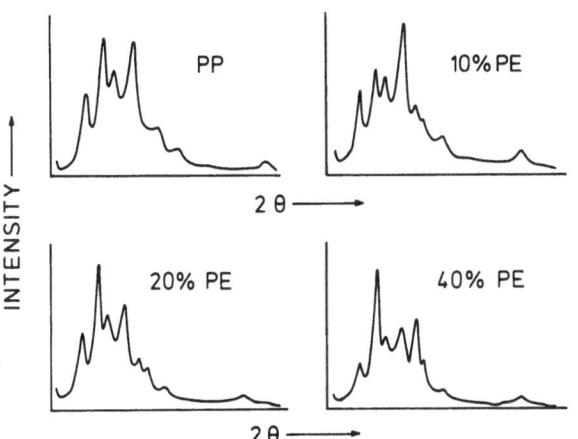

Fig. 7. Selection of measured wide angle X-ray scattering curves

Table 4. Crystal thicknesses and lattice distortions determined by the method of Hosemann and Wilke

PP/PE weight fraction/%	Polypropylene $\bar{l}_{110}/\text{Å}$	$g_{II}/\%$	Polyethylene $\bar{l}_{110}/\text{Å}$	$g_{II}/\%$
100/ 0	114 ± 2	6.30		
95/ 5	103 ± 5	6.33	(56 ± 1)	(2.92)
90/10	122 ± 8	6.99	54 ± 2	3.93
85/15	114 ± 6	6.25	44 ± 1	2.42
80/20	97 ± 4	5.90	60 ± 2	3.43
70/30	104 ± 5	5.35	63 ± 2	2.59
60/40	129 ± 8	6.47	75 ± 3	2.74

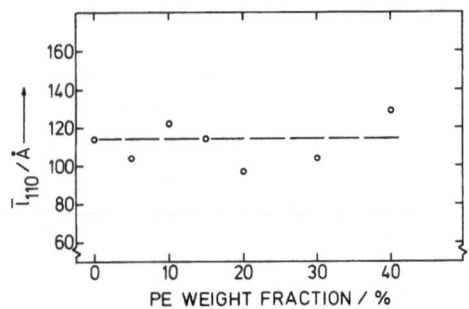

Fig. 8. Crystal thicknesses of the PP crystals in (110) direction

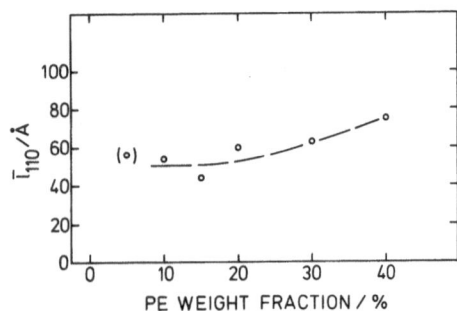

Fig. 9. Crystal thicknesses of the PE crystals in (110) direction

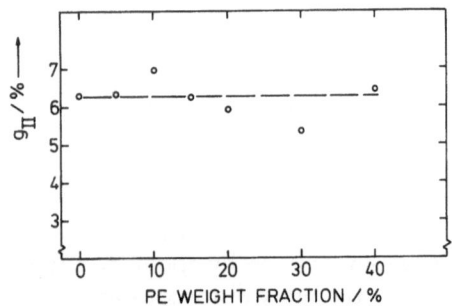

Fig. 10. Lattice distortions of the PP crystals

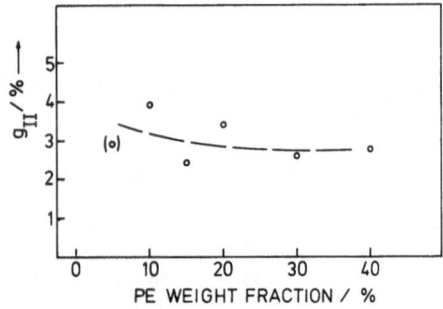

Fig. 11. Lattice distortions of the PE crystals

ture of the PE. These findings confirm that the PP crystal morphology is not altered by the presence of finely dispersed PE. Due to this dispersion, the polyethylene crystallization is hindered at small concentrations of the PE. All modifications of the morphology occur in the amorphous phase. The melt-blended system PP/ PE is strictly incompatible. All mechanical improvements have to be attributed to the particle size distribution of the polyethylene domains.

Technical products often contain polyethylene in a concentration of about 30%. In view of the results reported in this paper this becomes meaningful because this PE content assures a stable morphology under technical processing conditions.

Acknowledgment

This work has been financially supported by the Deutsche Forschungsgemeinschaft.

References

1. Galeski A, Pracella M, Martuscelli E (1984) J Polym Sci Phys 22:793
2. Zakin JL, Simha R, Hershey H (1966) J Appl Polym Sci 10:1455
3. Plochocki A (1966) Koll Z u Z Poylmere 208:168
4. Ogawa T, Tanaka S, Inaba T (1974) J Appl Polym Sci 18:1351
5. Kudlacek L, Kaplanova M, Knap F (1978) Faserforsch Textiltechn 29:286
6. Riess G, Schlienger M, Marti S (1980) J Macromol Sci Phys 17:335
7. Craig TO (1974) J Polym Sci Polym Chem Ed 12:2105
8. Karger-Kocsis J, Balajthy Z, Kollar L (1984) Kunststoffe 74:104
9. Karger-Kocsis J, Kallo A, Kuleznev VN (1984) Polymer 25:279
10. Dao KC (1984) Polymer 25:1527
11. Kryszewski M, Galeski A, Pakula T, Grebowicz J (1973) J Coll Interf Sci 44:85
12. Letz J (1979) J Polym Sci A–Z 8:1415
13. Ruland W (1977) Coll & Polym Sci 255:417
14. Stribeck M, Ruland W (1978) J Appl Cryst 11:535
15. Stribeck M (1980) Thesis, Univ Marburg
16. Wenig W, Schöller T (1985) Die Angew Makromol Chem 130:155
17. Wenig W, Meyer K (1980) Coll & Polym Sci 258:1009
18. Hosemann R, Wilke W (1964) Faserforsch Textiltechn 15:521
19. Wenig W, Hagenbeck G (1983) Die Angew Makromol Chem 119:1

Received May 13, 1985;
accepted August 15, 1985

Authors' address:

W. Wenig
Universität-GH-Duisburg
Laboratorium für Angewandte Physik
D-4100 Duisburg 1, F.R.G.

Progress in Colloid & Polymer Science   Progr Colloid & Polymer Sci 71:119–124 (1985)

Structure and properties of segmented polyamides*)**)

E. Bornschlegl, G. Goldbach and K. Meyer

Central Research and Development Department of Hüls AG, Marl, F.R.G.

Abstract: Segmented polyamides composed of polyether and polyamide structural units are industrially very interesting thermoplastic materials of construction, having a broad spectrum of properties. The principal reason for this broad spectrum is the polyphase structure of the materials and the possibility of influencing this structure by choice of the ratio of oligoether to oligoamide as well as choice of the mean block lengths.

In general, three phases can be detected:

1. an amorphous phase, generally forming the matrix, of oligolactam sequences and oligoether sequences. The oligoether component produces *internal plasticisation* (a marked lowering of the glass transition temperature),

2. a discontinuous amorphous phase which essentially consists of oligoether structural units and acts as an *internal toughener* (glass transition far below room temperature), and

3. a crystalline phase of sufficiently well ordered lactam sequences. Its melting range is drastically shifted towards lower temperatures with increasing oligoether content of the total material.

In addition, at relatively high oligoether contents, a crystalline oligoether phase is at times encountered to a minor degree.

Key words: Block copolyamides, polyphase structure, mixed phase formation, melting characteristics, glass transition temperature.

Introduction

Segmented polyamides (since the structural units are linked via an ester group, such systems are hereafter also referred to as polyether-ester-amides) of oligoether and oligolactam structural units [1–4] have the chain structure shown in figure 1. The technological properties can be varied in a controlled manner not only by using chemically different oligoethers and/or oligolactams but also by varying the (mean) block lengths. A suitable way of rapidly investigating the phase ratios is to combine differential scanning calorimeter examination with temperature-dependent torsional vibration tests. The former provide (through the enthalpy of melting and the position of the melting range) integral information on the crystalline regions while the latter (through the position of the glass transi-

tion range) permit characterisation of the amorphous components. (Such experiments of course do not provide information on the morphology).

Experimental

Concerning the preparation of the products, compare [1].

The differential scanning calorimetry experiments were carried out with a DSC 2 C apparatus from Perkin-Elmer (weight of sample about 5 mg). The samples were in each case heated, from the as-supplied (granular) state, to 220 °C and were then cooled to −50 °C and

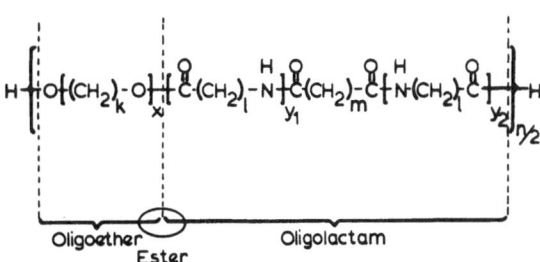

Fig. 1. Structure of polyether-ester-amides

---

*) In part given as a lecture at the Makromolekulares Kolloquium, Freiburg/Br. 28.02.–02.03.1985.

**) Dedicated to Prof. Dr. H.-G. Kilian on the occasion of his 60th birthday.

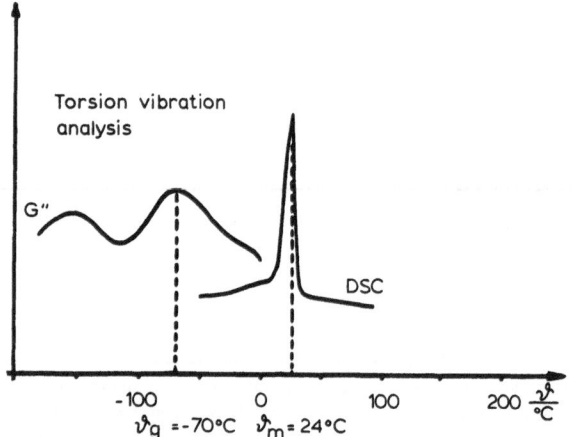

Fig. 2a) Loss modulus/temperature diagram and DSC melting thermogram oligotetrahydrofuran

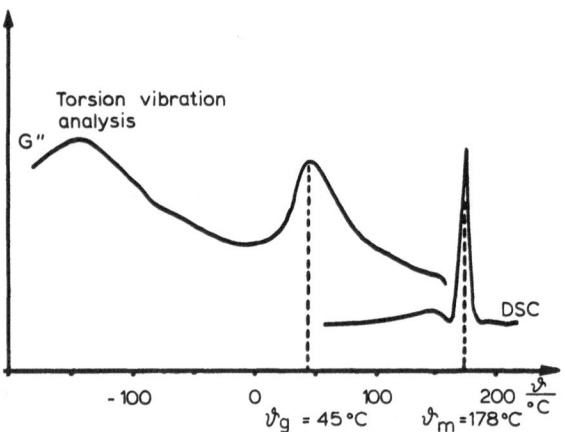

Fig. 2b) Loss modulus/temperature diagram and DSC melting thermogram of polyamide 12

and again heated (the rate of temperature change being in each 20 K. min⁻¹). The thermogram obtained on second melting was evaluated.

The torsional vibration analysis was carried out using an automatic torsional pendulum ATM III from Myrenne, at a measuring frequency of (1 ± 0.1) Hz. The test specimens were taken from compression-moulded sheets.

Results

Figure 2a and figure 2b show, on the same temperature axis, torsional vibration diagrams (loss moduli) and DSC melting thermograms for the polymers made from the two pure structural units (oligotetrahydrofuran and polyamide 12 respectively). The glass transition temperatures are −70 °C and +45 °C and the melting peak maxima are 24 °C and 178 °C respectively. In what follows, we first present results of DSC

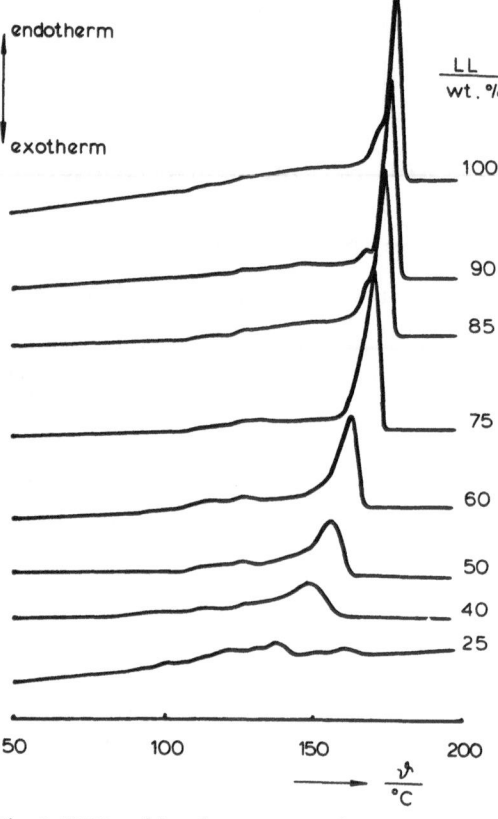

Fig. 3. DSC melting thermograms of a series of polyether-ester-amides of different content of oligolauryl-lactam structural units

investigations, followed by results of torsional vibration experiments.

Investigation of the crystalline components (DSC melting behaviour)

Figure 3 shows melting thermograms for a number of polyether-ester-amides, with the proportion of oligotetrahydrofuran increasing from top to bottom. The uppermost thermogram shown is that of the pure polyamide. The fact that it fits effortlessly into the series suggests that the effect involved is the melting of crystalline oligolauryl-lactam regions. This is also confirmed by, for example, X-ray wide-angle scattering.

A decrease in the heat of fusion and a lowering of the melting temperature can be observed. Since what is involved is the melting of oligolauryl-lactam crystallites, it seems sensible to relate the heat of fusion to the lauryl-lactam content. Doing so shows nearly the same value for all materials, the value being only slightly lower than that of the pure polyamide 12 (crystallinity 30–40 %). Accordingly, the decrease in heat of fusion is essentially a dilution effect.

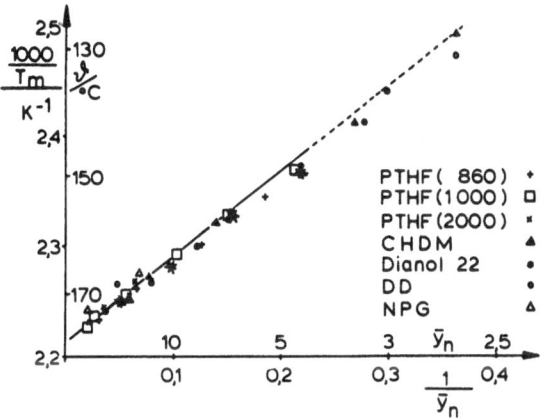

Fig. 4. Dependence of reciprocal absolute melting temperature (peak maximum) on the reciprocal average degree of condensation of oligoamid structural unites; different diol structural units and pure oligolauryl lactam-dicarboxylic acids (*), respectively

To interpret the lowering of the melting point of partially crystalline block copolymers, it has been proposed that an "oligomer melting" be assumed [5]. In fact, with the present polyether-ester-amides, when the composition was varied, the same oligotetrahydrofuran was used, namely the same soft phase. Under these circumstances, the proportion of hard phase can be lowered only by reducing its mean chain length. Accordingly, the series just illustrated concerns systems with a progressively decreasing mean chain length of the oligoamide segments. The term "oligomer melting" is simply intended to indicate that, at least as a rough approximation, the hard phase structural units melt and crystallize as if they were present as oli-

gomers and as if they were not linked to one another by the soft phase segments. This assumption leads to the model chosen below (fig. 4). A good straight line results, as was to be expected according to Flory [6]. In accordance with the concept of "oligomer melting", the same result is obtained if the mean molecular weight of the soft segment is varied or if other soft segments (table 1) [1] are used. The behaviour of the pure oligoamide-dicarboxylic acids, i. e. of the hard segments without connecting soft segments, also does not deviate from this pattern.

The chosen presentation (reciprocal melting point plotted against number-average chain length) would appear sensible if the lowering of the melting point was a consequence of reduced chain length only. However, in the case of the present oligoamide-dicarboxylic acids such is not the case. To vary the chain length, a dicarboxylic acid was employed. However, the latter does not necessarily become incorporated at the end of the chain. Accordingly, it not only lowers the melting point through varying the chain length but additionally lowers it through acting as a comonomer unit. The lowering of the melting point caused by the latter effect is in fact particularly great since the comonomer unit is incorporated in strict isolation: each oligoamide contains exactly one comonomer unit. This twofold lowering of the melting point has to be taken into account if, for example, it is desired to estimate the heat of fusion of the homopolyamide from the slope of a Flory graph.

The tetrahydrofuran constituents can also be present in a partially crystalline form. This is shown in figure 5, which represents the melting behaviour of a series based on an oligotetrahydrofuran of higher

Table 1. Soft segments analysed

Structure formula	Term	Abbreviation
$HO-[(CH_2)_4-O]_x H$	Oligotetrahydrofuran	PTHF ($\bar{M}_w$)
$HOCH_2-\langle H \rangle-CH_2OH$	Cyclohexanedimethanol	CHDM
$HO(CH_2)_2O-\langle O \rangle-\underset{\underset{CH_3}{\vert}}{\overset{\overset{CH_3}{\vert}}{C}}-\langle O \rangle-O(CH_2)_2OH$	Dianol 22	
$HO-(CH_2)_{12}-OH$	Dodecanediol	DD
$HOCH_2-\underset{\underset{CH_3}{\vert}}{\overset{\overset{CH_3}{\vert}}{C}}-CH_2OH$	Neopentyl glycol	NPG

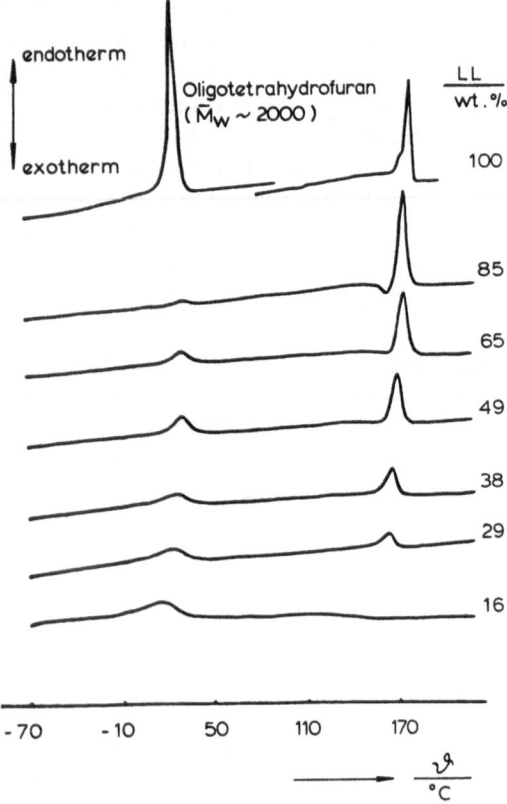

Fig. 5. DSC melting thermograms of a series of polyether-ester-amides of different content of oligolauryl-lactam structural units; oligotetrahydrofuran of higher molecular weight than in figure 3

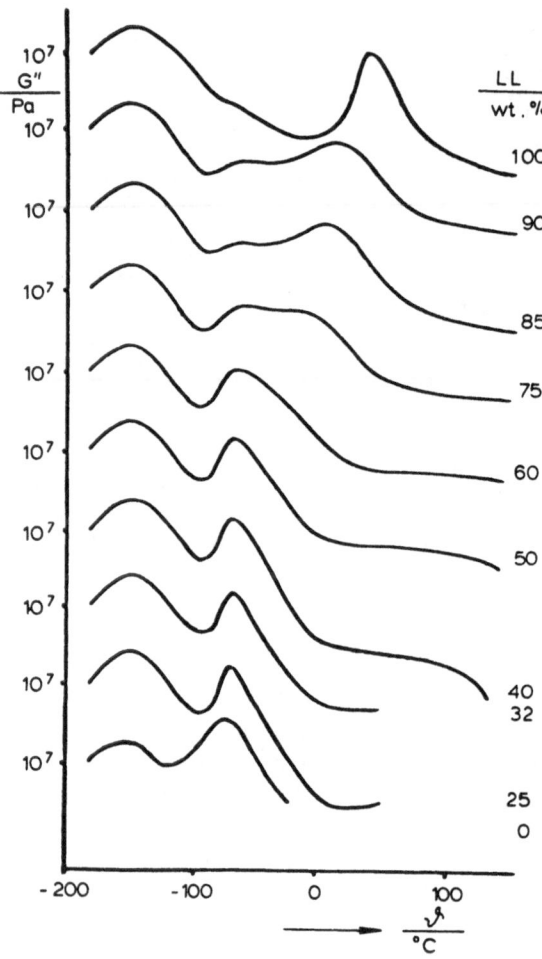

Fig. 6. Torsional vibration analysis; loss modulus of a series of polyetherester-amides of different content of oligolauryl-lactam structural units

molecular weight. If the melting signal of the oligotetrahydrofuran is in each case expressed on the basis of the oligotetrahydrofuran content, the same value is obtained in every case; it is about half that of the pure oligotetrahydrofuran, which has a very high crystallinity of 55–60 %. The effect on the position of the melting peak is striking: with increasing oligotetrahydrofuran content the melting point *decreases* and accordingly shifts away (at least at the highest oligotetrahydrofuran contents of the investigated range) from the melting point of the pure oligotetrahydrofuran. As yet, no substantiated interpretation of this phenomenon can be offered.

Investigation of the amorphous components (torsional vibration analysis)

Figure 6 shows the torsional vibration diagrams — loss modulus (G'') as a function of temperature — for the polyether-ester-amide series shown in figure 3. The two pure components, namely the oligotetrahydrofuran and polyamide 12, are also plotted. We do not propose to concern ourselves here with the low tem-

perature relaxation region (which does not vary significantly with composition). The two other relaxation regions show a clear dependence on composition. The relaxation region manifesting itself at the lower temperatures is obviously a consequence of the glass transition of the pure oligotetrahydrofuran and, as the composition is varied, so its conspicuousness changes markedly but its position only slightly. It is to be interpreted as a glass transition of a mixed phase of a large amount of amorphous oligotetrahydrofuran and a small proportion — changing only little — of amorphous oligolauryl-lactam. The changes in the second relaxation region, which is closely related to the glass transition of the pure polyamide 12, are distinctly more pronounced. What is involved is the glass transition of a mixed phase of a large amount of oligolauryl-lactam and a proportion of oligotetrahydrofuran which increases with increasing overall content of the latter.

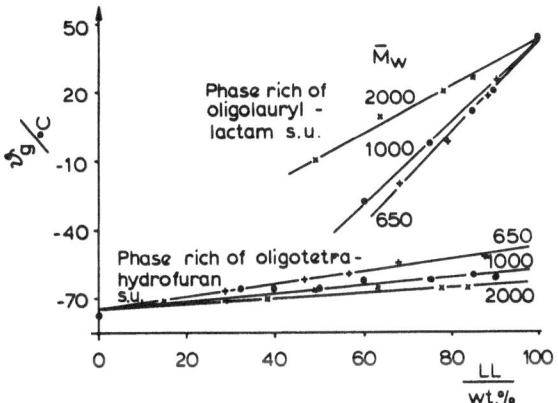

Fig. 7. Dependence of glass transition temperatures (temperature of maximum of loss modulus) on content of oligolauryl-lactam structural units; different molecular weights of oligotetrahydrofuran

Below a lauryl-lactam content of about 60% by weight, only one relaxation region is observed. We have not yet investigated whether this is due to the fact that the relaxation regions of the two amorphous phases which are still present cannot be resolved, or whether under these conditions only one amorphous phase is present.

Figure 7 shows the position of the two loss maxima as a function of the composition by weight of the products. The shift of the two glass transition towards lower temperatures shows that with increasing total oligotetrahydrofuran content the proportion of this component in the two amorphous phases also increases. To investigate what factors determine the distribution of the two amorphous species, namely the amorphous oligotetrahydrofuran and the amorphous oligoamide, between the two amorphous mixed phases, three series with oligotetrahydrofurans of different molecular weight were investigated.

As figure 7 shows, basically the same pattern is observed in all three cases. However, the shift in the glass transition is of differing conspicuousness depending on the chain length of the oligotetrahydrofuran used. For the same composition by weight, as the chain length of the soft segments increases, the glass transition of the phase rich in lauryl-lactam structural units (upper group of curves) moves to progressively higher values (corresponding to the decreasing content of oligotetrahydrofuran) and the glass transition of the phase rich in oligotetrahydrofuran moves to progressively lower values (corresponding to the decreasing proportion of polylauryl-lactam). This is in accordance with expectations; it simply means that with increasing chain length of the oligotetrahydrofuran

structural unit the tendency to form these two mixed phases diminishes.

Summary: Implications for technological properties

The polyether-ester-amides in general contain the following phases, formed by different portions of the same chains:

1. a crystalline oligolactam phase,

2. an amorphous homogeneously mixed phase of nearly the entire amorphous oligoamide and proportions of the amorphous oligotetrahydrofuran,

3. a second amorphous homogeneously mixed phase of small proportions of oligoamide and a relatively large amount of amorphous oligotetrahydrofuran, and

4. (at times, and only at temperatures below room temperature) a crystalline oligotetrahydrofuran phase.

The proportions and compositions of the phases depend on the weight ratio of oligolactam/oligotetrahydrofuran and on the mean molecular weight of the oligotetrahydrofuran employed.

From a technological point of view (cf. also [7, 8]) the amorphous mixed phase rich in oligolactam is to be regarded as an internally plasticised polyamide. The lowering of the glass transition temperature is due to oligotetrahydrofuran structural units. Since these are part of the chain, there is no danger of exudation of the "plasticiser". The crystalline oligoamide regions constitute thermo-reversible physical multiple crosslinking points. Below their melting range, they impart to the material the properties of a crosslinked and reinforced rubber, while at higher temperatures they permit processing as for thermoplastics. The components rich in oligotetrahydrofuran, which are elastomeric down to low temperatures, act as built-in toughening agents which further improve the mechanical properties.

References

1. Mumcu S, Burzin K, Feldmann R, Feinauer R (1978) Angew Makromol Chem 74:49
2. Deleens G, Foy P, Marechal E (1977) Eur Polym J 13:337, 343, 353
3. Castaldo L, Maglio G, Palumbo R (1978) J Polym Sci, Polymer Letter Ed 16:643
4. Sorta E, della Fortuna G (1980) Polymer 21:728
5. Schmid FG, Dröscher M (1983) Makromol Chem 184:2669
6. Flory PJ (1949) J Chem Phys 17:223
7. Horlbeck G, Mumcu S (1984) Plastverarb 35:38

8. della Fortuna G, Melis A, Perego G, Vitali R, Zotteri L (1979)
   Proc Int Rubber Conf, Milan, Italy, p 229

Received June 3, 1985;
accepted June 16, 1985

Authors' address:

Dr. Kurt Meyer
ZBFE/Angew. Phys. Chemie
Hüls AG
Postfach 1320
4370 Marl, F.R.G.

Progress in Colloid & Polymer Science　　　　Progr Colloid & Polymer Sci 71:125–133 (1985)

Structure and anisotropy in PC*)

II. WAXS, anisotropic heat conduction and birefringence in oriented samples

M. Pietralla[1]), H. R. Schubach[1]), M. Dettenmaier[2]), and B. Heise[1])

[1]) Abteilung Experimentelle Physik, University Ulm, Ulm, F.R.G.
[2]) MPI Polymerforschung, Mainz, F.R.G.

Abstract: Birefringence, anisotropic heat conduction, and wide angle X-ray scattering (WAXS) investigations have been performed on one set of uniaxially drawn samples of polycarbonate (PC). The orientation parameters from heat conduction and from WAXS are well correlated. The intrinsic birefringence is determined to $\Delta n_o = 0.106$ in agreement with the monomer polarizability calculated on the basis of model structures. The short range order as evidenced by WAXS does not depend on the degree of orientation in PC. Over small regions (≤ 1.5 nm) the molecular arrangement is similar to that in the crystalline state.

Key words: Polycarbonate, anisotropic heat conduction, intrinsic birefringence, wide angle X-ray scattering (WAXS), local structure.

Introduction

It is of interest to know the orientation of chain molecules within an anisotropic samples for many purposes. Knowledge of the second moments of the orientation distribution in the form of the generalized Legendre polynomials [1] is sufficient for all properties which are described by second rank material tensors. Fourth rank tensor properties such as the elastic tensor necessitate in addition the knowledge of the fourth moments. Wide angle X-ray scattering (WAXS) allows the determination of these moments if information on the short range structure is available. The preceding paper [17] by Schubach and Heise shows how the WAXS measurements can be evaluated. In contrast to X-ray analysis which requires a large amount of computer time, birefringence measurements are extremely rapid and simple. However, the orientation of the chains can only be determined if the intrinsic (i. e. monomer or segmental) birefringence Δn_o is known.

Unfortunately reliable values of Δn_0 are difficult to obtain. Recently, Biangardi [2] has reported a value of $\Delta n_o = 0.236$ for the monomer unit of PC. This value has been evaluated from X-ray measurements which will be discussed subsequently.

Anisotropy of heat conduction [3], though seldom measured, allows the determination of the upper and lower bounds of the chain orientation on the basis of very general assumptions [4].

In this paper we present measurements of the anisotropy of heat conduction obtained by an improved DeSénarmont method [5]. The orientation parameters deduced from the measurements are compared with those determined by WAXS. They allow only a small range of values of the intrinsic birefringence. The intrinsic birefringence is also calculated using polarizability data of the PC-monomer analogues published by Erman et al. [6] and structure models of the PC chain.

1. Experimental

Sample preparation

Sheets of polycarbonate with the trade name Makrolon 3200 (kindly supplied by Bayer AG, Leverkusen) were oriented above

---

*) Dedicated to Prof. Dr. H.-G. Kilian on the occasion of his 60th birthday.(B. Heise, M. Pietralla, H. R. Schubach): With gratitude for the vivid introduction into the science of polymers.

the glass transition temperature ($T_g = 149\,°C$) at $T = 160\,°C$ with a nominal strain rate of $\dot{\epsilon} = 0.03\ s^{-1}$. After drawing, the samples were rapidly cooled to room temperature in a stream of cold air whilst keeping their position in the clamps fixed. For comparison, two samples were oriented below the glass transition temperature at $T = 25\,°C$ and $129\,°C$, respectively with a nominal strain-rate of $\dot{\epsilon} = 6 \times 10^{-4}\ s^{-1}$.

At $25\,°C$ PC deforms by necking, at $129\,°C$ by formation of diffuse shear zones.

The birefringence of each sample was determined at a wavelength of 545 nm with a polarizing microscope, equipped with an Ehringhaus compensator. The measurements were carried out by compensating large retardations with a wedge-shaped piece of PC and small ones with the Ehringhaus compensator.

Anisotropic heat conduction

The anisotropy of heat conduction is closely related to the orientation of chain segments. If one is only interested in the orientation of chain segments the measurements of absolute values are dispensable. In this case the anisotropy ratio $A = k''/k^\perp$ where k'' and $k^\perp$ denote the thermal conductivities in stretching direction and perpendicular to it, contains the information needed [3]. This ratio can be determined directly with the so-called DeSénarmont method [3, 5] which is a transient point source method. This method and the equipment used are described in [5]. The measurements have been carried out at $30\,°C$. The results are plotted in figure 1 against the birefringence. The magnitude of the anisotropy compares well to values derived from measurements of Hennig [7] (At $\lambda \sim 1.8$, $A \sim 1.8$). PC exhibits the highest anisotropy of all amorphous polymers investigated hitherto.

2. Orientation and anisotropic heat conduction

In amorphous polymers the heat conduction mechanism is essentially that of a polymeric liquid even in the glassy state. In the normal liquids the contribution from convective heat transport is negligible (less than 1%) [8]. Thus only the collisional part is operative, yielding a constant mean free path l in the kinetic equation [9]

$$k = (1/3)\ C\ v\ l. \qquad (1)$$

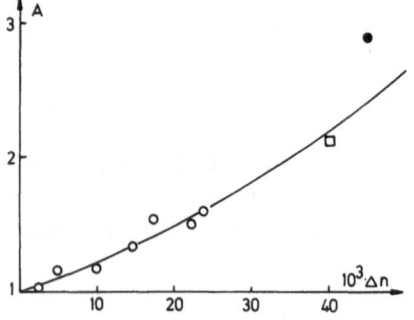

Fig. 1. Heat conduction anisotropy A versus birefringence Δn at $T = 22\,°C$. $\bigcirc$ sample stretched at $T_s = 160\,°C$; $\square$ sample stretched at $T_s = 25\,°C$; $\bullet$ sample stretched at $T_s = 129\,°C$; —— calculated with $\Delta n_o = 0.106$ and $A_o = 10$

The temperature dependence of the conductivity k is given by that of the specific heat C from T_g down to ~ 25 K since the "sound velocity" v is only slightly temperature dependent. The small deflection observed when passing through T_g to higher temperatures [10] is mainly due to the step in the thermal expansion coefficient [11] (C is the specific heat per volume at constant pressure) and to the step in the sound velocity gradient.

In this picture the anisotropy in thermal conductivity must be attributed to the anisotropy in the collision process which results from the presence of van der Waals and covalent bonds. The collisions due to the van der Waals forces result in the intrinsic perpendicular contribution $k_o^\perp$ and the collisions due to the primary bonds result in the intrinsic parallel contribution k_o''. If these contributions are averaged like the macroscopic ones, i. e. like second rank tensors, then the conductivity (parallel) and the resistivity (series) averaging pose the limits of the macroscopic conductivities. Because of the small mean free path (~ 0.15 nm [9]) it is assumed that the conductivity averaging will be the most suitable for the collisional model. The magnitude of the intrinsic anisotropy of thermal conductivity has been estimated from the conductivities of ordinary liquids (van der Waals bonds) and of glasses (covalent bonds) to be in the range [12]

$$5 \lesssim A_o \lesssim 25 \qquad (2)$$

with the most probable value $A_o \sim 10$.

The macroscopic anisotropy of thermal conductivity is a function of orientation and intrinsic anisotropy and is given by

$$A = \frac{1 + 2q_c P_2}{1 - q_c P_2}; \quad q_c = \frac{A_o - 1}{A_o + 2}. \qquad (2.1)$$

From this equation the orientation parameter P_2 obtained after conductivity averaging can be determined via

$$P_2^c = \frac{A - 1}{A + 2} \cdot \frac{A_o + 2}{A_o - 1} \qquad (2.2)$$

provided the intrinsic anisotropy is known. All the amorphous polymers we have measured so far [4] exhibit values of $A_0 \sim 10$. In view of the high anisotropies measured we assume $10 \lesssim A_0 \lesssim 16$ for the PC chain. The upper limit has been estimated roughly from the van der Waals type description of real networks [13a] applied to the glassy state [13b]. An

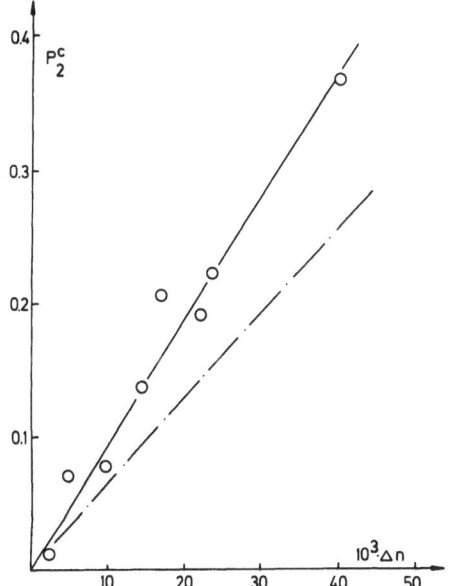

Fig. 2. Orientation parameter P_2^c (conductivity average eq. (2.2) versus birefringence Δn calculated with the intrinsic heat conduction anisotropoy $A_o = 10$. The solid line yields $\Delta n_o = 0.106$. The dash-dotted line is the lower limit ($A_o = \infty$) of the orientation derived from heat conduction anisotropy. (The last data point of fig. 1 has been omitted)

intrinsic anisotropy lower than $A_o \sim 10$ is unreasonable because the orientation parameter for the samples with the highest anisotropies measured will then exceed values of $P_2 = 0.5$. This value is unrealistic for network systems.

The orientation parameter determined in this manner is plotted versus birefringence in figure 2. The data fit well on a straight line from which the intrinsic birefringence is calculated from the slope of the straight line $\Delta n = \Delta n_0 P_2$ to $\Delta n_0 = 0.106$.

It is possible to estimate the upper limit of the intrinsic birefringence of the units producing the heat conduction anisotropy. Assuming that the intrinsic anisotropy in heat conduction is infinitely large, i. e. $A_o \to \infty$, the minimum orientation is needed to produce the macroscopic anisotropy of the sample. This requires the maximum intrinsic birefringence $\Delta n_o = 0.13$. This value is well below that estimated by Biangardi [2]. The general problem in this context is the assignment of the unit or segment involved in the orientation process because the different experimental methods may correspond to units of different size. In the case of heat conduction the picture of the collisional transport leads to the conclusion that this unit must be closely related to the statistical equivalent seg-

ment, the Kuhn segment. The unflexible part of a monomer or segment will act as a whole, thus both the size of the statistical segment and that of the adjacent flexibel parts will determine the anisotropy of the collision process. For the PC chain we regard the monomer unit to be this segment because of the high flexibility of the carbonate group. From heat conduction anisotropy the range allowed for the intrinsic anisotropy of this segment is given by

$$0.10 \leq \Delta n_o \leq 0.13 \,.$$

3. Orientation of the chain axis revealed by WAXS

The structure of amorphous materials may be described by the density correlation function (DCF) $\hat{\varrho}(\vec{R})$ [14, 15]. The DCF gives the probability to find an atom a vector distance $\vec{R}$ apart from another atom. Assuming cylindrical symmetry (as is the case for uniaxial stretched materials) one can calculate the orientation distribution function (ODF) [14]

$$D(R, \alpha) = \frac{1}{2} \frac{\hat{\varrho}(R, \alpha)}{\hat{\varrho}(R)} \tag{3.1}$$

which describes the probability that an interatomic distance vector $\vec{R}$ makes an angle α with the stretching direction (see fig. 3).

$$\hat{\varrho}(R) = \frac{1}{2} \int_0^\pi \hat{\varrho}(R, \alpha) \sin \alpha \, d\alpha \tag{3.2}$$

is the isotropic mean value of the cylindric symmetric DCF $\hat{\varrho}(R, \alpha)$. Because of the symmetry one can expand the ODF in the even Legendre polynominals $P_{2n}(\cos \alpha)$

$$D(R, \alpha) = \sum_{n=0}^{\infty} D_{2n}(R) P_{2n}(\cos \alpha) \tag{3.3}$$

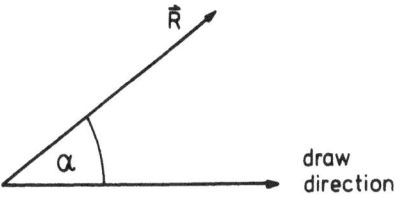

Fig. 3. Angle α between draw direction and distance vector $\vec{R}$

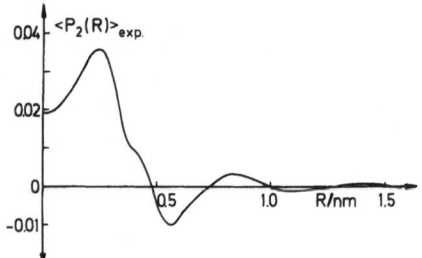

Fig. 4. Orientation parameter $\langle P_2(R)\rangle_{\text{exp}}$ derived by WAXS

with

$$D_{2n}(R) = \frac{4n+1}{2} \int\limits_0^\pi D(R,\alpha)\, P_{2n}(\cos\alpha)\, \sin\alpha\, d\alpha\,. \tag{3.4}$$

The Herman orientation parameters $\langle P_{2n}(R)\rangle$ are very closely related to the Legendre polynominal expansion coefficients of the ODF.

$$\langle P_{2n}(R)\rangle = \frac{2}{4n+1}\, D_{2n}(R)\,.$$

The orientation parameters $\langle P_{2n}(R)\rangle$ are still functions of the distances. Figure 4 shows the orientation parameter $\langle P_2(R)\rangle$ for a specimen of polycarbonate with a draw ratio of $\lambda = 2.5$. From this orientation parameter one cannot deduce directly the orientation of a chain segment. The orientation of distance vectors within a chain segment differ from the orientation of the chain axis because of the molecular architecture. Provided the orientation of the interatomic distance vectors within a chain segment with respect to its axis is known one can calculate the orientation parameters from [14,16]

$$\langle P_{2n}(R)\rangle_{\text{exp}} = \langle P_{2n}(R)\rangle_{\text{intra}}\, \langle P_{2n}\rangle_{\text{chain}} \tag{3.5}$$

$\langle P_{2n}(R)\rangle_{\text{intra}}$ are the orientation parameters of the intrachain distance vectors with respect to the chain axis.

$\langle P_{2n}\rangle_{\text{chain}}$ is the orientation parameter of the chain segment with respect to the draw direction. $\langle P_{2n}(R)\rangle_{\text{exp}}$ is the orientation parameter determined by WAXS. The chain orientation is then given by

$$\langle P_{2n}\rangle_{\text{chain}} = \frac{\langle P_{2n}(R)\rangle_{\text{exp}}}{\langle P_{2n}(R)\rangle_{\text{intra}}}\,. \tag{3.6}$$

In order to calculate the orientation parameter $\langle P_{2n}(R)\rangle_{\text{intra}}$ we calculated the DCF $\hat{\varrho}_{\text{intra}}(\vec{R})$ for a short

chain segment of two monomer units [14]. The configuration of this chain segment is similar to that of figure 10 with the exception that we now use an angle of $110\,^\circ$ between C_{carbonat}-O-C_{phenyl} in order to build up a stretched chain. The structure is not exactly the one we deduced from WAXS and from birefringence calculations, but it serves as a first approximation. Figure 5 shows the isotropic mean value of the DCF derived from WAXS [13,14,17] and derived from the model. Because the model contains only distances within the chain, $\hat{\varrho}(\vec{R})_{\text{intra}}$ approximates zero for higher R. There-

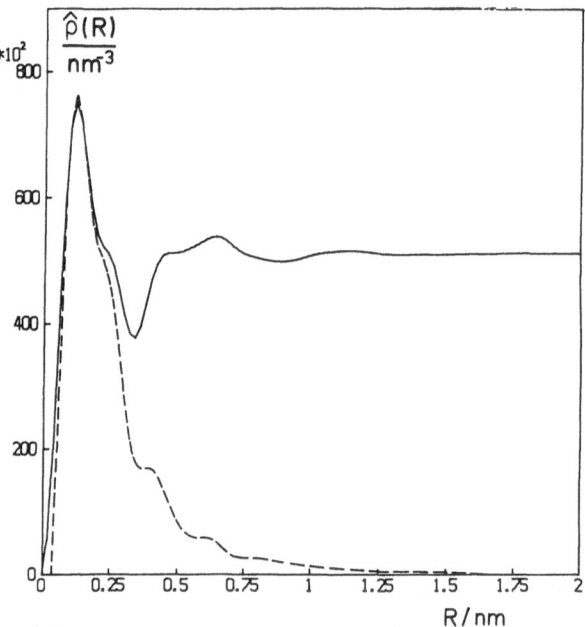

Fig. 5. Isotropic mean value of the density correlation function. —— experiment; — — — model

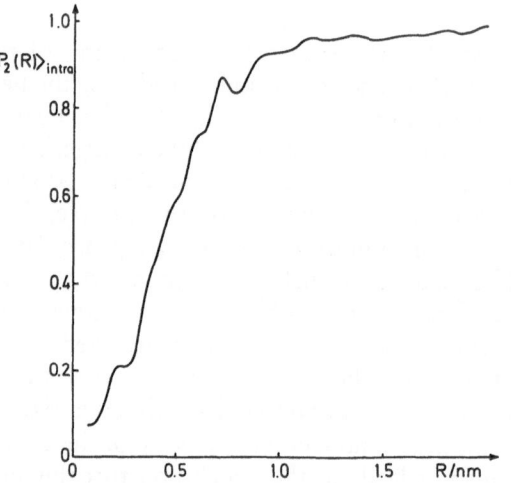

Fig. 6. Orientation parameter $\langle P_2(R)\rangle_{\text{intra}}$ of the chain model

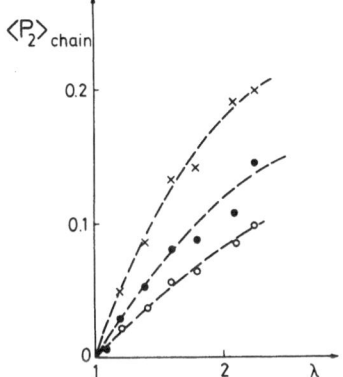

Fig. 7. Orientation parameter $\langle P_2 \rangle_{chain}$ for ● derived at the distance of 0.14 nm (next neighbours); ○ derived at the distance of 0.25 nm (next but one neighbours); × derived from the orientation of the main halo in the scattering intensity (see text)

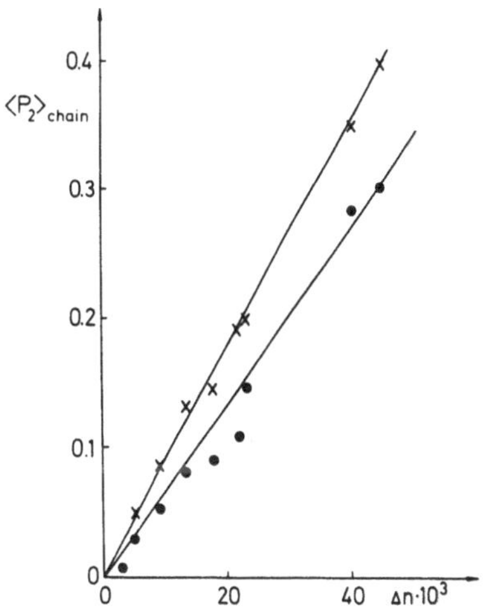

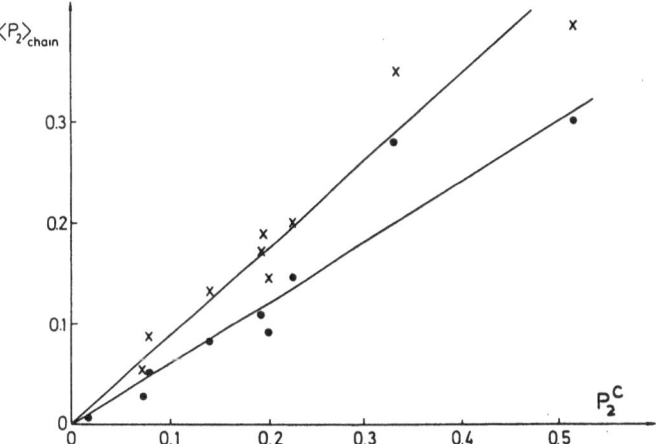

Fig. 8. Birefringence and chain orientation derived by WAXS. ● $\langle P_2 \rangle_{chain}$ at a distance of 0.14 nm; × from main halo of the scattering intensity

fore the calculations of the chain segment orientation are limited up to 0.3 nm.

Figure 6 shows $\langle P_2(R) \rangle_{intra}$ for the model. $\langle P_2(R) \rangle_{intra}$ approximates unity at higher R because the larger distance vectors ar more aligned to the chain axis. Figure 7 shows the chain orientation parameter $\langle P_2 \rangle_{chain}$ calculated at distances of next neighbours ($R = 0.14$ nm) and of next but one neighbours ($R = 0.25$ nm). The chain orientation deduced from the orientation of the X-ray scattering halo at the scattering vector $Q = 12.1$ nm^{-1} is included. This halo is due to the correlations of next neighbour chains [14, 17].

If one takes the values of crystalline polycarbonate from Bonart [18], who suggested that neighbouring chains are shifted by $c/8$ against each other[1]), one gets $\langle P_2(Q = 12.1 \, \text{nm}^{-1}) \rangle_{intra} = -0.27$ (see fig. 12). Provided the model structure reproduces the structure within the material, the points should match each other. But this cannot be expected because of the rough estimation of this model. The values of $\langle P_2 \rangle_{chain}$ derived at distances of next but one neighbours represent the lower limit of the chain segment orientation. The fact that we assume a somewhat inclined conformation of the phenyl rings (see section IV) decreases the value of $\langle P_2(R) \rangle_{intra}$ for next but one neighbour distances and as a consequence the corresponding values $\langle P_2 \rangle_{chain}$ increase. Figure 8 shows a plot of the chain axis orientation versus birefringence. In figure 9 P_2^c derived from

[1]) The *c*-axis is parallel to the chain axis and the repeat unit consists of two monomer units.

Fig. 9. P_2 derived from the anisotropy of heat conduction and WAXS. Notation see figure 8

the anisotropy of heat conduction is compared with $\langle P_2 \rangle_{chain}$.

4. Calculation of monomer birefringence

For the following calculation we generally assume that the average orientation of the monomers with respect to the chain axis has rotational symmetry. This is an obvious assumption for glassy polymers which is not neccessarily valid for the crystalline state. On this assumption it is sufficient to calculate only the uniaxial part of the birefringence. It is given for sufficient small birefringences by[2])

$$\Delta n_o = \frac{2\pi(n^2 + 2)^2}{9n} N \, \Delta\alpha_o \qquad (4.1)$$

where n is the mean refractive index, N the number density of monomers and $\Delta\alpha_o$ the uniaxially (transversely) averaged polarizability difference of the monomer. The polarizability tensor of a molecule is frequently calculated by assuming the additivity scheme of bond polarizabilities to be valid (citations in [19]). In addition the bond angles of the molecule are needed. In many cases this method yields good approximative values but may fail in others (even in the sign of $\Delta\alpha_o$) because of internal field effects. It could be shown [4] that the additivity scheme is only applicable if the Lorentz-Lorenz (LL) equation is valid. In fact, the derivation of the Lorentz-Lorenz equation leads to equation (4.1).

The LL equation is derived on the assumption of an isotropic structure contribution to the internal field.[3])

If local order occurs (intra and intermolecular) the internal field becomes anisotropic resulting in deviations from the additivity scheme. The calculations according to (4.1) can be considered to be much more reliable if the polarizability difference $\Delta\alpha_o$ has been determined for the monomer or part of it from other experimental methods again under the assumption of the LL equation.

In this case the severest contribution of the internal field due to the molecular constitution is incorporated in the polarizability difference derived. Using this value again in the LL equation or its equivalent (4.1) thus embraces the internal field contribution inherent in the measurements of $\Delta\alpha_o$. This would not hold true if the polarizability difference would have been calculated from bond polarizabilities. Fortunately such measurements of $\Delta\alpha_o$ exist for the monomer of the Bisphenol-A-PC. Erman and coworkers [6] have evaluated the polarizability tensors of the phenyl and of the carbonate group within the PC configuration. From measurements of depolarized Rayleigh scattering (DRS), electric (Kerr effect), and magnetic (Cotton-Mouton effect) birefringence. These measurements have been conducted on the monomer analogues Diphenylcarbonate (DPC) and Diphenylpropane (DPP) and on the polymer. For the former the configurational energy has been calculated as well, thus yielding the most probable bond angles. The polarizability tensors of the constituent molecular configurations determined in this manner embrace the internal field effect. They differ significantly from those derived on the assumption of tensor additivity. The fully substituted C-atom with the two methyl groups can be assumed to be isotropic [6] or to give only a negligible contribution compared to the other groups. The values taken from [6] are listed in table 1.

The reference coordinate system of the molecule has been attached to the carbonate group, see figure 10. $\Delta\alpha_o^\phi$ denotes the uniaxial part of the polarizability difference of the phenyl group, $\Delta\alpha_o^C$ that of the carbonate

Table 1. Polarizability data of the monomer constituents. These data of the anisotropic part of the polarizability tensor are taken from reference [6]. For comparison it must be noticed that the coordinate system used in this paper differ from those used in reference [6]. The anisotropic part is given by $\hat{\underline{\alpha}} = \underline{\alpha} - \frac{1}{3}$ trace $\underline{\alpha}$ and can be written in terms of its uniaxial $\Delta\alpha_o$ and transverse $\Delta\alpha_+$ components as $\hat{\underline{\alpha}} = \Delta\alpha_o$ diag $(2/3, -1/3, -1/3) + \Delta\alpha_+$ diag $(0, 1/2, -1/2)$

	$\hat{\alpha}_{xx}$	$\hat{\alpha}_{yy}$	$\hat{\alpha}_{zz}$	$\Delta\alpha_o$	$\Delta\alpha_+$
carbonate group	−1.21	1.04	0.17	1.56	−1.38
phenyl group	−3.9	−1.0	4.9	7.35	−2.9
			10^{-24} cm^3		

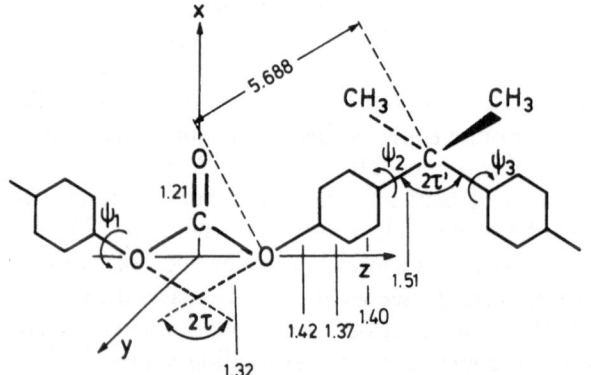

Fig. 10. The polycarbonate monomer according to [6]. The angles τ and τ' are 67° and 56° respectively thus allowing no planar chain like that in the figure. The effective tilting angle θ_{eff} is that of the O-phenyl-C configuration with respect to the coordinate system shown. Hence it is a combiantion of τ, τ' and a rotation around the C-O bonds of the carbonate group. The torsion angles ψ_i can have the same or opposite sign.

---

[2]) The equation is given in *cgs* units because the *SI* units are very inconvenient in this case. For going over to the *SI* system the substitution $\alpha/4\pi \in_o \rightarrow \alpha$ must be made converting the units to $4\pi \in_o 10^{-30}$ m^3 (*SI*) $\cong 10^{-24}$ cm^3 (*cgs*)

[3]) The field itself is anisotropic on account of the anisotropic molecular polarizability. An anisotropic structure contribution results in an anisotropic internal field even for an isotropic molecular polarizability.

group, $\Delta\alpha_o^M$ that of the monomer and $\Delta\alpha_+^\phi$ is the transverse part of the polarizability difference of the phenyl group (for definitions see table 1). Then the uniaxial part of the monomer can be written as

$$\Delta\alpha_o^M = 2\Delta\alpha_o^\phi(\theta, \psi) + \Delta\alpha_o^C$$
$$= 2P_{200}\Delta\alpha_o^\phi + 6P_{202}\Delta\alpha_+^\phi + \Delta\alpha_o^C \qquad (4.2)$$

where

$$P_{200} = \frac{1}{2}\left(3\left\langle\cos^2\theta\right\rangle - 1\right); P_{202} = \frac{1}{4}\left\langle\sin^2\theta\cos 2\psi\right\rangle$$

are the generalized orientation parameters [1] and apply to specific as well as to averaged configurations[4]). The angles θ, ψ occurring in (4.2) are indicated in figure 10 and belong to the set of Euler angles describing the orientation.

We assume that the bond angles 2τ, $2\tau'$ determined for the monomer analogues will not change when transferring to the polymer. On the basis of this assumption it must be concluded that monomer groups in PC cannot build up a planar zig-zag chain since $2\tau = 134°$ and $2\tau' = 112°$. Hence rotations most probable at the C-O bond of the carbonate group must accomodate the chain to a common linear chain axis. This explains the fact that the crystalline *c*-axis embraces two monomers [18]. These rotations and the bond angle can be lumped together into an effective tilting angle θ_{eff} of the O-phenyl-C configuration within the chain which is the one needed in equation (4.2). This effective tilting angle can be estimated from the crystallografic data of Bonart [18] who determined $c = 2.08$ nm in the chain direction. Assuming the carbonate groups to be parallel to the chain axis, the length of 1.653 nm is left for the four phenyl-group configurations having a linear length of 2.267 nm. The resulting tilting angle is $\theta_{\text{eff}} = 43.2°$.

The torsion angles ψ of the phenyl group are $\pm 46°$ for the monomer analogues and about $50°$ for the polymer [6]. The latter value represents an average which may be deduced in principle from $\langle\cos^2\psi\rangle$.

Since both the torsion angles ψ_1, ψ_2 published in [6] are very close to the isotropy angle $\psi_{is} = 45°$[5]), they are lumped together. The results of our calculation is depicted in figure 11. In the upper part the dependence

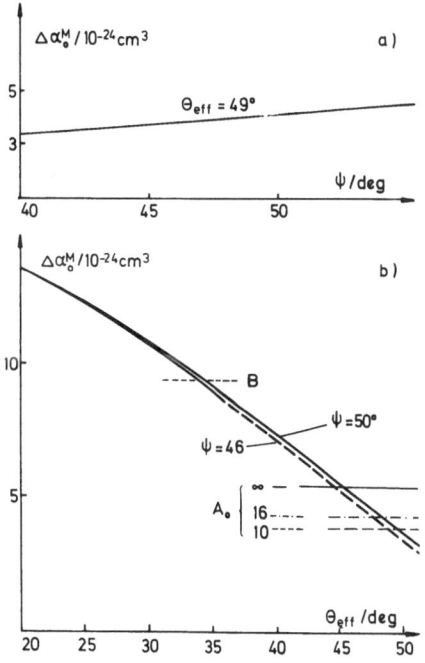

Fig. 11. The polarizability anisotropy $\Delta\alpha_o^M$ of the polycarbonate monomer versus the configurational angles. The curves have been calculated with equation (4.2). a) $\Delta\alpha_o^M$ versus the torsion angle ψ with fixed tilting angle θ_{eff}; b) $\Delta\alpha_o^M$ versus tilting angle θ_{eff} with fixed torsion angle ψ. The range allowed from heat conduction anisotropy is indicated together with the corresponding intrinsic anisotropy A_o. B marks the polarizability value derived from the intrinsic birefringence given by Biangardi [2]

of the torsion angle is plotted for a fixed tilting angle. It is evident that the torsion angle has only little influence within the range discussed. Since this range is close to the isotropy angle $\psi_{is} = 45°$ where P_{202} vanishes we must conclude that a broad near random distribution can give the same orientation parameter and cannot be excluded for the glassy polymer. Thus the torsion angle should be interpreted only as a representative mean value. Adopting this value to be $\psi = 50°$ [6] the dependence on the tilting angle θ_{eff} is plotted in the lower part of figure 11.

With the amorphous density $\varrho = 1.2$ gcm^{-3} the mean refractive index $n = 1.59$ [9,18] the molar mass $m = 254$ g mol^{-1} and the intrinsic (monomer) birefringence from heat conduction

$$0.10 \leq \Delta n_o \leq 0.13$$

the possible range of the monomers anisotropic polarizability is calculated to fall into the range

$$3.9 \leq \Delta\alpha_o \leq 4.4 \; (\times 10^{-24} \; \text{cm}^3).$$

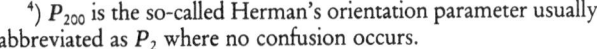

[4]) P_{200} is the so-called Herman's orientation parameter usually abbreviated as P_2 where no confusion occurs.

[5]) At this specific angle P_{202} vanishes as in the isotropic case which is characterized by a random distribution of ψ.

These values are also indicated in the lower part of figure 11. The result is highly satisfactory due to the fact that the angles θ_{eff} and ψ derived from crystallografic data [18] and from configurational energy calculations [6] fit very closely to the allowed range derived from heat conduction anisotropy. For the most probable range of heat conduction anisotropy $10 \lesssim A_o \lesssim 16$ the effective tilting angle is only enhanced to

$$47.5° \lesssim \theta_{eff} \lesssim 48.5°$$

for fixed $\psi = 50°$. Thus it appears that the average monomer configuration within the glassy PC is not far from that in the crystalline state. A 4°–5° higher average tilting angle θ_{eff} obviously suffices to build up the perturbed conformations of the amorphous phase.

Compared to the value $\Delta n_o = 0.236$ ($\Delta \alpha_o = 9.26 \times 10^{-24}$ cm^3) derived by Biangardi [2] the above values are lower by a factor of two. It is worthwhile to trace the reason for this difference. The upper limit from heat conduction ansiotropy is $\Delta n_o = 0.13$. Hence, we conclude that the simplified X-ray treatment of Biangardi [2] obviously does not determine directly the monomer orientation.

The published value itself leads to an effective tilting angle of 35°. This would correspond nicely to a coordinate system with the z-axis joining the two phenyl groups. The exact angle of 34° of the phenyl groups enhances the anisotropy by a small amount of 0.4 to 9.6 which is counterbalanced by the now rotated carbonate group ($\lesssim 20°$) to the value of 9.2. If this interpretation is legitimate, the method of Biangardi [2] would obviously be sensitive to the diphenyl correlations of neighbouring chains.

5. Discussion

The data presented give a clear picture of the orientational state in PC. It is now possible to compare the WAXS data of the orientated samples with those of isotropic ones by orientational averaging of the former. The resulting RDDFs are identical to those of isotropic samples [17]. That means that the local conformations and the neighbourhood of a monomer unit within the range of ≤ 1.5 nm remains constant irrespective of the macroscopic orientation. The difference in local structure in the range > 1.5 nm leading to the oriented states cannot clearly be detected by WAXS. This result legitimates the assumption made for the determination of orientation from heat conduction and from birefringence. The idea of an intrinsic birefringence is only possible with the assumption of a

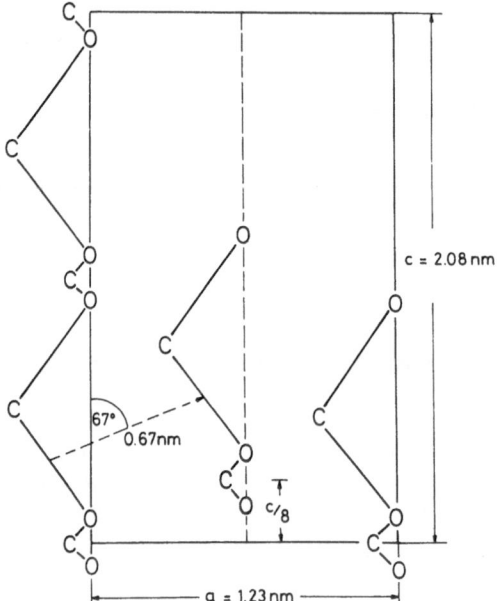

Fig. 12. Projection of the unit cell on the ac-plane according to Bonart [18]. The shift of $c/8$ produces the chain-chain correlation indicated by the dashed arrow. This correlation is assumed to produce the main halo

constant internal field which in turn requires a constant nearest neighbour structure on the average. The linear relationship between birefringence and the orientation parameter $\langle P_2 \rangle_{chain}$ from WAXS additionally counts for the constancy of the local structure. On the basis of these findings it is easily understood that the anisotropic heat conduction can also be described by a constant intrinsic anisotropy. With respect to the birefringence the local field is obviously mainly determined by the monomer configuration. The use of the well known additivity rule would have led to markedly different results. This demonstrates the need to use polarizability data corrected by the internal field when calculating the birefringence of polymer segments. An illustrative example is polyisobutylene which should have zero birefringence when using the additivity scheme, but in reality has a birefringence comparable to that of polyethylene [19]. However, the additivity scheme may be used to calculate the mean refractive index.

On a very local scale the chain conformation found is not very different from that in the crystalline one. This can be concluded not only by the similarity of the average torsion and tilting angles. Evaluating $\langle P_2 \rangle_{chain}$ from the WAXS data of the amorphous halo yields a quantitative agreement with that from heat conduction (fig. 9) provided the chains are assumed to be staggered with a shift of $c/8$ (in terms of the unit cell) as in the crystalline state (see fig. 12). This treatment of the amorphous halo is in contrast to that of Biangardi [2] who assumed the wide angle X-ray scattering of the main halo to be normal to the chains. The intensity of the halo is a sum over different contributions. The main contribution can be represented by that of the distance vectors of equal units. If no staggering is assumed the results of Biangardi will be reproduced. The tilting angle $\theta_{eff} \sim 35°$ derived from his data as discussed in section IV must be regarded as fortuitous. It cannot be interpreted as the tilting angle of the phenyl rings. The essential quantity, the value of $\langle P_2 \rangle_{intra} = -0.27$ is determined mainly by the ratio of the chain distances to the shift of $c/8$ and is only slightly dependent on the monomer configuration assuming a parallel chain packing. Biangardi assumed implicitly $\langle P_2 \rangle_{intra} = -0.5$ which explains the factor of about two compared to our results.

Our findings about the short range order of PC are not in contradiction to light scattering experiments [21] where no indications of liquid crystalline ordering has been found. In fact, the orientation correlations obsered by WAXS occur on a very local scale (≤ 1.5 nm). The apparent constancy of the depolarization parameter δ_o when going from solutions to the bulk corresponds to our conclusion that the internal field contribution is mainly intramolecular.

The measurement of the chain orientation in amorphous polymers is a routine method in principle. The evaluation of the data however requires a sufficiently accurate knowledge of the local structure like the monomer configuration and the arrangement of neighboured chains. Any simple minded interpretation of the amorphous halo may introduce serious errors. The X-ray scattering is sensitive to the atomic arrangements discussed above. Both birefringence and the anisotropic heat conduction are average values and are thus more directly coupled to the axially averaged monomer properties and hence to the chain axis provided the assumption of the constancy of the intrinsic parameters is fulfilled. The orientation parameters found by us in the uniaxially drawn PC are high and up to the limits of a deformed network.

Acknowledgement

We gratefully acknowledge the assistance of Mrs. B. Groner in performing the heat conduction experiments. This work has kindly been supported by the Deutsche Forschungsgemeinschaft (DFG).

References

1. Jarvis DA, Hutchinson IJ, Bower DI, Ward IM (1980) Polymer 21:41
2. Biangardi HJ (1980) (ed) Käufer H, Kunststoff-Forschung, Vol 1, TU Berlin
3. Pietralla M (1981) Coll & Polym Sci 259:111
4. Pietralla M (1982) Habilitationsschrift Ulm
5. Blum K, Kilian HG, Pietralla M (1983) J Phys E Sci Instr 16:807
6. a) Erman B, Wu D, Irvine PA, Marivn DC, Flory PJ (1982) Macromol 15:670
6. b) Erman B, Marvin DC, Irvine PA, Flory PJ (1982) ibid 15:664
7. a) Hennig J (1967) Kunststoffe 57:385
7. b) Hellwege KH, Hennig J, Knappe W (1963) Kolloid Z u Z Polym 188:121
8. McLaughlin (1969) (ed) Tye RP, Thermal Conductivity, Academic Press, London, p 42
9. Greig D (1982) (ed) Ward IM, Developments in Oriented Polymers 1, Applied Sci Pub, London, p 84
10. Van Krevelen DW (1976) Properties of Polymers, Elsevier Sci Pub Comp, Amsterdam, tables 4.8 and 10.2
11. Dietz W (1977) Coll & Polym Sci 255:755
12. Eiermann K (1964) Kolloid Z u Z Polym 198:5
13. a) Kilian HG (1981) Polymer 22:209; (1983) Kautschuk + Gummi 36:959
13. b) Kilian HG (1984) private communication
14. Schubach HR (1984) Dissertation, Ulm
15. Wright AC (1974) Adv Struct Res Diffr Meth 5:1
16. Mitchell GR, Windle AH (1983) Polymer 24:1513
17. Schubach HR, Heise B, to be published in Coll & Polym Sci
18. Bonart R (1966) Die Makromol Chem 92:149
19. Pietralla M (1980) J Polym Sci PPE 18:1717
20. Liberman MH, Debolt LC, Flory PJ (1974) J Polym Sci PPE 12:187
21. Dettenmaier M, Kausch HH (1981) Coll & Polym Sci 259:209

Received June 21, 1985;
accepted August 9, 1985

Authors' address:

M. Pietralla
Universität Ulm
Abt. Experimentelle Physik
Oberer Eselsberg
D-7900 Ulm, F.R.G.

Progress in Colloid & Polymer Science   Progr Colloid & Polymer Sci 71:134–139 (1985)

Orientation relaxation of fluorescent molecules in uniaxially drawn PVC-films during annealing*)**)

R. Neuert, H. Springer, and G. Hinrichsen

Technische Universität Berlin, Institut für Nichtmetallische Werkstoffe, Polymerphysik, Berlin, F.R.G.

Abstract: PVC-films doped with fluorescent molecules of different shapes were uniaxially drawn at 65 and 80 °C and subsequently annealed with free and fixed ends. The orientation behaviour of the fluorescent molecules and the polymer segments is investigated by UV-dichroism and fluorescence polarization and by birefringence respectively. The disorientation process of the fluorescent probes is shown to depend on drawing temperature and length of the probes. Longer fluorescent molecules lose their orientation more rapidly and shorter ones more slowly than the polymer segments on an average.

Key words: Orientation relaxation, fluorescence polarization, dichroism, birefringence, polyvinylchloride.

Introduction

As shown in several papers [1–3], probe orientation during uniaxial stretching of polymer films depends — although not exclusively — on the shape of the probe molecules, especailly on the length or the ratio of length to diameter. This has been found also for PVC-films [4] and may be explained in a first approximation by the hypothesis that longer probe molecules are better aligned to the polymer chains than shorter ones. But this hypothesis does not explain in a sufficient way that the longest probes are higher orientated than the polymer segments. To understand this phenomenon one may assume a preferred alignment of the longer probes to locally higher orientated segments. Such an assumption is in accordance with experimental results described in [5] and [6], which suggest that the segmental orientation is dependent on chain conformation.

To get more insight into the orientation and alignment behaviour of the probes, the present paper deals with the orientation change of the probes during annealing. In [5] IR-dichroic measurements are used to follow conformationally sensitive relaxation in drawn PVC-samples annealed with free and fixed ends. Unfortunately with the IR-spectra of our PVC-samples we do not see any chance for separating and evaluating the relevant bands in a straightforward manner. Only curve analysis as in [7] may lead to reliable results, but a number of assumptions about peak positions and widths have to be made and thus makes the procedure tedious and questionable. Therefore in this paper the segmental orientation is only characterized by an average value taken from birefringence measurements.

Experimental

Sample preparation

Preparation of the doped samples is described elsewhere [4]. The fluorescent molecules used for this report are presented in table 1. The lengths of the probes relative to the length of the DPS-molecule are also indicated.

The samples were drawn in a temperature cabinet using a tensile testing machine (Instron 1026). Drawing temperatures were set at 65 and 80 °C and the drawing rate at 20 %/min.

---

*) Poster presented during the Spring Conference of the Deutsche Physikalische Gesellschaft, March 18–20, 1985 in Lausanne.

**) Dedicated to Prof. Dr. H.-G. Kilian on the occasion of his 60th birthday.

Table 1. Fluorescent molecules used as probes

Probe name	Abbre-viation	Formula	L/L_{DPS}
1,4-diphenyl-1,3-butadiene	DPB		0.68
1,6-diphenyl-1,3,5-hexa-triene	DPH		0.81
1,8-diphenyl-1,3,5,7-octat-etraene	DPO		0.94
trans-4,4'-diphenyl-stilbene	DPS		1.00
1,4-bis(o-methylstyryl) benzene	MSB		0.91
4,4-bis(2buty-loctyloxy)-p-quaterphenyl	BIBUQ		1.35

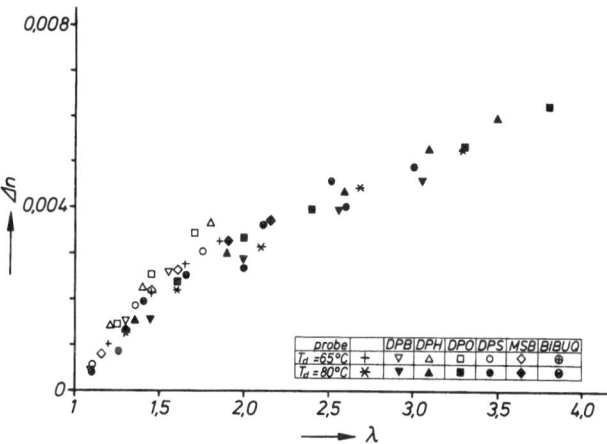

Fig. 1. Birefringence Δn plotted against draw ratio. Parameters: T_d, probes

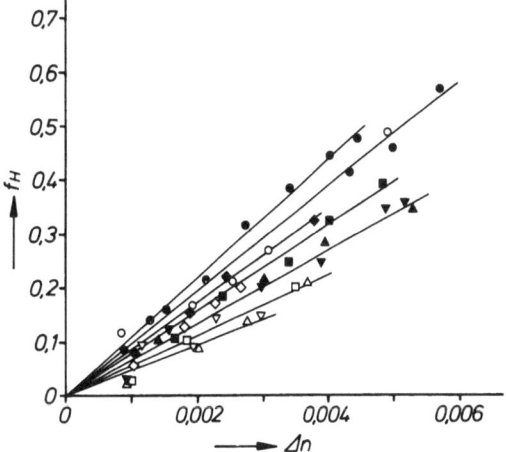

Fig. 2. Hermans' orientation factor f_H plotted against birefringence Δn. Symbols and parameters as in figure 1

Birefringence, dichroism and fluorescence polarization

These were measured as described in [8].

IR-dichroism

IR-dichroic measurements were performed on a Nicolet Fourier transform infrared spectrometer (MX-1) using a grid polarizer.

Annealing procedure

Drawn samples exhibiting a relatively high draw ratio were chosen for relaxation. The annealing temperatures lay between 80 and 87 °C. Annealing was carried out in an air oven for five minutes at each temperature.

Results

Characterization of the drawn samples

Figure 1 shows the birefringence Δn of the uniaxially stretched samples as a function of draw ratio λ. The maximum attainable draw ratio depends on the drawing temperature (T_d) and amounts to 1.8 for $T_d = 65$ °C and 3.8 for $T_d = 80$ °C. The birefringence values of samples drawn at 65 °C are slightly higher than those of the samples drawn at 80 °C, indicating temperature depending orientation relaxation during the drawing process. Doping has no significant influence on the Δn-λ-relation.

The interrelation between the orientation of fluorescent molecules and polymer segments is seen from figure 2, in which Hermans' orientation factor f_H evaluated from fluorescence polarization is plotted against birefringence. As in preceeding papers [3–5] the measured values can be approximated by straight lines whereby the longer molecules obviously orientate to a higher degree than the shorter ones. We consider only the length of the probes, because the cross section diameters are nearly equal. The dependence of f_H on the relative length L/L_{DPS} of the probes at a constant segmental orientation ($\Delta n = 0.004$) is graphed in figure 3. Adopting for the intrinsic birefringence of PVC $\Delta n_o = 0.013$ (cf. [3]) Hermans' orientation factor of the poly-

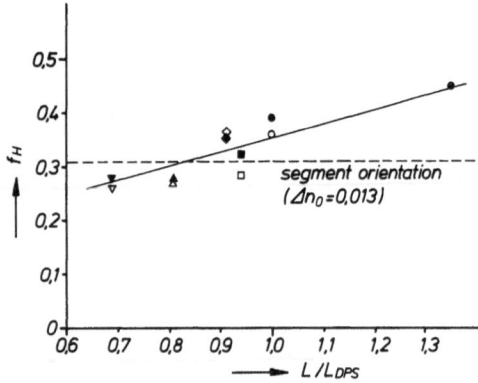

Fig. 3. Hermans' orientation factor f_H of probes in PVC-samples with $\Delta n = 0.004$. Symbols and parameters as in figure 1. The dashed line indicates the segmental orientation

Table 2. Orientation data of samples chosen for annealing experiments

Probe	$T_d/°C$	Sample ends	λ	$\Delta n \cdot 10^3$	f_H
DPB	65	Free	1.7	3.2	0.19
DPH			1.75	3.4	0.18
DPO			1.7	3.3	0.21
DPS			1.75	3.2	0.29
MSB			1.6	3.2	0.26
DPB		Fixed	1.6	2.7	0.17
DPH			1.8	3.5	0.19
DPO			1.8	3.2	0.22
DPS			1.75	3.2	0.29
MSB			1.6	3.0	0.24
DPB	80	Free	3.05	4.8	0.35
DPH			2.6	4.8	0.36
DPO			3.0	5.2	0.42
DPS			2.55	5.1	0.48
MSB			2.1	3.8	0.35
BIBUQ			2.05	3.4	0.41
DPB		Fixed	3.05	4.9	0.35
DPH			2.5	4.7	0.36
DPO			3.0	5.3	0.42
DPS			2.55	5.1	0.49
MSB			2.1	4.0	0.37
BIBUQ			2.05	4.0	0.47

mer segments can also be calculated. For comparison this value is drawn in figure 3 as a dashed line.

Orientation relaxation of the annealed samples

Table 2 yields the orientation data of samples chosen for annealing. For samples drawn at 65 °C the draw ratio λ lies between 1.6 and 1.8 and for samples drawn at 80 °C λ varies between 2 and 3.

To simplify a comparison of samples exhibiting different orientation values before annealing, the measured quantities of the annealed samples are divided by the initial value of the corresponding quantity. Such reduced quantities are indexed in the following with "rel".

The relaxation behaviour of the polymer segments for the samples of table 2 is outlined in figure 4. Doping

obviously plays no significant role for the segmental relaxation.

The disorientation of the fluorescent molecules as a function of annealing temperature can be gathered from figures 5 and 6 for annealing with fixed and free ends respectively. The relaxation behaviour depends

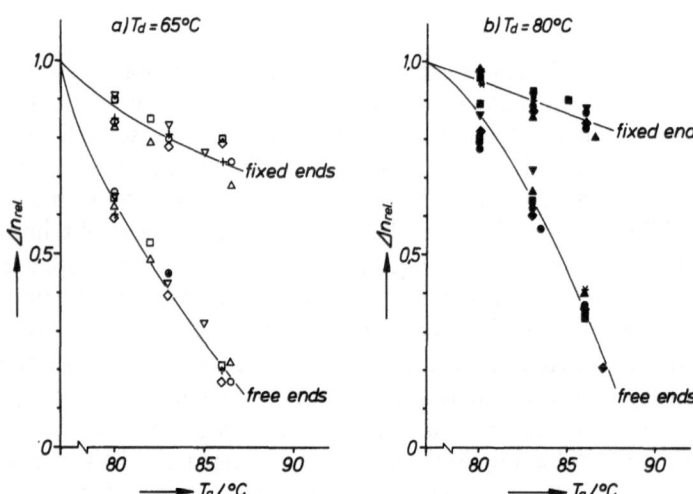

Fig. 4. Relative birefringence Δn_{rel} as a function of annealing temperature T_a. Symbols as in figure 1. Parameter: annealing with free or fixed ends. Drawing temperature: a) 65 °C, b) 80 °C

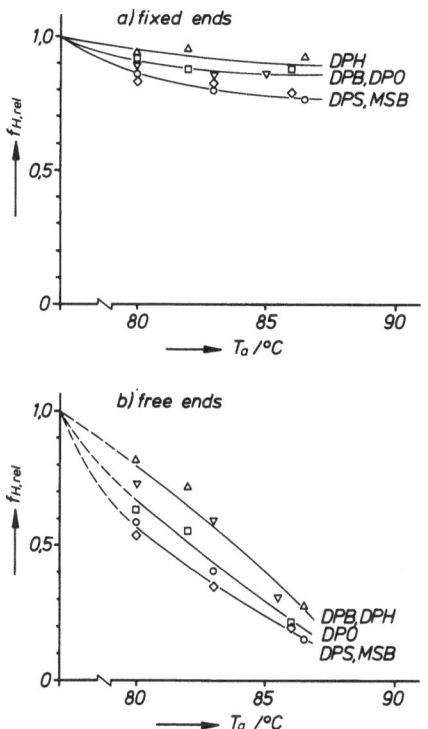

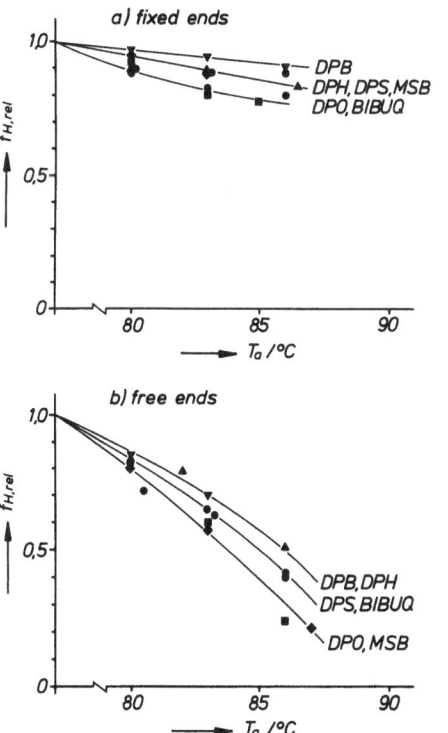

Fig. 5. Hermans' orientation factor f_H of probes in PVC-samples drawn at 65 °C as a function of annealing temperature T_a. Symbols as in figure 1. Annealing a) with fixed ends and b) with free ends

Fig. 6. Hermans' orientation factor f_H of probes in PVC-samples drawn at 80 °C as a function of annealing temperature T_a. Symbols as in figure 1. Annealing a) with fixed ends and b) with free ends

on the length of the probes, especially when the samples have been drawn at 65 °C and the annealing is performed with free ends.

Discussion

To compare segment and probe orientation relaxation directly, in figure 7 Δn_{rel} is plotted against $f_{H, rel}$. A clear separation of the measured values according to the fluorescent probes is only possible for samples drawn at the lower temperature. As with the f_H-Δn-relation, for samples stretched to different draw ratios, the length of the probes seems to be mainly responsible for the observed differences. To elucidate this, in figure 8 the orientation factor of the various probes, observed at the moment when the segmental orientation has decreased to half the initial value (Δn_{rel} = 0.5), is shown as a function of the relative probe length. Obviously the orientation of the short probes diminishes more slowly and that of the long probes more strongly than that of an "average chain segment".

Considering the orientation and disorientation processes together, one can achieve two different "orientation hysteresis" curves for a short and a long

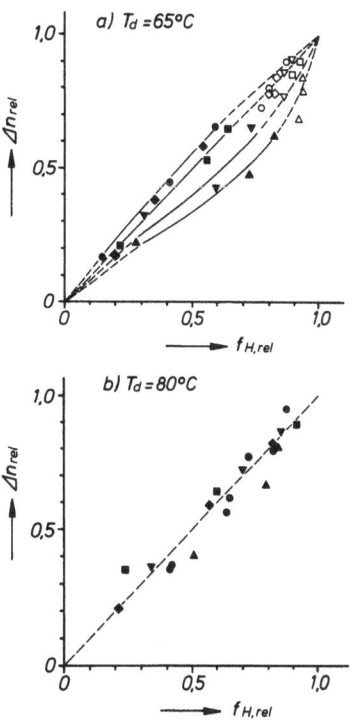

Fig. 7. Comparison of segment and probe disorientation for samples drawn at a) 65 °C and b) 80 °C. Symbols as in figure 1, but empty and filled symbols indicate annealing with fixed and free ends respectively

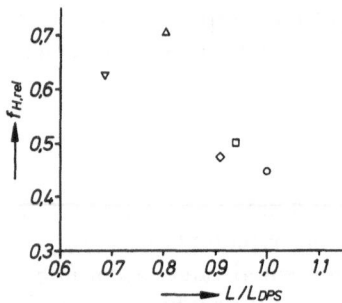

Fig. 8. Relative orientation factor $f_{H,\text{rel}}$ for probes of samples drawn at 65 °C and annealed with free ends to $\Delta n_{\text{rel}} = 0.5$ as a function of relative probe length. Symbols as in figure 1

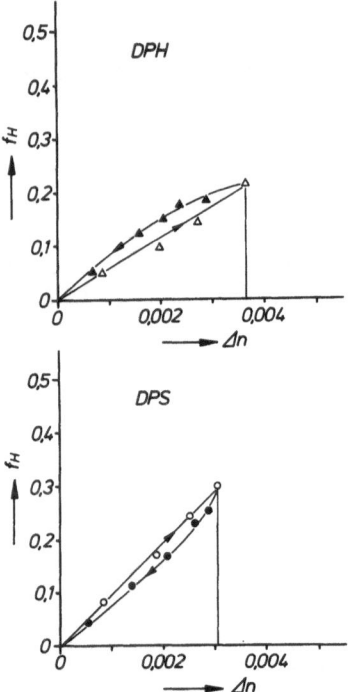

Fig. 9. Hermans' orientation factor f_H for a) DPH and b) DPS in PVC-samples as a function of Δn. The empty and filled symbols give the values during drawing at 65 °C and annealing at various temperatures respectively

probe molecule respectively. This is demonstrated in figure 9 for DPH (short probe) and DPS (long probe). These observations are consistent with the following model:

During uniaxial drawing the polymer segments orientate in a locally and/or conformationally inhomogeneous way. The latter effect also has been suggested in [6] in interpreting IR-dichroic measure-

ments. The longer probe molecules tend to align preferentially to the higher orientated sequences of segments (e.g. trans-sequences), whereas the shorter probes rather adopt the orientation of shorter sequences. Orientation relaxation first causes disorientation of the higher orientated sequences of segments, thus leading to a relatively quick decrease of the orientation of the longer fluorescent molecules and a slower disorientation fo the shorter ones.

This model clearly holds only for the samples drawn at 65 °C. This can be understood if one takes into account that at $T_d = 80$ °C the local and conformational inhomogeneity is already reduced by relaxation processes during stretching.

Finally it should be mentioned that the observed phenomena cannot be explained by a simple model which only assumes that the longer probes align better to the polymer chain axis than the shorter ones. This model would hold for the drawing behaviour as long as the probes are less orientated than the segments, but fail in describing the relaxation process, because the hysteresis effects are incompatible with this model.

Conclusion

The orientation and disorientation behaviour of polymer segments and rodlike fluorescent probes in PVC-films shows the following features:

Depending on drawing temperature a local and/or conformational inhomogeneity of segmental orientation exists. For samples drawn at temperatures significantly below T_g, this leads during annealing to an "orientation hysteresis", which depends on the length of the fluorescent molecules. Longer probes disorientate faster and shorter ones more slowly than an average polymer segment.

Acknowledgement

Financial support by the Deutsche Forschungsgemeinschaft is gratefully acknowledged.

References

1. Thulstrup EW, Michl J (1982) J Am Chem Soc 104:5594
2. Hennecke M (1981) Thesis, Kaiserslautern
3. Springer H, Neuert R, Müller FD, Hinrichsen G (1983) Coll & Polym Sci 261:800
4. Springer H, Kussi J, Richter HJ, Hinrichsen G (1981) Coll & Polym Sci 259:911
5. Gabarayeva AD, Shishkin NI (1973) Vysokomol Soyed A 15:2769
6. Shindo Y, Read BE, Stein RS (1968) Makromol Chem 118:272
7. Theodorou M, Jasse B (1983) J Polym Sci Polym Phys Ed 21:2263

8. Springer H, Neuert R, Müller FD, Hinrichsen G (1984) Coll &
Polym Sci 262:46

Received May 17, 1985;
accepted August 9, 1985

Authors' address:

Dr. H. Springer
Institut für Nichtmetallische Werkstoffe
Polymerphysik
TU Berlin
Englische Str. 20
D-1000 Berlin 12, F.R.G.

Progress in Colloid & Polymer Science　　　　Progr Colloid & Polymer Sci 71:140–144 (1985)

Preferred orientation of the internal structure of carbon layers in carbon fibers *)

R. Plaetschke and W. Ruland

Fachbereich Physikalische Chemie, Bereich Polymere, Philipps-Universität, Marburg/Lahn, F.R.G.

Abstract: In addition to the preferred orientation of the carbon layer normals, carbon fibers can show a preferred orientation of the internal structure of the carbon layers. The degree of orientation is a function of the heat-treatment temperature, the type of orientation depends on the starting material. The origin of the orientation is probably a preferred direction of growth during the first stages of pyrolysis.

Key words: X-ray scattering, carbon fibers, *a*-axis orientation, general fiber symmetry.

1. Introduction

The preferred orientation of the carbon layer planes in carbon fibers with respect to the fiber axis has been studied in detail [1], notably in connection with the relationship between structure and mechanical properties [2–5], in relation to the micropore orientation [6,7], the intercalation [8] and to fluorination [9]. In all these studies it was assumed that no preferred orientation exists for the internal structure of the carbon layers, i. e. the orientation of the two-dimensional hexagonal structure of these layers was considered random with respect to the layer normal.

Preliminary X-ray studies on carbon fibers from mesophase pitch carried out in this laboratory indicated that a preferred orientation of the internal layer structure is observable in certain cases[1]. The aim of the work reported here was to investigate this effect in detail.

2. Theoretical

In the case of a random orientation of the internal structure of the carbon layers with respect to the layer normals, the orientation distributions (pole figures) of all reflections are completely defined by the orientation distribution of the layer normals (simple fiber

symmetry). The latter is experimentally obtained from the intensity distributions of the $(00l)$ reflections for constant absolute values of the reciprocal space vector s_{00l} ($s = 2 \sin \theta/\lambda$). It has been shown [10] that the intensity distribution of the (hk) interferences of nongraphitic carbons are related to the orientation distribution of the layer normals by the function

$$F_c (\sigma, \phi) = 2 \int_0^\pi g_c (\beta) \, d\eta \qquad (1)$$

where g_c is the orientation distribution of the layer normals (c-axis orientation), $\sin \sigma = s_{hk}/s$, and $\cos \beta = \cos \phi \cos \sigma + \sin \phi \sin \sigma \cos \eta$. Figure 1 shows the directions and angles involved in this calculation.

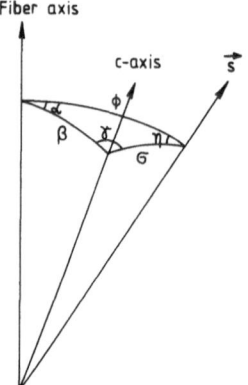

Fig. 1. Geometry of orientation angles and directions

---

*) Dedicated to Prof. Dr. H.-G. Kilian on the occasion of his 60th birthday.
[1]) For PAN-base carbonfibers, such orientations have been discussed earlier [15,16].

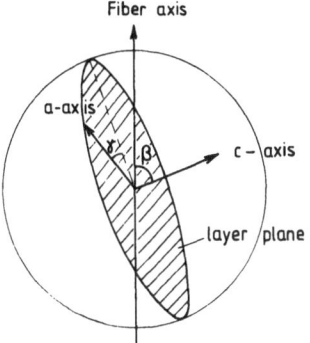

Fig. 2. Orientation of a carbon layer with respect to the fiber axis

In order to extend this treatment to include a preferred orientation of the internal structure of the carbon layers, let us assume that the orientation of the layer normals (c-axis) and of the internal layer structure (a-axis) are not correlated and that the latter is defined by the orientation distribution $g_a(\gamma)$. Figure 2 shows a schematic presentation of the geometry involved. β is the angle between the c-axis and the fiber axis, γ is the angle between the a-axis and the section of the layer plane and the plane including the fiber axis and the c-axis.

The angle γ is shown in the spherical triangle defined in figure 1 which corresponds to that already defined in the earlier work [10]. Inspection of the treatment given in the earlier work reveals that the effect of an a-axis orientation changes equation (1) to

$$F_{a,c}(\sigma, \phi) = 2 \int_0^\pi g_a(\gamma)\, g_c(\beta)\, d\eta \qquad (2)$$

where γ is related to β, σ and ϕ by

$$\cos\phi = \cos\beta \cos\sigma + \sin\beta \sin\sigma \cos\gamma.$$

Substituting η by β in equation (2) one obtains

$$F_{a,c}(\sigma, \phi) = \int \frac{g_a(\gamma)\, g_c(\beta)\, \sin\beta\, d\beta}{\sqrt{\sin^2\phi \sin^2\sigma - (\cos\beta - \cos\phi \cos\sigma)^2}}.$$

The limits of the integral are defined by the range of real values for the denominator.

Considering the intensity distribution in ϕ for $s = s_{hk}$, i. e. on the maximum of the (hk) interferences, one obtains

$$F_{a,c}\left(\frac{\pi}{2}, \phi\right) = \int_{\frac{\pi}{2} - \phi}^{\frac{\pi}{2} + \phi} \frac{g_a(\gamma)\, g_c(\beta)\, \sin\beta\, d\beta}{\sqrt{\sin^2\phi - \cos^2\beta}}. \qquad (3)$$

For carbon fibers with a high preferred orientation of the layer normals, $g_c(\beta)$ is non-zero only in a narrow range of β values in the vicinity of $\pi/2$. In that case, equation (3) can be approximated by

$$F_{a,c}\left(\frac{\pi}{2}, \phi\right) \simeq g_a(\phi)\, F_c\left(\frac{\pi}{2}, \phi\right). \qquad (4)$$

Due to the hexagonal structure of the carbon layers, g_a is periodic

$$g_a(\phi) = g_a\left(\phi - \frac{\pi}{3}\right).$$

Furthermore, if one considers the intensity distribution of the (10) interference to be determined by $g_a(\phi)$, that of the (11) interference is determined by $g_a\left(\phi - \frac{\pi}{6}\right)$, i.e. maxima in $g_a(\phi)$ correspond to minima in $g_a\left(\phi - \frac{\pi}{6}\right)$ (see fig. 3).

A simplified determination of $g_a(\phi)$ consists in measuring the intensity distribution as a function of ϕ for constant $s = s_{10}$ and $s = s_{11}$. The ratio of the two curves is given by

$$\frac{I(s_{10}, \phi)}{I(s_{11}, \phi)} \sim \frac{g_a(\phi)}{g_a\left(\phi - \frac{\pi}{6}\right)} = J(\phi) \qquad (5)$$

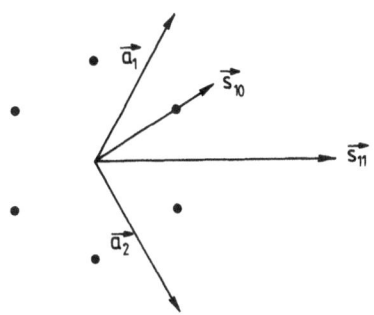

Fig. 3. Orientation of the hexagonal layer structure with respect to the unit cell vectors a_1 and a_2 and the reciprocal lattice vectors s_{10} and s_{11}

provided $g_c(\beta)$ is sufficiently narrow and that no correlation exists between preferred orientation and other structural parameters, e. g. graphitization.

Taking into account the multiplicity of (10) the function $g_a(\phi)$ can be approximated by

$$
g_a(\phi) = \frac{1}{3}\sum_{m=0}^{2} \frac{1-q_a^2}{1+q_a^2-2q_a\cos 2\left(\phi-\pi\,\dfrac{m}{3}\right)}
$$
$$
= \frac{1-p^2}{1+p^2-2p\cos 6\phi} \qquad (6)
$$

where q_a is the orientation parameter for the a-axis and $p = q_a^3$. The range of q_a and p values is -1 to 1. $q_a = p = 0$ is random orientation.

For positive values of q_a, the functions $g_a(\phi)$ and $J(\phi)$ show maxima at $0°$, $60°$ etc., i. e. the carbon hexagons are oriented preferrentially with the corner in the direction of the fiber axis (see fig. 3).

For negative values of q_a the functions $g_a(\phi)$ and $J(\phi)$ show maxima at $30°$, $90°$ etc., i. e. the carbon hexagons are oriented preferrentially with the side in the direction of the fiber axis (see fig. 3).

According to the above expressions the ratio Q of the maximum and the minimum value of $J(\phi)$ is given by

$$
Q = \left(\frac{1+|p|}{1-|p|}\right)^4 = \left(\frac{1+|q_a|^3}{1-|q_a|^3}\right)^4
$$

i. e.

$$
|q_a| = \left(\frac{Q^{1/4}-1}{Q^{1/4}+1}\right)^{1/3}. \qquad (7)
$$

The function $F_c\left(\dfrac{\pi}{2},\phi\right)$ can be approximated by [10]

$$
F_c\left(\frac{\pi}{2},\phi\right) \sim [(1+q_c)^2-4q_c\sin^2\phi]^{-1/2} \qquad (8)
$$

where q_c is the orientation parameter for the c-axis (q_c is identical to the parameter q in the earlier papers).

3. Experimental

Four samples of mesophase pitch carbon fibers, five samples of PAN-base carbon fibers and four samples of cellulose-base carbon fibers were chosen for the studies[2]). Designation and characteristics of the sample are listed in table 1.

[2]) The authors are indebted to Dr. L. S. Singer from Union Carbide Parma Research Center for providing the majority of the samples and to Dr. Böder from Sigri, Meitingen, for the samples P4 and P6.

Table 1. Designation and characteristics of the samples

Sample	Type	Heat treatment temperature (°C)	Tensile modulus (GPa)
M1	mesophase pitch	(thermoset pitch)	
M2	"	2500	607
M3	"	2750	586
M4	"	3000	634
P1	PAN	2500	400
P2	"	2750	434
P3	"	3000	483
P4	"	1350	
P6	"	2700	
C1	Cellulose		290
C2	"		359
C3	"		455
C4	"		524

X-ray scattering intensities were obtained by photofilm and by counter techniques. CuK$_\alpha$ radiation was used in both cases. The diffractometer set-up for the counter measurements consits of a curved quartz crystal as focussing monochromator with the sample in symmeterical transmission. Parallel bundles of fibers were mounted on a frame in such a way that the absorption factor is independent of the angle of rotation ϕ within the plane of the frame. The intensity was measured with a xenon-filled proportional counter and pulse-height discrimination. The relationships between the Bragg angle θ, the angle of rotation ϕ and the reciprocal space vector s ($s = 2\sin\theta/\lambda$) as well as various correction factors have been discussed in detail elsewhere [11].

4. Results and discussion

Figure 4a shows a plot of the intensity distribution $I(s_{10}, \phi)$ for the sample M4, i. e. the variation of the intensity with ϕ on the maximum of the (10) interference ring. For comparison, figure 4b shows a plot of the function $F_{a,c}\left(\dfrac{\pi}{2},\phi\right)$ calculated with the use of equations (4), (6) and (8) and the parameters $q_a = 0.4$ and $q_c = -0.8$.

Figure 5 shows the evaluation of the $I(s_{10}, \phi)$ and $I(s_{11}, \phi)$ curves in terms of equation (5) for the sample M3 in comparison with a theoretical curve for $g_a(\phi)/g_a\left(\phi-\dfrac{\pi}{6}\right)$ using equation (6) for $q_a = 0.33$. Systematic deviations from the theoretical curve occur for $\phi < 15°$ and $\phi > 75°$. The former are due to the fact that the calculation neglects the influence of the finite width of the (hk) rods on $I(s_{hk}, \phi)$, the latter are due to the influence of the tails of the $(00l)$ lines. A positive value of q_a

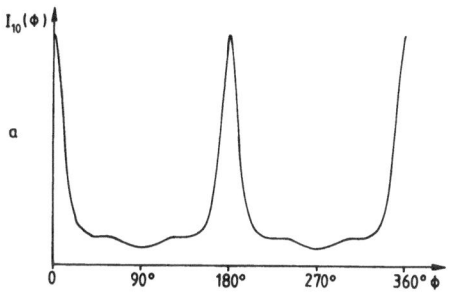

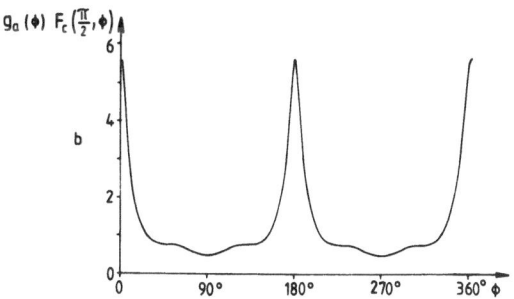

Fig. 4. Intensity distribution on the (10) interference ring as a function of ϕ. a) Observed (sample M4); b) calculated

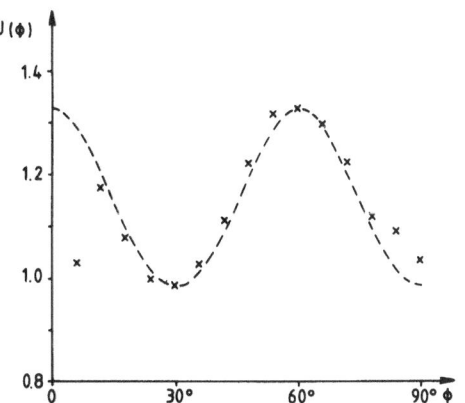

Fig. 5. Intensity ratio $J(\phi)$ for sample M3. $\times \times \times$ observed; $---$ calculated ($q_a = 0.33$)

Table 2. a-axis and c-axis orientation parameters and type of orientation

Sample	q_a	$|q_c|$	Type of orientation
M1	0	0.49	—
M2	0.30	0.90	◯
M3	0.33	0.92	◯
M4	0.41	0.93	◯
P1	−0.30	0.82	◯
P2	−0.27	0.83	◯
P3	−0.19	0.87	◯
P4	0	0.62	—
P5	−0.27	0.82	◯
C1	0	0.85	—
C2	0	0.86	—
C3	0.30	0.89	◯
C4	0.31	0.91	◯

is indicative for a preferred orientation of the carbon hexagons with the corner in the direction of the fiber axis.

Table 2 summarizes the results obtained from the intensity measurements on the (002), (10) and (11) interference rings.

The q_a-values listed in table 2 indicate that mesophase pitch and cellulose-base fibers show a-axis orientation with the corner of the hexagon in the direction of the fiber axis, whereas PAN-base fibers show a-axis orientation with the side of the hexagon in the direction of the fiber axis. In the case of mesophase pitch and cellulose-base fibers the a-axis orientation increases with increasing heat treatment temperature (HTT), no a-axis orientation is present at lower HTT.

In PAN-base fibers the a-axis orientation apparently goes through a maximum with increasing HTT.

The type of preferred orientation of the a-axis orientation resembles the expected orientation of the intermediate stages of pyrolysis in the case of PAN-fibers [12, 13] ("black orlon") and of cellulose fibers [14] ("longitudinal polymerization"). This orientation is not, however, detectable at lower HTT. This may be due to the fact that the orientation of the carbon hexagons formed in the early stages of pyrolysis is primarily with respect to the direction of growth of the ribbon-shaped carbon layers and that the increase of the c-axis orientation is correlated with an increase of the orientation of the long axis of the ribbons in the direction of the fiber axis. This does not, however, explain the decrease of the a-axis orientation at higher HTT in the case of PAN-base fibers.

An explanation for the a-axis orientation in the case of mesophase pitch fibers would be the preponderance of aromatic molecules in the mesophase pitch with a shape in favor of a corner-on orientation of the carbon hexagons on uni-axial deformation.

First results of a quantitative evaluation of the intensity distribution of general (hkl) interferences reveal that the description and explanation of the a-axis orientation may be more complicated than that given in this paper. The results of these studies will be presented later.

Acknowledgement

The authors are indebted to the Deutsche Forschungsgemeinschaft for supporting this work.

References

1. Ruland W (1967) J Appl Phys 38:3585
2. Ruland W (1969) Appl Polymer Symposia 9:293
3. Fourdeux A, Perret R, Ruland W (1971) International Conference on Carbon Fibres, their Composites and Applications, London, Prepr No 9
4. Perret R, Ruland W (1973) Kolloid Z u Z Polym 251:34
5. Fischer L, Ruland W (1979) Coll & Polym Sci 257:449
6. Perret R, Ruland W (1969) J Appl Cryst 2:209
7. Perret R, Ruland W (1970) J Appl Cryst 3:525
8. Hérinckx C, Perret R, Ruland W (1972) Carbon 10:711
9. Fischer L, Siemann U, Ruland W (1983) Coll & Polym Sci 261:744
10. Ruland W, Tompa H (1968) Acta Cryst A 24:93
11. Ruland W (1967) Norelco Reporter 14:12
12. Houtz RC (1950) Text Res J 20:765
13. Grassie N, Hay JN (1962) J Polym Sci 56:189
14. Bacon R, Tang MM (1964) Carbon 2:221
15. Ergun S (1972) Nat Phys Sci 238:137
16. Stuart T, Zubzanda O, Feughelmen M (1973) Nat Phys Sci 242:42

Received January 15, 1985;
accepted February 12, 1985

Authors' address:

Prof. Dr. W. Ruland
Dipl.-Phys. R. Plaetschke
Fachbereich Physikalische Chemie, Bereich Polymere
Philipps-Universität Marburg/Lahn
Hans-Meerwein-Str., Geb. H
D-3350 Marburg/Lahn 1, F.R.G.

Progress in Colloid & Polymer Science

Progr Colloid & Polymer Sci 71:145–153 (1985)

Density, energy and entropy of defects in the crystalline regions of crosslinked polyethylene*)**)

E. Jäger, J. Müller, and B.-J. Jungnickel

Deutsches Kunststoff-Institut, Darmstadt, F.R.G.

Abstract: By comparison of small angle X-ray scattering, thermal analytical and electron microscopic investigations, it is shown that crosslinked polyethylene crystallizes according to the usual folded chain scheme up to defect densities of 0.9 % and crystallizes micellarly for higher ones. The energy needed for the insertion of a crosslink into the lattice was estimated to be 3.3 eV, indicating a strong deterioration of the crystal. Similarly, the calculated entropy change due to crosslinks in the crystalline regions, 7 meV/K per crosslink, is rather high. This value can be explained by assuming that crosslinks are incorporated into the crystals in a variety of conformational and configurational different ways.

Key words: Crosslinked polyethylene, crystallization kinetics, energy and entropy of crystalline distortion points, chain folded and micellar crystallization.

1. Introduction

Crystallizable polymers can contain non-crystallizable elements of different kinds such as copolymer units, branches or crosslinks. The crystallization behavior of polymers which contain such defects is a matter of controversy. Especially the problem whether such defects are excluded from the growing crystal [1, 2], whether they are distributed statistically homogeneously in a partially crystalline material [3] or whether they are incorporated into the crystals to a level lying between these two limiting cases [4] is not definitely solved.

The extent to which defects influence the crystallization behavior and the amount to which — if at all — they are incorporated into the crystal can have kinetic, enthalpic or entropic causes. They lead to a shift of the melting temperature, to a decrease of the melting enthalpy and they influence both size and shape of the crystalline lamellae depending on the overall defect content. In this connection one has to consider that a defect incorporated into a crystal has to be characterized not only with respect to its energy requirement and the structural distortions which it induces but also with respect to its different conformational and configurational realization possibilities. The last issues strongly affect the changes in entropy. Beside this, as usual the entropy is influenced by a possible defect demixing and the distribution of the defects within the crystalline regions.

Therefore, there are many parameters which are not contained in and which can hardly be incorporated into the mentioned theories [1–4]. Therefore, the problem probably cannot be answered generally and each case has to be dealt with separately. Thus, it is surely not allowed to transmit results which are valid for the crystallization behavior of branched polyethylene (PE) (e. g. [5]) to that of crosslinked polyethylene and vice versa. For homogeneously crosslinked PE we concluded from wide angle X-ray measurements (WAXS) in a former paper [6] that the crosslink density within the crystalline region (X_c) increases up to an overall crosslink density (X) of some one percent, then decreases suddenly until reaching zero and again

---

*) Presented at the spring meeting of the Fachausschuß Polymerphysik of the Deutsche Physikalische Gesellschaft, Lausanne/Switzerland, March 18th–20th, 1985.

**) Dedicated with compliments to Prof. Dr. H.-G. Kilian on the occasion of his 60th birthday.

increases slightly. This behavior can be described by a thermodynamic model that predicts a chain folded crystallization scheme for $X < 1\%$ and a fringed micellar one for higher X-values. This is consistent with theoretical considerations by Kilian et al. [7], which described the crystallization behavior of moderately and more distorted polymers as fringed micellar crystallization with defect exclusion by eutectic demixing. Indeed, the crystallization of densely crosslinked PE can be understood and described quantitatively on this basis [8].

Nevertheless, the critical concentration for which chain folded crystallization is replaced by micellar crystallization, and the crystallization behavior in the vicinity of this critical concentration must be thoroughly examined, particularly because of the paucity of reports on the crystallization of homogeneously crosslinked PE. Detailed investigations are performed only for PE crosslinked by irradiation at room temperature [9,10], for branched PE [5] and for PE containing small amounts of copolymer units [11,12]. In this paper we report thermal analytical, electron microscopic and small angle X-ray scattering investigations on PE crosslinked by irradiation with fast electrons in the melt. The present results shed light on the questions mentioned above with reference to crosslinked PE. In addition, they allow not only to quantify the relation between crosslink density in the crystalline regions and the crosslink density in the crystallizing melt, but also to estimate the entropic and enthalpic changes after insertion of crosslinks into the crystals. The results are related closely to those reported earlier on WAXS of the same material [6] and with the findings on the crystallization behavior of PE with higher crosslink content [8].

2. Experimental

2.1 Sample material and preparation

Test bars were prepared from highly linear PE Lupolen[®1]) 5260 Z with a weight mean molecular weight of $M_w = 400\,000$. The samples were inserted into an irradition chamber, which was then evacuated. They were molten at (453 ± 1) K, homogenized for 2 h and then irradiated with electrons of 1 MeV at this temperature. Radiation was delivered by a Van-de-Graaf generator. The irradiation power was 3.3 Gysec^{-1} and the irradiation time varied between 0.5 min and 4 h. Under the thermal conditions chosen, crosslinking occurs almost exclusively with radiation chemistry, chain scission and creation of double bonds being negligible [13]. By subsequent suitable structure investigations, the crosslinking was shown to be homogeneous. After irradiation, the samples were cooled at (403 ± 1) K, stored at this temperature for 15 h and finally brought to room temperature.

2.2 Estimation of crosslink density

The measurements were twofold. On the one hand, the crosslink density was estimated by measuring the strain modulus [14]. The drawing was performed at 440 K, i. e. far above the equilibrium melting temperature ($T_m^o = 414$ K) to prevent stress induced crystallization. A slow crosshead speed of 0.5 mmsec^{-1} ensured that the number of effective network chains was not essentially influenced by entanglements. By simultaneous stress relaxation measurements the strain was proved to be purely elastic.

The crosslink density was also estimated by swelling measurements. For this purpose, the samples were swollen in xylene at 403 K for 24 h. From the equilibrium swelling, the crosslink density was calculated by use of the Flory-Rehner equation [15]. For the interaction parameter, the value $\chi = 0.31$ was chosen, this value being measured for linear PE at 373 K [16].

2.3 Small angle X-ray scattering (SAXS)

The SAXS measurements were performed with a Kratky camera (entrance slit width: 60 μm; detector slit width: 150 μm; 2ϑ detection range: (0.07–5)°; Cu-K$_\alpha$ radiation). The scattering curves were de-smeared [17] and Fourier transformed to obtain the autocorrelation function. This function was subsequently analysed to obtain the invariant, the specific internal surface and the amorphous and crystalline thicknesses respectively, assuming a partially crystalline lamellar two-phase system [18]. From these values, two independent SAXS degrees of crystallinity could be calculated, both of which were equal within the limits of error.

2.4 Electron microscopy (EM)

Sample pieces measuring $5 \times 3 \times 1$ mm^3 were treated with uranyle acetate according to a method described by Kanig [19]. Microtome cuts of (50–100) μm thickness were then taken at 215 K and finally investigated in the EM. On the electron micrographs, the crystalline lamellae are bright strips or spots. From these, the distances between adjacent lamellae (long periods) and the amorphous thicknesses were measured. Only lamellae oriented perpendicular to the cut surface were measured.

2.5 Thermal analysis

The melting behavior was analysed by DSC measurements (sample mass: $\approx$ 8 mg; heat rate: 2.5 Kmin^{-1}; sensitivity: 0.4 mJsec^{-1}). From the DSC traces, the specific melting enthalpy Δh (area between base line and measuring curve, normalized for the sample mass) and the melting temperature T_m (junction point between base line and the tangent through the point of inflection of the back side of the melting peak) were estimated. From the Δh-values, by $\alpha_{DSC} = \Delta h / \Delta h_o$ with $\Delta h_o = 287$ Jg^{-1} [20] a DSC degree of crystallinity was calculated.

3. Results

3.1 Crosslink density

In figure 1 the crosslink density X is shown as a function of the irradiation dose D. X is defined to be the number of C-atoms, bonded in crosslinks

---

[1]) [®]Lupolen is a trade name for PE of the BASF AG/West Germany.

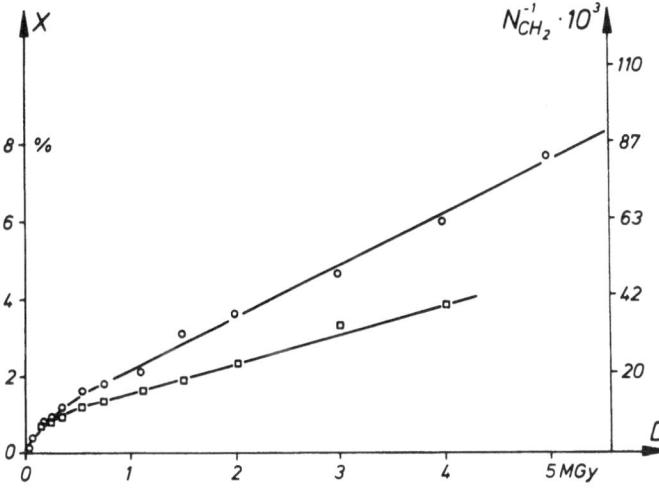

Fig. 1. Crosslink densities X versus irradiating doses D (–□–□–: by stress-strain measurements, –O–O–: by swelling measurements). N_{CH_2}: mean number of CH_2 units of the network chains

($>$CH–CH$<$), relative to their overall number. For the sake of perspicuity, the amount of CH_2-units between two crosslinks is also shown. The values which are estimated by modulus and by swelling measurements coincide well for doses smaller than 1 MGy. For higher irradiation doses, the true crosslink density is systematically higher than that calculated, since the basic assumption of the Flory-Rehner and Treloar theories, that is Gaussian behavior of the network chains, is not fullfilled. It will be shown later, however, that the range of crosslink densities which is important

for the question dealt with here is covered by irradiation doses lower than 1 MGy.

3.2 Small angle X-ray scattering data

The variation of the degree of crystallinity α_X as a function of the crosslink density is shown in figure 2. α_X was calculated according to

$$\alpha_X = d_c/(d_a + d_c) \qquad (1)$$

and

$$J_o = 2\pi^2 \alpha_X (1 - \alpha_X) (\Delta\eta)^2 \qquad (2)$$

(J_o: invariant, $J_o = \int J(s)\, s^2\, ds$; $\Delta\eta$: electron density difference between amorphous and crystalline regions; d_c, d_a: thicknesses of the crystalline and amorphous layers respectively). Figure 3 shows the dependence of the specific internal surface a_s on the crosslink density.

3.3 Electron-microscopy observations

Some EM pictures of the melt-irradiated PEs are displayed in figures 4 a–d. Because of the contrasting technique, the crystalline regions are displayed as bright strips or spots, the amorphous layers in between by dark ones. The non-irradiated sample (fig. 4 a) has the lamellar structure which is typical for PE. The lateral dimension of the lamellae is of some microns, whereas the lamellar thicknesses d_c are about 30 nm. It can be concluded from the series of pictures that with increasing irradiation dose the thicknesses, as well as the lateral dimensions of the lamellae decrease. For the sample irradiated with 0.5 MGy, both have comparable values, that is, only cubelike crystal bricks occur. Especially, for irradiation doses between 0.2 MGy and 0.5 MGy the lateral dimensions of the lamellae decrease rapidly. In samples irradiated with doses above 1.1 MGy, no crystal structures were distinguishable.

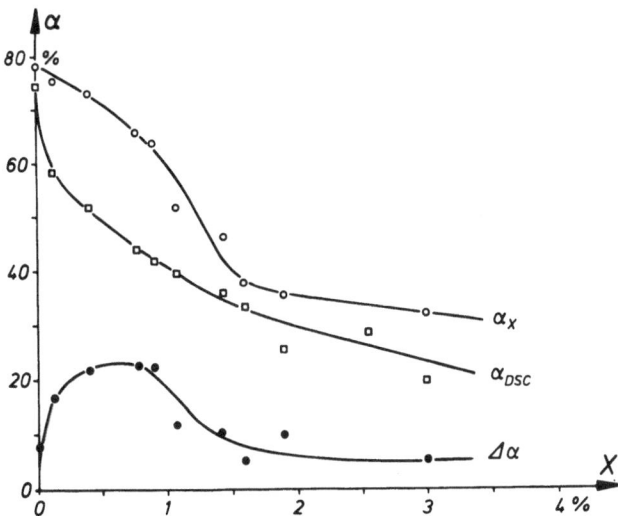

Fig. 2. SAXS (–O–O–, α_X) and DSC crystallinities (–□–□–, α_{DSC}) versus crosslink densities X. $\Delta\alpha = \alpha_X - \alpha_{DSC}$

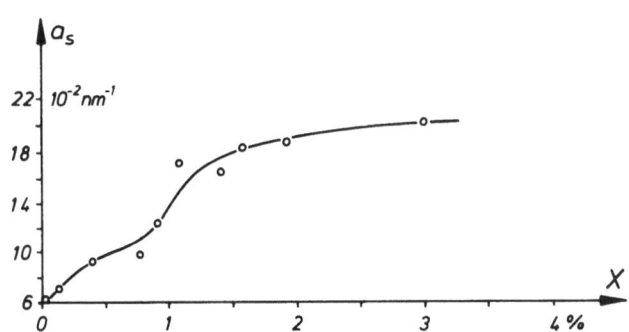

Fig. 3. Specific internal surface a_s as a function of crosslink density X

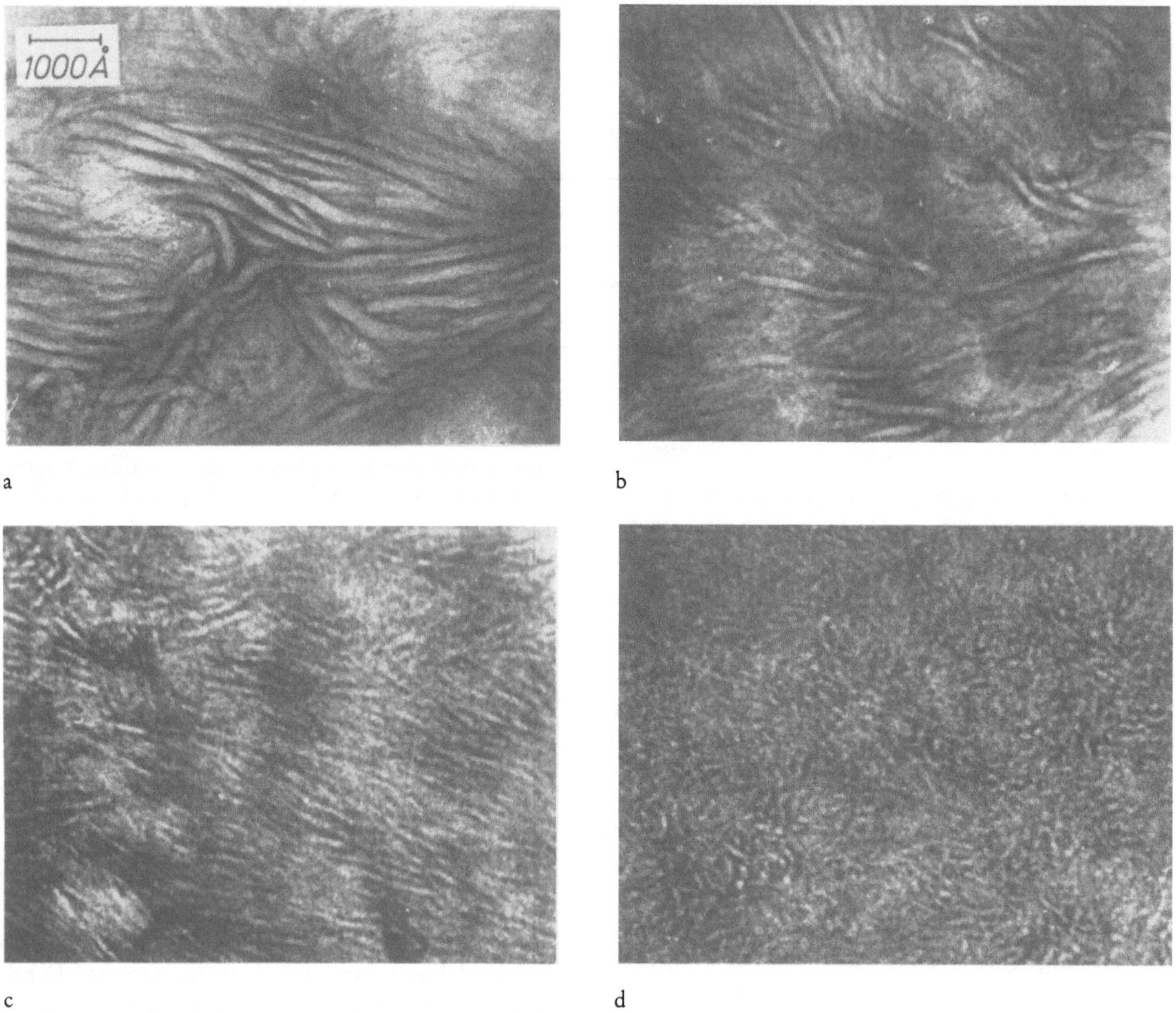

Progress in Colloid & Polymer Science, Vol. 71 (1985)

148

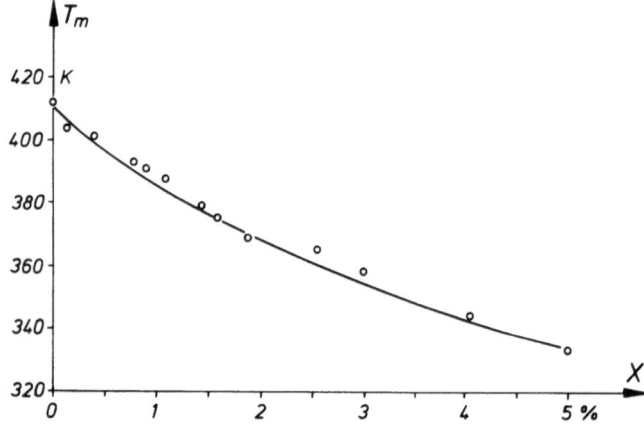

Fig. 4. Electron micrographs of the supermolecular structure of the melt-irradiated cyrstallized PE. a) $X = 0.00\%$; b) $X = 0.55\%$; c) $X = 0.93\%$; d) $X = 1.41\%$

Fig. 5. Melting temperatures T_m as a function of crosslink densities X

3.4. Thermal analysis

Figures 2 and 5 illustrate the thermal analytically estimated degree of crystallinity and the melting temperature as a function of the crosslink density.

4. Discussion

4.1 Distribution and energy of crosslinks

Firstly, it is remarkable that samples containing network chains of only 30 CH$_2$ groups, that is, with lengths comparable to the long periods, still have degrees of crystallinity of some 30%. These values are astonishingly high. Further important information is obtained by a comparison of the degrees of crystallin-

ity found by SAXS and by DSC. The following discussion is based on the assumption that α_X immediately reflects the mass content of the crystalline phase. In contrast, the melting enthalpy Δh and, therefore, α_{DSC} too, is influenced by[2])
– the crystalline mass content: Δh_m;
– defects in the crystalline regions: Δh_d;
– contributions of surface energies: Δh_s.

Δh_s depends on the size of the specific internal surface a_s and the specific surface energy σ:

$$\Delta h_s = \frac{1}{\varrho} \left(\sigma_s a_{ss} + \sigma_e a_{se} \right) \tag{3}$$

(ϱ: density). In this equation, distinction is made between σ_e, the σ-value of the end surface of the lamellae, and σ_s, the side contribution to the specific surface energy, the latter surface being parallel to the chains. Usually, $\sigma_e \approx 10\,\sigma_s$. Similarly, one has to distinguish between the contributions of the side and of the end faces of the lamellae to the specific internal surface (a_{ss}, a_{se}).

$\Delta h_d'$ can be written as

$$\Delta h_d' = \varepsilon_d \alpha_X X_c A M^{-1} \tag{4}$$

ε_d is the energy requirement for the insertion of a crosslink into the crystalline regions and X_c is the concentration of crosslinks within the crystals. The molecular weight of a CH_2-unit is designated as M and A is the Avogadro number. By α_X it is considered that Δh relates to the whole mass, whereas $\Delta h_d'$ refers only to the crystalline mass content. Summarizing, α_{DSC} can be written as

$$\alpha_{DSC} = \alpha_X - \alpha_d - \alpha_s; \quad \alpha_d = \Delta h_d'/\Delta h_0;$$
$$\alpha_s = \Delta h_s/\Delta h_0. \tag{5}$$

Therefore, the difference $\Delta \alpha = \alpha_X - \alpha_{DSC}$ (fig. 2) reflects the contributions of the crystal surfaces and defects to the DSC degrees of crystallinity.

Division of the specific internal surface a_s by the density and by the (true) degree of crystallinity α_X of the respective sample gives the amount of the crystalline surface relative to a crystalline mass unit. This value is displayed in figure 6 as a function of the overall

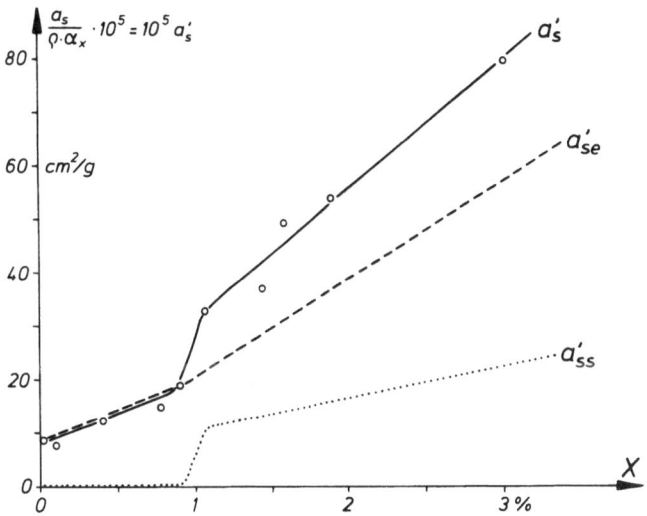

Fig. 6. Contributions of lamellar side (a_{ss}') and end faces (a_{se}') to the specific internal surface a_s'

crosslink density X. At an X-value of about 0.9 % a remarkable step can be seen. This sudden increase of the crystal surface occurs at the same crosslink density at which in the EM pictures a change from laterally largely extended lamellae to cube-like crystal bricks is observed. Therefore, the step in figure 6 could be attributed to a sudden increase of the contribution of the side faces of the lamellae. Taking this into consideration, a division of the specific internal surface into contributions of the end and side faces is made in figure 6.

For the calculation of Δh_s, one needs the values of σ_s and σ_e, the latter for the different kinds of end faces, that is, for lamellar and micellar ones. In the literature, values are given for σ_e by Wunderlich [20], Mandelkern [21], Zachmann [22] and Kilian et al. [7]. In the following, we use the values given by Kilian which refer especially to crosslinked PE. Therefore, we put $\sigma_{ef} = 115 \times 10^{-7}\,Jcm^{-2}$ and $\sigma_{em} = 78 \times 10^{-7}\,Jcm^{-2}$ for a folded and a micellar surface respectively. For σ_s, the value $\sigma_s = 10 \times 10^{-7}\,Jcm^{-2}$, given by Wunderlich [20], is used.

As already mentioned, the EM pictures revealed that samples with crosslink densities up to $X \approx 0.9\,\%$ exhibited crystalline lamellae of large lateral dimensions. As will be shown below, under these conditions chain folded crystallization occurred. For higher crosslink densities, the crystalline regions are cube-like and micellar crystallization takes place. However, the assumptions concerning the crystallization schemes in the two crosslink density ranges which define the σ_e-

---

[2]) Throughout this paper, large letters denote absolute values small letters specific ones, that is, values related to a mass or to a volume unit respectively. Dashed symbols (e', a', . . .) refer to the crystal content, undashed ones to the whole mass.

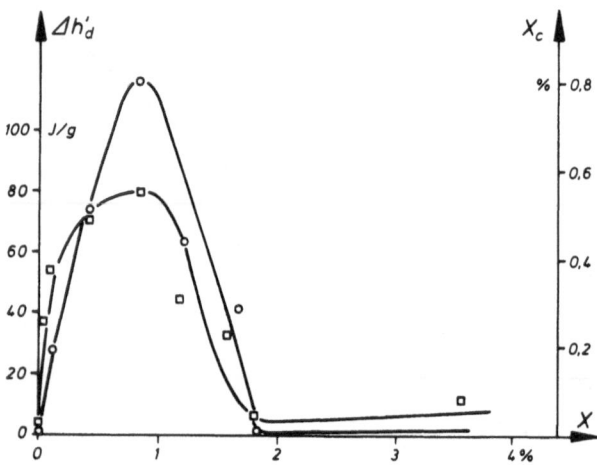

Fig. 7. Specific defect energy $\Delta h'_d$ (-□-□-, left) of crosslinks and crosslink density (-O-O-, right) in the crystalline regions X_c as a function of the overall crosslink density X

values in the latter calculations do not play an important role in the separation of Δh_s from Δh since $\Delta h'_d \gg \Delta h_s$ in the investigated X-range.

Using all these considerations, the specific surface energy can first be calculated and eliminated from the values of Δh. Subsequently, the specific defect energy $\Delta h'_d$ can be calculated, the course of which, dependent on the crosslink density X, is shown in figure 7.

On the basis of pure statistical considerations, Balta-Calleja et al. [23] derived some equations which relate the degree of crystallinity (α_X), the overall crosslink density (X), the lamellar thickness (d_c) and the crystalline crosslink density (X_c). Using these equations, we could estimate the connection between X_c and X. The results are also displayed in figure 7. Obviously, $\Delta h'_d(X)$ and $X_c(X)$ behave similarly. For small X-values, both $\Delta h'_d$ and X_c increase approximately proportional to X. For $0.9\% < X < 1.8\%$, both values decrease until becoming about zero and then remain at zero for higher crosslink densities. In the same X-range, that is for $0.9\% < X < 1.8\%$, a change from laterally largely dimensioned lamellar crystals to a lot of smaller, cube-like crystals (EM pictures) and a sudden increase of the specific crystal surface (SAXS) is observed.

Using the values of $\Delta h'_d$ and X_c, the defect energy of a single crystalline network point (ε_d) could be estimated. We obtained constant ε_d-values of $\varepsilon_d \approx 3.3$ eV/crosslink in the investigated X-range. This constancy had to be expected for our small X_c-values, for which the crosslinks in the crystals are so distant that they

cannot interact energetically. But nevertheless, this result is a hint that the different theories and measurements which were used for the evaluation are mutually consistent, and is evidence for the reliability of the ε_d-value itself. It is rather high, comparable to the valence energy of a C-C-bond and thus indicates a strong deterioration of the crystal lattice. Nevertheless, it seems to be reasonable, if one considers that one crosslink can influence structural changes in a large crystal volume.

The observation that X_c decreases and a_s simultaneously increases, instantly at a certain X value, can be explained by a change in the crystallization scheme. For a certain crystallization scheme "C_i" the amount of crosslinks incorporated into the crystals depends on the overall crosslink density in a specific manner (fig. 8). However, the insertion of crosslinks into the lattice and the crystallization itself are possible only up to a critical crosslink density X_i^o. Either crystals with $X_c > X_{ci}^o$ are thermodynamically unstable or they are not kinetically realizable. Therefore, any material with $X > X_i^o$ cannot crystallize according to scheme C_i. It may be, however, that there are different kinds of crystallization schemes (C_1, C_2, ...) which differ for a given material and for given crystallization conditions by the probability of their occurrence and by their specific functions $X_{ci}(X)$. In particular, polymers can crystallize according to a chain-folded as well as to a micellar procedure. We associate these procedures to the curves for C_1 and C_2 in figure 8. Usually, process C_1, that is chain folding, is preferred. Compared to a micellar crystal growth, it results in much larger X_c values since in contrast to a pure alignment of chains, crosslinks are led back to the growth face with folding. Therefore, the relation $X_1^o < X_2^o$ must hold. Consequently, if $X <$

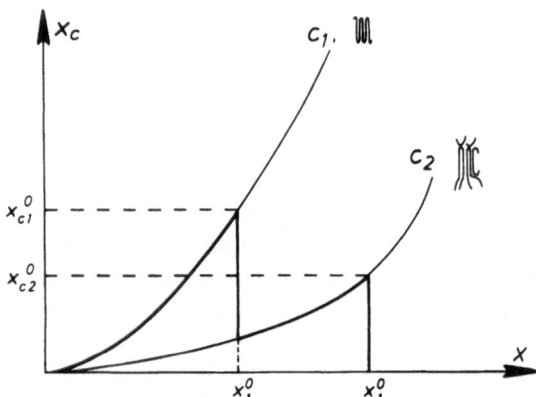

Fig. 8. Dependence of the amount of crystalline crosslinks X_c on the overall defect density X for different crystallization schemes (schematically, cf. text)

X_1^o chain folding occurs and if $X_1^o < X < X_2^o$ micellar crystallization takes place. The overall function $X_c(X)$ is then given by the thick curve in figure 8.

The model satisfactorily describes the behavior found here. On the one hand we observe for X values smaller than 0.9% crystalline regions with large lateral dimensions (lamellae). This is a hint for a chain folded crystallization. One the other hand, in the same X region to some extent $X \approx X_c$. Obviously, for $X \approx 0.9\%$ the limit for chain folded crystallization is reached. For higher X values, the penetration of crosslinks into the crystals strongly decreases and the shape of the crystals changes to cube-like. We conclude that at this crosslink density crystallization changes to a micellar one.

For $X \approx 0.9\%$, this being the limit fo the occurrence of chain folded crystallization, the specific defect energy amounts approximately to $82\,\mathrm{Jg}^{-1}$. This value is less than one third of the melting enthalpy of the ideal crystal ($\Delta h_o = 287\,\mathrm{Jg}^{-1}$). Chain folded crystallization with insertion of crosslinks for higher X values would also give a further gain of enthalpy. Therefore, the assumed change in the crystallization procedure has not only equilibrium thermodynamically reasons but must have kinetical reasons as well.

The comparison of the network chain lengths in an all-trans conformation (l_c) with the crystal thicknesses (d_c, fig. 9) gives an additional hint for the occurrence of micellar crystallization. For a simple alignment of chains, the relations $l_c \approx d_c$ must hold. Indeed, in figure 9 it can be seen that the curves for l_c and d_c converge for higher X values. Therefore, for higher crosslink densities the crystal thickness is controlled by the network chain lenght, thus indicating a micellar crystallization procedure.

4.2 Entropic changes

As is well known, the melting temperature T_m is given by

$$T_m = \Delta h / \Delta s. \tag{6}$$

Δ designates differences between the melt and the crystalline phase, that is, $\Delta h = h_m - h_c$ and $\Delta s = s_m - s_c$. We define

$$\delta h_m(X, a) = h_m(0, 0) - h_m(X, a). \tag{7}$$

Similar equations are defined with respect to the crystalline phase and to the entropy (cf. fig. 11). Therefore, δ describes changes due to the existence of crosslinks in the respective phase. Additionally, δh_c considers the influence of crystalline surface enthalpy. Consequently,

$$T_m = T_m^0 \left[\frac{1 - \dfrac{\delta h_m - \delta h_c}{\Delta h_0}}{1 - T_m^0 \dfrac{\delta s_m - \delta s_c}{\Delta h_0}} \right] \tag{8}$$

with

$$T_m^0 = \Delta h_0 / \Delta s_0 \tag{9}$$

T_m^o being the melting temperature of defect-free crystals with negligible surface enthalpy contributions. We put $\delta h_m = 0$, assuming that the insertion of crosslinks into the melt does not change the energy content of this phase. Using the α_{DSC} values of figure 2 and the equation $\Delta h = \Delta h_o\, \alpha_{\mathrm{DSC}}$ (cf. paragraph 2.5.) and considering the arguments in connection with equation (5), it can be concluded at δh_c. Then by means of equations (6)–(9), from the melting temperatures one can estimate $\Delta s = \Delta s_o - \delta s_m + \delta s_c$ (cf. fig. 11 and eq. (7)). In figure 10 the values of $\Delta s = \Delta s' = \Delta h/(T_m \alpha_X) = \Delta h'/T_m$ are given as a function of the overall crosslink density. For the discussion below, the values of X_c are given too.

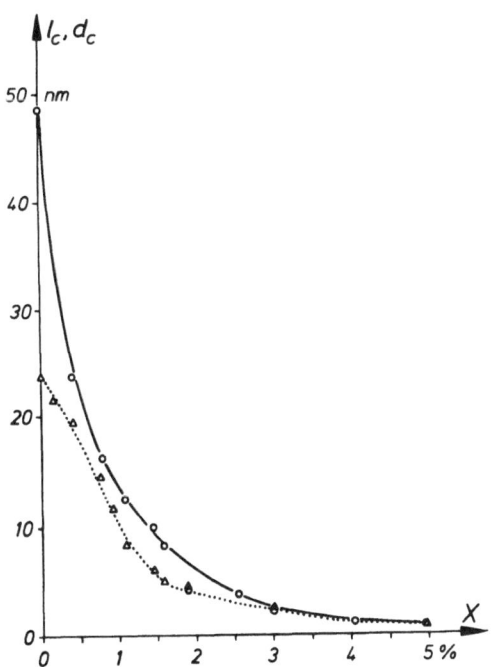

Fig. 9. Network chain lengths l_c (–O–O–) in all-trans conformation and lamellar thicknesses d_c ($\cdots\triangle\cdots\triangle\cdots$) versus crosslink densities X

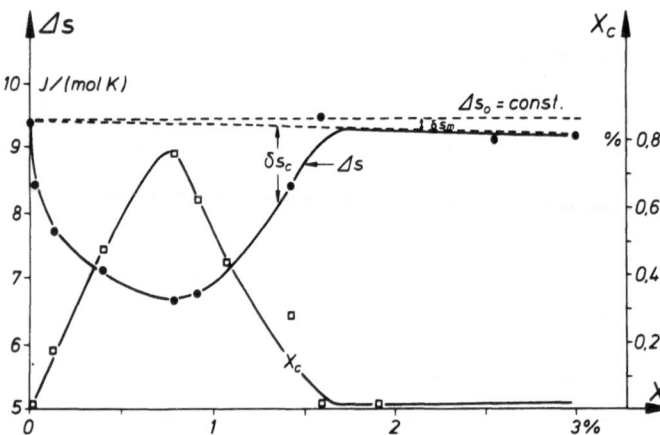

Fig. 10. Melt entropy Δs as a function of the crosslink density X and the contributions of the crystalline phase and the melt

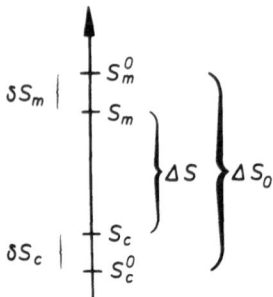

Fig. 11. Entropy levels in linear and crosslinked PE (schematically, cf. text)

The melting entropy $\Delta s(o) \approx 9.4$ J/(mol K) of the non-irradiated sample agrees well with the value of $\Delta s_o = 9.78$ J/(mol K) reported by Mandelkern [21]. The difference is due to lattice distortions already in the uncrosslinked material which cannot be prevented. With increasing X, Δs decreases, reaching a minimum value of $\Delta s = 6.7$ J/(mol K) at the same X value, at which X_c reaches its maximum. With then decreasing insertion of crosslinks, the entropy increases and approximately reassumes its starting value for those X, for which $X_c \approx 0$.

Δs consists of three contributions, that is Δs_o, δs_m and δs_c. Δs_o, that is the melting entropy of the ideal crystal, should not depend on the crosslink density. In figure 10 this is drawn as a straight line ($\Delta s_o = const = 9.4$ J/(mol K)). Obviously, the difference $\Delta s^* = \Delta s_o - \Delta s = \delta s_m - \delta s_c$ is directly correlated to the crystalline crosslink density.

The values of $\Delta s^* = \delta s_m - \delta s_c$ cannot be explained quantitatively by equations given for the entropy

change with crystallization in the literature. For example, some authors have tried to describe the crystallization behavior of crosslinked polymers by a theory of Flory [1], which deals with the crystallization behavior of statistical copolymers [21]. The theory assumes exclusion of non-crystallizable co-units ("B") from the crystals of component "A", this "demixing" of the co-units leading to an entropy change of

$$\Delta s_E^* = - R \ln X_A. \tag{10}$$

X_A being the relative molar content of A-units. This contribution to the entropy change with crystallization always leads to a melting point depression. For our samples, $\Delta s_E^* \approx 0.3$ J/(mol K), this is only some percent of the observed entropy change. If the defects are included partially, according to Helfand and Lauritzen [4]

$$\Delta s_I^* = - R \ln [X_A + X_B \exp (- \varepsilon_d/kT)] \tag{11}$$

(cf. eq. (4)). By using this expression, the observed entropy changes cannot be explained either. This is also true if ε_d changes by an order of magnitude.

By insertion of crosslinks into the crystals, the entropy of the crystalline phase should increase, that is $\delta s_c < 0$. In the melt, two competitive effects occur: the insertion of defects into the melt should increase the melt entropy, but the reduction of the conformal realization possibilities of the network chains due to the insertion of crosslinks should decrease the melt entropy. The corresponding entropy levels are given in figure 11. A negative value is chosen arbitrarily for δs_m.

As a first approximation, the absolute value of δs_m should depend linearly on the crosslink density for small X values. Similarly, δs_c should reflect the amount of the crystalline crosslink density. Especially, $\delta s_c \approx 0$ if $X_c \approx 0$. Therefore, as is done in figure 10, by a linear extrapolation of the course of Δs for large X values to the value of $\Delta s(0)$, a separation of the contributions of δs_m and δs_c can be performed.

Concerning δs_m, a value of $\delta s_m \approx 0.5$ meV/K per crosslink can be deduced. Therefore, the entropy of the crosslinked melt is lower than that of the linear one and, hence, the loss of entropy by diminution of the conformational and configurational realization possibilities of the network chains must exceed the gain in entropy by mixing with the crosslinks.

For the crystals, $\delta s_c = - 7$ meV/K per crosslink was calculated. This rather high value cannot be explained by the mixing entropy of a single kind of distortion point. Therefore, it must be assumed that the cross-links are incorporated into the crystal in many different conformational and configurational ways.

References

1. Flory PJ (1955) Trans Faraday Soc 51:848
2. Kilian HG (1965) Kolloid Z u Z Polym 202:97
3. Sanchez JC, Eby RK (1973) Res Nat Bur Stand 77A:353
4. Helfand E, Lauritzen JI (1973) Macromol 6:631
5. Balta-Calleja FJ, Gonzales-Ortega JC, Martinez de Salazar J (1978) Polymer 19:1094
6. Gielenz G, Jungnickel BJ (1982) Coll & Polym Sci 260:742
7. Heise B, Kilian HG, Schmidt H (1981) Coll & Polym Sci 259:611
8. Kilian HG, Unseld K, Jäger E, Müller J, Jungnickel BJ, Coll & Polym Sci, in press
9. Zoepfle FJ. Markovic V, Silverman J (1984) J Polym Sci, Polym Chem Ed 22:2017
10. Ungar G, Grupp DT, Keller A (1980) Polymer 21:1273; ibid 1278, ibid 1284
11. Kortleve G, Tuijnman CAF, Vonk CG (1972) J Polym Sci (A-2) 10:123
12. Martuscelli E (1975) J Macromol Sci (B) 11:1
13. Charlesby A (1960) Atomic Radiation and Polymers, Pergamon Press, Oxford
14. Treloar LRG (1958) The Physics of Rubber Elasticity, Clarendon Press, Oxford
15. Flory PJ, Rehner J (1943) J Chem Phys 11:521
16. Tung LH (1957) J Polym Sci 24:333
17. Strobl GR (1970) Acta Cryst A 26:367
18. Strobl GR, Schneider M (1980) J Poylm Sci 18:1343
19. Kanig G (1975) Progr Coll & Polym Sci 57:176
20. Wunderlich B (1973) Macromolecular Physics II, Academic Press, New York–London
21. Mandelkern L (1964) Crystallization of Polymers, Mc Graw Hill, New York
22. Zachmann HG (1967) Kolloid Z u Z Polym 216–217:180
23. Martinez Salazar J, Balta-Calleja FJ (1980) Polymer Bulletin 2:163

Received April 23, 1985;
accepted July 3, 1985

Authors' address:

B.-J. Jungnickel
Deutsches Kunststoff-Institut
Schloßgartenstr. 6 R
D-6100 Darmstadt, F.R.G.

Progress in Colloid & Polymer Science   Progr Colloid & Polymer Sci 71:154–163 (1985)

Investigation of the deformation and relaxation of various polyethylenes by X-ray diffraction*)

E.-M. Reck, H. Schenk, and W. Wilke

Abteilung Experimentelle Physik, Universität Ulm, Ulm, F.R.G.

Abstract: Three different PE samples (linear, branched, very high molecular weight) are investigated for different draw ratios (uniaxial deformation) by wide and small angle X-ray diffraction.

Crystallite sizes and lattice distortions show qualitatively the same behaviour, but quantitative differences occur. There is a remarkable difference in the superstructure between "normal" and very high molecular weight samples. The relaxation at different temperatures is investigated for linear PE. The degree of rebuilding of the original structure depends on the draw ratio and relaxation temperature (by fixed relaxation time).

The results are discussed within the framework of dislocation concepts.

Key words: Deformation of PE, crystallite sizes, lattice distortions, dislocations and paracystallinity.

Introduction

During deformation of semicrystalline polymers the superstructure as well as the crystallite size and perfection changes. For linear and branched PE the superstructure changes from lamellar cluster structures to a fibrillar structure, as is well known. But the behaviour is different for very high molecular PE (GUR); this follows from SAXS experiments. From WAXS experiments the change of the crystallite sizes and the lattice distortions can be calculated. These values are compared for the above mentioned three types of PE. For linear PE the reverse process, the relaxation of deformed samples at different temperatures, was investigated. The question was whether the original structure is rebuilt by recovery and recrystallization processes.

Theory

The integral width of the wide angle X-ray reflections is determined (after correction of instrumental

and sample thickness broadening) by the average crystallite shape and size and the lattice distortions. The analysis of the experimental widths was possible by using lattice distortions of the paracrystalline type. A possible explanation for this fact will be discussed later.

The integral width of the reflection with Miller indices $h = (h_1, h_2, h_3)$ is given by

$$\delta\beta_h = \frac{1}{D_h} + \frac{\pi^2 g_s^2}{d_{\hat{h}}} \cdot p^2 = \delta\beta_h(S^2) + \delta\beta_h(Z) \quad (1)$$

with

$$g_s^2 = \frac{\sum\limits_{i,k=1}^{3} \hat{h}_i^2 \cdot (\hat{h}_k/\bar{a}_k^2) \cdot g_{ki}^2}{\sum\limits_{i=1}^{3} \frac{\hat{h}_i^2}{\bar{a}_i^2}} \quad (2)$$

$\bar{a}_i$ are the lattice parameters, $d_{\hat{h}}$ is the net plane distance ($\hat{h}$ the Miller indices without common measure), p the order of the reflection. The g_{ik} are the principal values of the paracrystal fluctuation tensors (fig. 1). The first term on the right side of equation (1) describes the contribution of the crystallite size and is determined by the average diameter of the crystallites, perpendicular to the netplanes with Miller indices h.

---

*) Dedicated to Prof. Dr. H.-G. Kilian on the occasion of his 60th birthday.

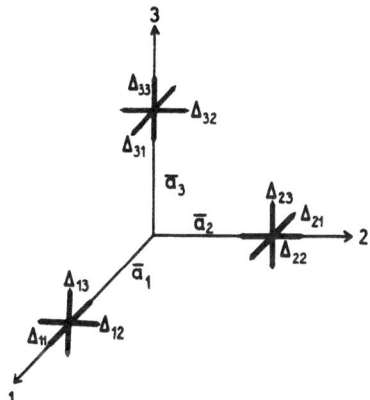

Fig. 1. Fluctuation tensors of paracrystalline distortions

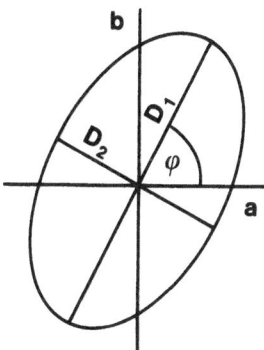

Fig. 2. Shape ellipsoid (a–b plane), D_3 perpendicular to the plane of paper (c-axis direction)

The crystallite shape is approximated by an ellipsoid with principal axes $\bar{D}_1, \bar{D}_2, \bar{D}_3$, oriented in such a way that the axis $\bar{D}_3$ is parallel to the crystallographic a_3- axis (chain direction) (fig. 2). For details of the sketched method see [1, 2]. From a set of measured integral widths $\delta\beta_h$ the values $\bar{D}_i$ and g_{ik} are calculated by a least-square program. The volume V and surface S of the crystallites are then given by

$$V = \frac{\pi}{6} \cdot \bar{D}_1 \cdot \bar{D}_2 \cdot \bar{D}_3 \tag{3}$$

$$S \approx \frac{\pi \bar{D}_3}{2} \left[\frac{3}{2} (\bar{D}_1 + \bar{D}_2) - \sqrt{\bar{D}_1 \cdot \bar{D}_2} \right]. \tag{4}$$

During deformation of the sample the crystallites break up and new internal surfaces are formed. The volume of the new crystallites decreases to V_λ (draw ratio λ) and their number increases to V_1/V_λ. The relative increase of the internal surface follows from

$$S_G = S_\lambda \cdot \frac{V_1}{V_\lambda}. \tag{5}$$

The opposite process takes place during relaxation of the samples in conjunction with recovery and recrystallization processes.

The distortion of the crystallites, described by the g_{ik}-values, can be characterized by the averaged paracrystalline fluctuations

$$\bar{g}_{ik} = \frac{1}{9} \sum_{i,k=1}^{3} g_{ik} \tag{6}$$

$$\bar{g} = \frac{1}{3} \sum_{i=1}^{3} g_{ii} \tag{7}$$

$\bar{g}$ contains only the distance fluctuations (g_{ii}) and no angle fluctuations (g_{ik}, $i \neq k$) (see fig. 1).

The paracrystalline distortions may be produced by statistically distributed dislocations [3], as will be shown in the following. The line profile is within a constant factor given by [4]

$$i(b) = \int y(t) \exp(-2\pi i b t) \, dt \tag{8}$$

with

$$\begin{aligned} y(t) &= F^2 \left\langle \exp(2\pi i b L_t) \right\rangle \\ &\approx F^2 \exp(-2\pi^2 b^2 \langle L_t^2 \rangle) \\ &= F^2 \exp(-2\pi^2 b^2 \Delta L^2(t)) \end{aligned} \tag{9}$$

L_t = (length of a column of m unit cells (parallel to $b_{\hat{h}}$) of the distorted crystal) $- m \, d_{\hat{h}}$.
The direction of t is perpendicular to the netplanes with Miller indices $\hat{h}$.
F^2 = structure factor.

For the paracrystal we have [5]

$$\langle L_t^2 \rangle = \Delta L^2(t) = \frac{|t|}{\langle d_{\hat{h}} \rangle} \cdot \Delta_{\hat{h}}^2 \tag{10}$$

$$\Delta_{\hat{h}}^2 = \langle d_{\hat{h}}^2 \rangle - \langle d_{\hat{h}} \rangle^2 \tag{11}$$

$$g_s^2 = \frac{\Delta_{\hat{h}}^2}{\langle d_{\hat{h}} \rangle^2} \tag{12}$$

and

$$b = \frac{p}{\langle d_{\hat{h}} \rangle} \quad (13)$$

for a reflection of order p.

The integral width is then given by

$$\delta\beta_h = \frac{\int i_h(b)\, db}{i_h(o)} = \frac{y(o)}{\int y(t)\, dt}$$

$$= \left[2 \int_0^\infty \exp\left[-2\pi^2 p^2 \left(g_s^2/\langle d_{\hat{h}}\rangle\right) t\right] dt\right]^{-1}$$

$$= \frac{\pi^2 g_s^2}{\langle d_{\hat{h}} \rangle} \cdot p^2. \quad (14)$$

In the case of an elastic deformed lattice follows [4]

$$\Delta L^2(t) = t^2 \langle \varepsilon_t^2 \rangle. \quad (15)$$

To calculate $\langle \varepsilon_t^2 \rangle$ for noninteracting dislocations, one has to average the inhomogeneous strain field

$$\varepsilon(\hat{h}) = \frac{C(\hat{h})}{r} \quad (16)$$

about the distance t at the position r from the dislocation line and about r between the core-radius nb_B (b_B = magnitude of the Burgers vector) and a cut-off radius r^x [6]:

$$\langle \varepsilon_t(\hat{h})^2 \rangle = \frac{1}{r^x - nb_B} \int_{nb_B}^{r^x} \left[\frac{1}{t} \int_r^{r+t} \frac{C^2(\hat{h})}{r^2}\, dr\right] dr \quad (17)$$

$$= \frac{C^2(\hat{h})}{t(r^x - nb_B)} \ln \frac{r^x(nb_B + t)}{nb_B(r^x + t)} \quad (18)$$

$C(\hat{h})$ depends on the reflection $\hat{h}$, the dislocation density, the Burgers vector and the elastic constants.

The t-dependence of $\langle \varepsilon_t^2(\hat{h}) \rangle$ is determined mainly by the factor t^{-1} in (18). It therefore follows in good approximation, using (14), (15) and (18),

$$\delta\beta_h = \{2 \int_0^\infty \exp\left[-2\pi^2 \left(p^2/\langle d_{\hat{h}}\rangle^2\right)\right.$$

$$\left. \cdot \left(C^2(\hat{h})/(r^x - nb_B)\right) \cdot t\right] dt\}^{-1} \quad (19)$$

$$= \frac{\pi^2 C^2(\hat{h})}{\langle d_{\hat{h}}\rangle^2 (r^x - nb_B)} \cdot p^2$$

$$\cong \frac{\pi^2}{\langle d_{\hat{h}}\rangle} g_s^2 \cdot p^2.$$

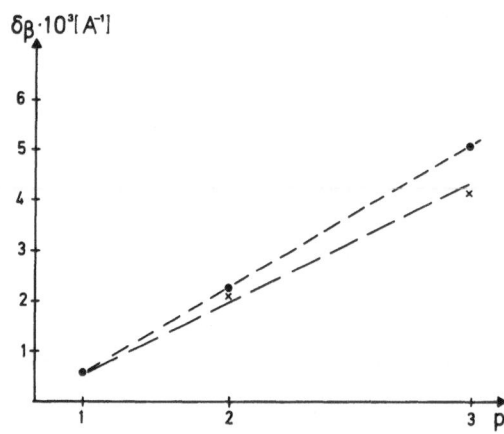

Fig. 3. Integral width $\delta\beta$ for different orders of reflection p, calculated with (19) ($\bullet\bullet\bullet$) and including the ln term in (18) ($\times\times\times$). Parameters: $g_s = 10^{-2}$, $d = 0.5$ nm, $r^x = 15$ nm, $nb_B = 3 \cdot 0.25$ nm $= 0.75$ nm. The curves are adjusted for $p = 1$

The characteristic p^2-dependence is the same as in (14), Q.E.D. To see the influence of the logarithmic factor in (18), for a representative set of parameters the integral width was calculated by inserting (18) in (14) and numerical integration. The result is shown in figure 3. The deviation from p^2-dependence is small and within the experimental errors.

Experimental

Two types of polyethylene, linear L 6041 D and very high molecular GUR, were examined over their whole range of drawing and then compared with a branched polyethylene (L 1840 D) [7, 8]. In addition relaxed samples of linear PE were examined (see table 1).

The original material was melted twice at 200 °C under pressure. The foil of GUR, examined at small angles was only melted once (fig. 4). All foils of L 6041 D were annealed for 3½ hours at 130 °C. Samples were then punched out of the foils of polyethylene and drawn to several draw ratios (table 1). After drawing they were fixed in length. In the case of L 6041 D relaxed samples (one end free) were also produced. There the draw ratios could not be determined

Table 1. Characteristics of the compared PE samples

	GUR	L 6041 D	L 1840 D
Molecular weight	$(3.5-4) \cdot 10^6$	$2 \cdot 10^5$	10^5
CH$_3$/100 monomers	7	1	28
Crystallinity/%	60–62	84–89	40–42
Original material	fine powder	plate (6 mm)	granulate
Drawing rate $\frac{\Delta \cdot i}{l_o}\left|\frac{\%}{\text{min}}\right.$	1.8	6.6	0.7
Drawing temperature/°C	80	80	32/21

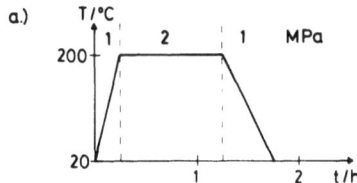

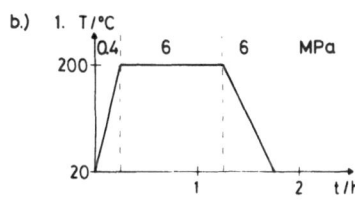
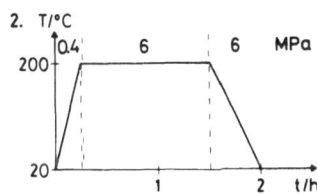

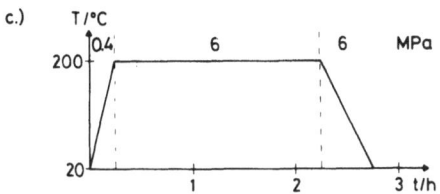

Fig. 4. Temperature curves during pressing. a) L 6041 D first and second melting; b) GUR 1. first and 2. second melting; c) GUR (small angle experiments)

Table 2. Examined samples (d: thickness in µm, λ: draw ratio) at wide angles

GUR	λ	1	1.4	1.8	3.1	3.9	5.5	6.8
	d	120	120	91	145	139	132	188

L 6041 D	unrelaxed	λ	1		2.1	4.5	7.8
		d	130		105	110	65
	relaxed at 130 °C	λ	1.9–2.2		4.4–4.6		7.8
		d	170		120		75
	relaxed at 135 °C	λ	2.0–2.2		4.3–4.6		7.7–7.9
		d	190		170		165
	relaxed at 139 °C	λ	1.9–2.3		4.3–4.7		7.7–8.1
		d	230		210		150

exactly, therefore a tolerance of drawing ranges is given in table 2. For the relaxation process three temperatures were used (130 °C, 135 °C, 139 °C) and the relaxation time always took three hours from the moment of heating (starting at 80 °C).

The dynamical SAXS experiments with the GUR sample (thickness of the sample: 1 mm) during drawing were made with synchrotron radiation (DESY, Hamburg) and received with a vidicon system.

For the wide angle X-ray experiments, Guinier cameras were used (Huber, Germany; XDC-700, Incentive Research & Development AB, Sweden). The films (Kodak Sb-X-Ray film, developer Kodak LX 24, fixing bath Kodak AL4) were exposed only in the linear range. The intensity curves (microdensitometer Mark III CS,

Joyce, Loebl & Co. Ltd.) were separated into scattering from crystalline regions, amorphous regions and background scattering. Considerably overlapping reflexes were separated by a computer program.

The integral widths were determined as mean values from five independent exposures and then used to search numerically the minimum value of the following sum:

$$\sum_{i=1}^{N} \{[\delta\beta_h^i - \delta\beta_h^i(Z) - \delta\beta_h^i(S^2)] \, G_i\}^2 \cdot \frac{100}{(\delta\beta_h^i)^2}.$$

As a result of this minimalization the crystallite-parameters discussed in the theoretical introduction were obtained.

Experimental results

1. Uniaxial deformation

a) Change of crystallite size

From the three types of PE examined, the mean crystallite sizes $\bar{D}_i$ and their change during drawing are shown in the following figures (fig. 5). The volume V of the crystallites, given in figure 6 was calculated with (3).

b) Change of distortions

As an integral value for the paracrystalline distortions the values $\bar{g}_{ik}$ and $\bar{g}$ ((6) and (7)) were used (fig. 7).

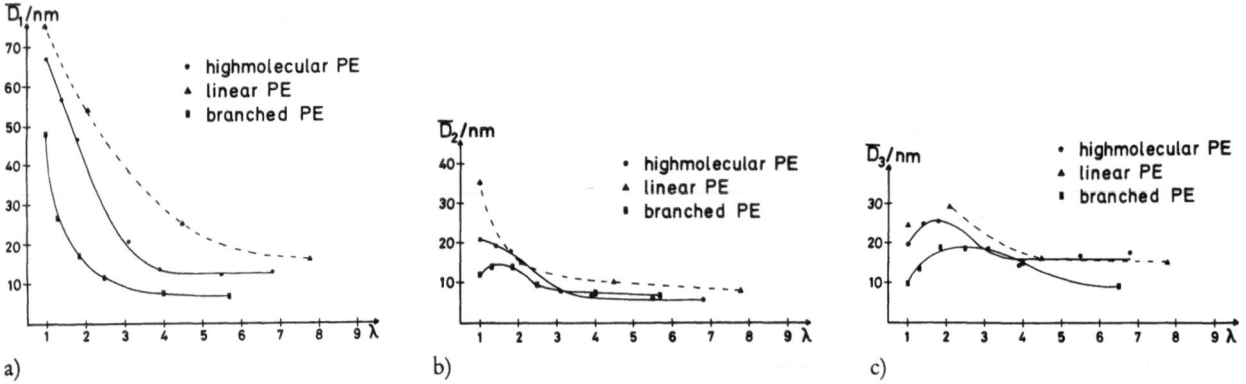

Fig. 5. Mean crystallite sizes of different PE. a), b) lateral dimensions; c) extension in chain direction

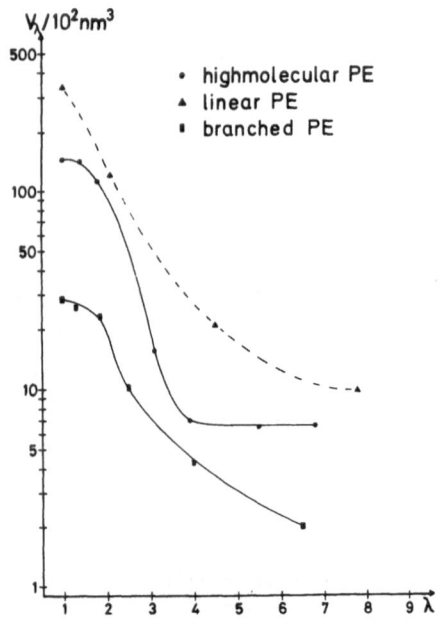

Fig. 6. Volume of the crystallites

2. Dynamical measurements (GUR) with synchrotron radiation

The pictures in figure 8 show the registered SAXS patterns of GUR at different draw ratios (drawing temperature 100 °C). The direction of drawing is vertical (the sample is faded in on the right side).

3. Relaxation behaviour (linear PE)

In figure 9 the mean crystallite sizes $\bar{D}_i$ of the samples relaxed at different temperatures (RT) are given. The volume of these crystallites is shown in figure 10. A summary of the results is given in table 3. Figure 11 shows the effect of relaxation on the superstructure and figure 12 that on the orientation for certain samples. The lattice distortion $\bar{g}_{ik}^2$ and the angle φ of the shape ellipsoid are given in table 3.

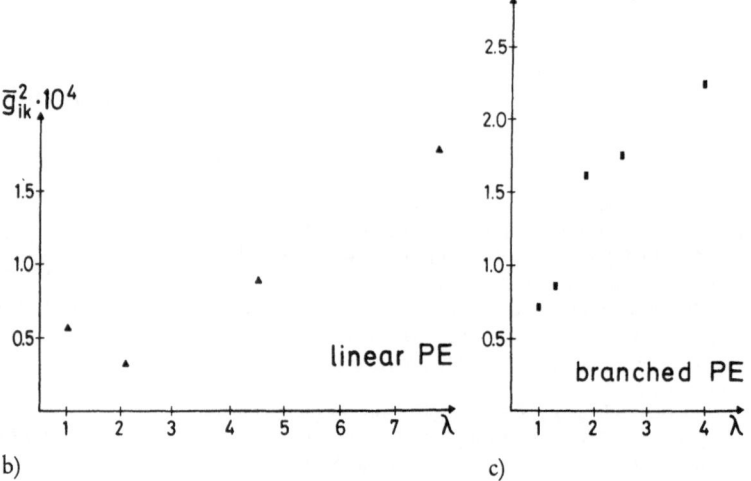

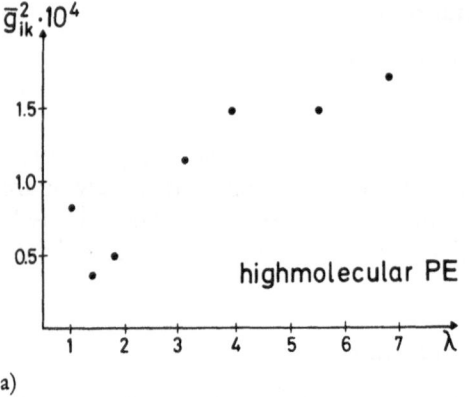

Fig. 7. Paracrystalline distortions. a) very high molecular PE (GUR); b) linear PE; c) branched PE

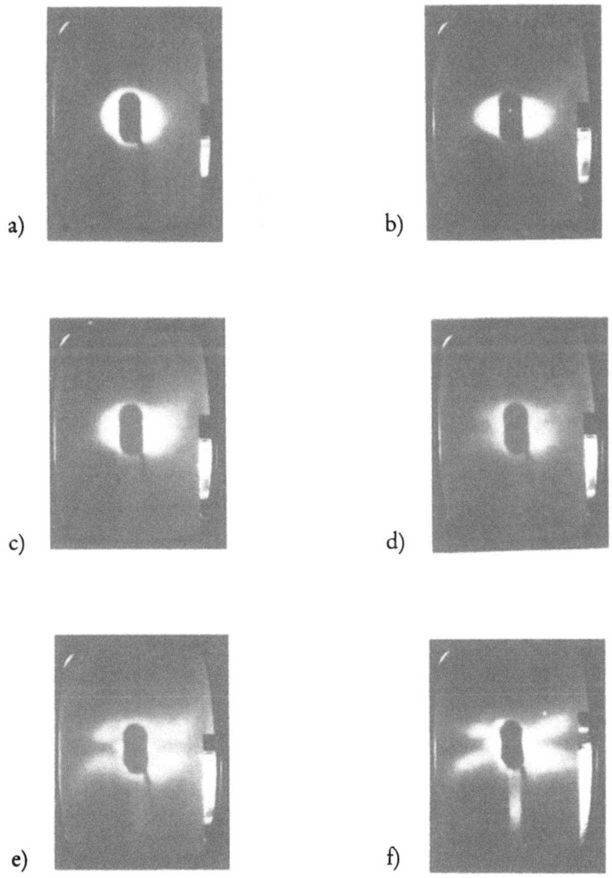

Fig. 8. SAXS-patterns of GUR, registered during drawing (at $T = 100\,°C$) [8] with synchrotron radiation. a) $\lambda = 1$; b) $\lambda \approx 1.5$–2; c) $\lambda \approx 2.5$–3; d) $\lambda \approx 3$; e) $\lambda \approx 4$; f) $\lambda \approx 7$

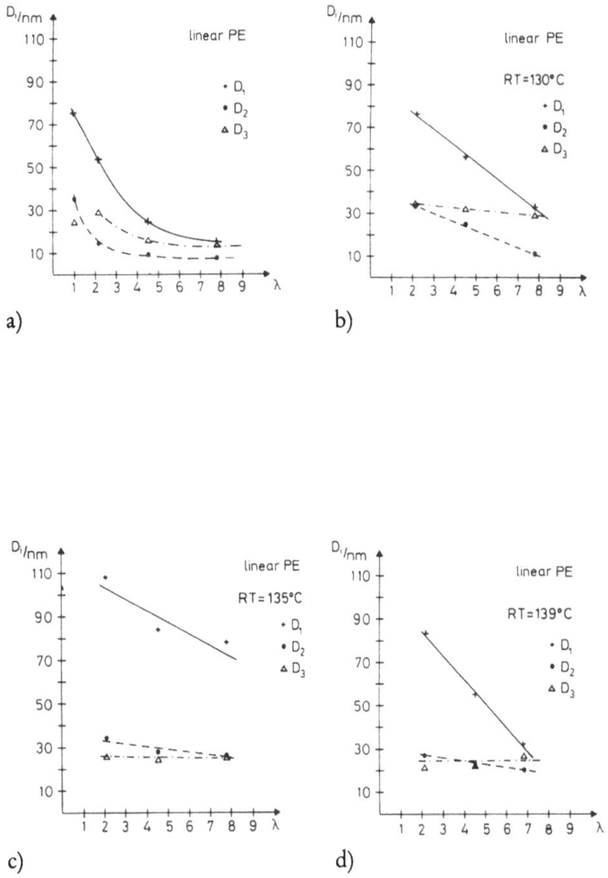

Fig. 9. Mean crystallite sizes of linear PE. a) unrelaxed; b) relaxed at $RT = 130\,°C$; c) relaxed at $RT = 135\,°C$; d) relaxed at $RT = 139\,°C$

Table 3. a) Lattice distortion $\bar{g}_{ik}^2 \cdot 10^4$

$RT^{\backslash \lambda}$	2.1	4.5	7.8
130 °C	0.69	0.45	0.79
135 °C	0.45	0.71	0.52
139 °C	0.61	0.81	0.24

b) Angle φ

$RT^{\backslash \lambda}$	2.1	4.5	7.8
130 °C	47	48	43
135 °C	44	44	45
139 °C	42	46	42

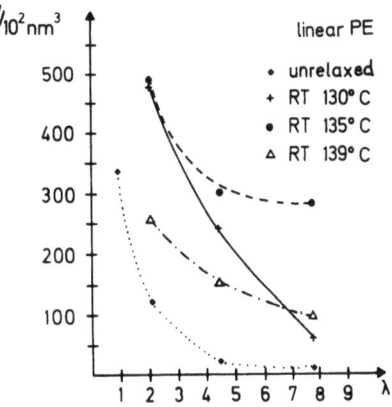

Fig. 10. Volume of the crystallites before and after relaxation

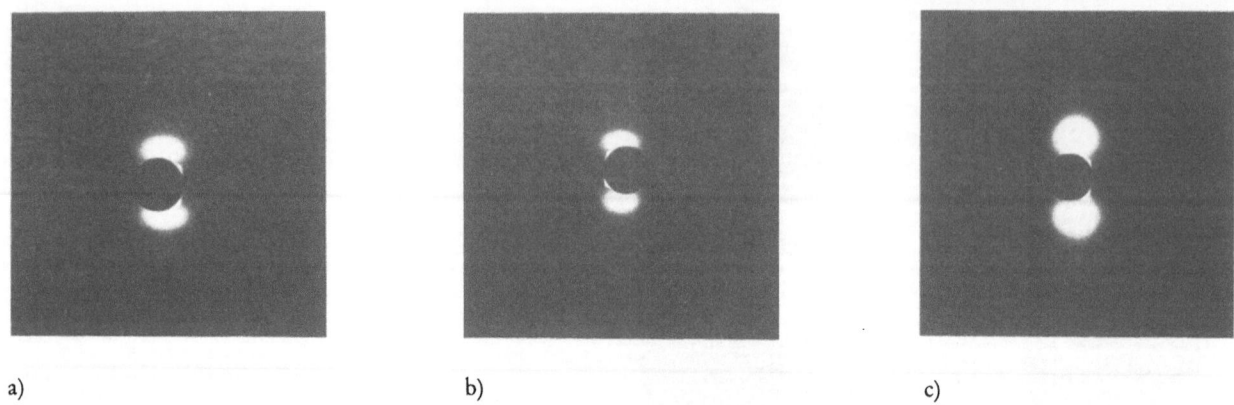

a)   b)   c)

Fig. 11. Small angle scattering from relaxed samples. a) $\lambda = 4.5\ RT = 130\,°C$; b) $\lambda = 7.8\ RT = 130\,°C$; c) $\lambda = 7.8\ RT = 135\,°C$

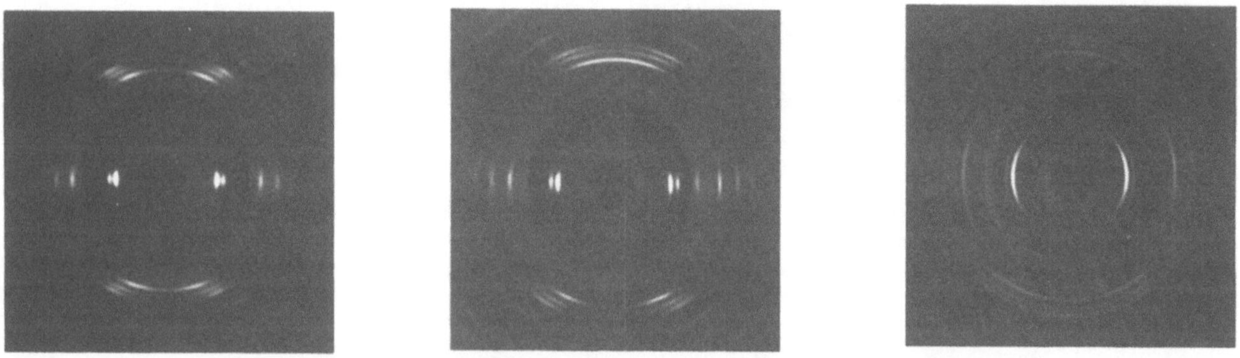

a)   b)   c)

d)   e)

Fig. 12. Wide angle scattering from relaxed samples. a) $\lambda = 7.8$; b) $\lambda = 7.8\ RT = 130\,°C$; c) $\lambda = 7.8\ RT = 135\,°C$; d) $\lambda = 4.5$; e) $\lambda = 4.5\ RT = 130\,°C$

Conclusions

In the theoretical part it was shown that it is possible to explain the paracrystalline line broadening by strain fields of dislocations. Therefore we will assume that $\bar{g}_{ik}$ and $\bar{g}$ is a measure of the dislocation density. For further experimental evidence for the existence of dislocations in polmyers see [17], for annealing effects [18]. Other defects may exist (kinks, Reneker defects...) with a point defect strain field (no remarkable contribution to the line width) which are not discussed here. The discussion about the importance of dislocation mechanisms in polymer crystal deformation has been continuing for a long time [9–16], and there are some open questions, but it seems a reasonable working hypothesis for a qualitative discussion. Some general results from the solid state physics of metals (and other low molecular weight crystals) are used for the interpretation of our measurements.

The following discussion of the results is divided into the two main parts of deformation and relaxation and within these sections the different drawing ranges are distinguished.

I. Uniaxial deformation

1. $1 \leq \lambda \leq 3\text{-}4$

In this range we found a lamellar superstructure to start with, which is destroyed during stretching, and then a fibrillar superstructure is formed. This process is accompanied by a decrease in crystallite size (figs. 5, 6) and strong changes in lattice distortions (fig. 7). The initial maximum for $\bar{D}_i$ can be explained as a selective orientation effect, because only the best oriented part of the crystallite ensemble contributes to the line width (as a result of the measuring method). Different effects contribute to the observed behaviour of the crystallites:

a) Deformation twinning with generation of twin boundaries [19].

b) In the course of the deformation process some crystallites come to be thermodynamically unstable and local melting and recrystallization occurs [20], resulting in a change of the crystallite size distribution.

c) Plastic deformation of the crystals with increasing dislocation density. Probably the multiplication by conventional Frank-Read sources does not work because of the smallness of the crystals. But there are suggestions for other mechanisms of dislocation generation by Frank-Read like sources [13] or thermal initiation coupled with an applied stress [14–16]. Another possibility is given if dislocations pile up at the grain boundaries, where the stress concentrations cause dislocation sources operating in neighbouring crystallites [21]. The increase in dislocation density leads to the formation of new grain boundaries and consequently to the observed decrease in crystallite size. The changes of the lattice distortions are also caused by the dislocation density. The initial decrease of $\bar{g}_{ik}$ is explained by the fact that small, strongly distorted crystallites become unstable first and disappear. This effect may also contribute to the development of the maxima of $\bar{D}_2$ and $\bar{D}_3$.

2. $3\text{-}4 \leq \lambda \leq \lambda_{max}$

There is a fibrillar superstructure (linear and branched PE) or a "V"-superstructure (GUR, see below). No remarkable change of crystallite sizes is observed, but a further increase of lattice distortions. This means nearly unchanged grain boundaries and increasing dislocation density in the interior of the crystallites. The deformation process is mainly determined by grain boundary slip. The crystallites are cross-linked by tie-molecules and therefore stress induction and subsequently dislocation generation inside the crystallites occur during deformation. The continuous increasing dislocation density explains the rise of $\bar{g}_{ik}$ and $\bar{g}$ with λ up to λ_{max}.

The cross-linking of the crystallites is enhanced in the case of very high molecular PE (GUR), such that a

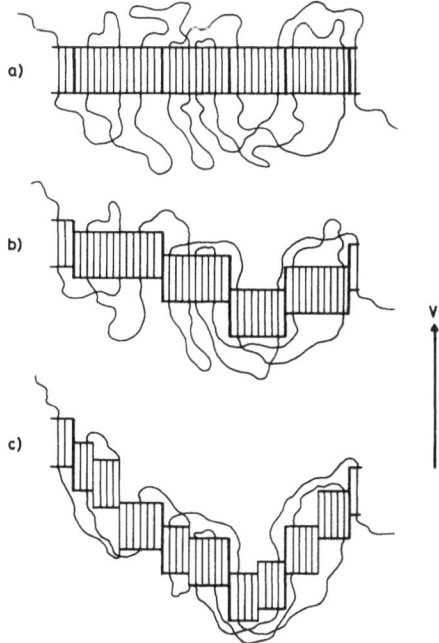

Fig. 13. Schematic picture of the "V" structure, built during drawing

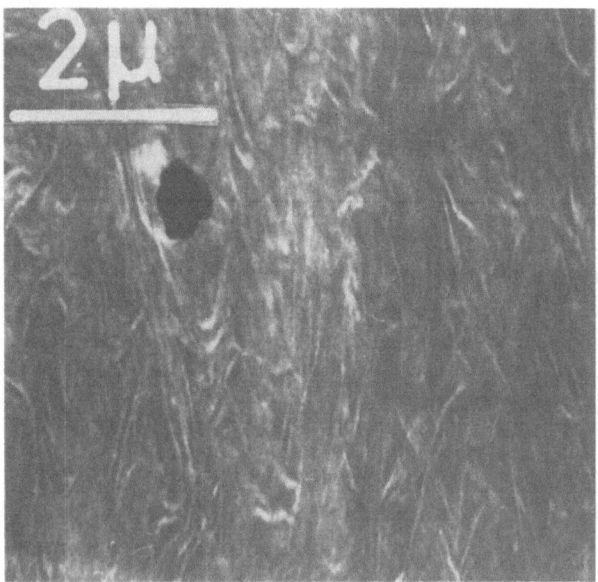

Fig. 14. Electromicrograph of uniaxially deformed GUR ($\lambda \approx 5$), showing the "V" structure [22]

complete slip to a fibrillar structure is suppressed and a "V" shaped superstructure arises. This is drawn schematically in figure 13, and figure 14 shows the EM picture of the "V" structure [22]. Additionally a chemical cross-linking during deformation is possible in this case [23].

3. $\lambda \sim 2.5$

In the middle range of the draw ratio λ a discontinuity in the increase of $\bar{g}_{ik}$ and $\bar{g}$ is indicated. (To ensure the effect, more experiments in the range $\lambda \sim 2$–3 are needed.) In this range the superstructure is probably completely destroyed and afterwards rebuilt. Strong evidence for this opinion comes from dynamical SAXS experiments with synchrotron radiation [8]: the discontinuous scattering intensity vanishes, indicating a destroyed superstructure. Static measurements with relaxed samples always yield discontinuous small angle scattering [24].

II. Relaxation

1. $\lambda < 2.5$

a) According to the concept of network memory [25] the superstructure and orientation are practically completely restored (isotropic lamellar superstructure) (figs. 11, 12).

b) The crystallite sizes $\bar{D}_i$ grow at about the values of the original isotropic sample (fig. 9). The driving mechanisms are remelting and recrystallization. For the relaxation temperature of 130 °C the value $\bar{D}_3$ (relaxed) $> \bar{D}_3$ (isotropic) is an annealing effect, favoured by chain orientation (chains parallel 3-direction). $\bar{D}_1$ (relaxed) $> \bar{D}_1$ (isotropic) at relaxation temperature 135 °C is explained by melting of small crystallites, therefore an increase of $\bar{D}_1$ and (110)-growing of the remaining crystallites during cooling. The measured angle φ (table 3) of the shape ellipsoid (fig. 2) is in accordance with this conception.

c) The lattice distortions $\bar{g}_{ik}$ after relaxation are nearly the same as those of the initial sample. Under the influence of thermal lattice vibrations a recovery process occurs with a decrease of dislocation density [21]. The low angle boundaries formed after recovery have no long range strain fields and give no remarkable contribution to $\bar{g}_{ik}$.

2. $\lambda > 2.5$

The crystallite volume V after relaxation is smaller than V of the original isotropic sample (fig. 10), which indicates that the structure is not completely restored. Probably the complete destruction of the superstructure during deformation at $\lambda \sim 2.5$ is of importance for the explanation of this behaviour. Because of the two-phase (crystalline, amorphous) structure of PE a full recrystallization as in metals is more difficult.

In some cases we found $\bar{g}_{ik}$ (relaxed) $< \bar{g}_{ik}$ (isotropic), which points to enhanced defect healing with increasing relaxation temperature and draw ratio. The results for the three different relaxation temperatures (RT) are summarized as follows (compare figs. 9–12).

a) RT 130 °C: No complete rebuilding of the superstructure, we found a fibrillar structure also after relaxation. The orientation also is not restored, there is a residual anisotropy. As for $\lambda < 2.5$ the finding $\bar{D}_3$ (relaxed) $> \bar{D}_3$ (isotropic) is an annealing effect, favoured by chain orientation.

b) RT 135 °C: For $\lambda = 4.5$ the isotropic lamellar superstructure is restored. The sample with $\lambda = 7.5$ showed a higher degree of orientation before relaxation, compared with the sample $\lambda = 4.5$ and after relaxation a fibrillar structure and anisotropic orientation. Perhaps a longer relaxation time could restore the original superstructure.

c) $RT = 139$ °C: At this relaxation temperature most of the crystallites are molten and a high chain mobility exists. Therefore true relaxation occurs only partly and the predominant process is "crystallisation from the melt" during cooling to room temperature. Because there is no annealing effect (the isotropic sample was annealed at 130 °C for 3 hours), the values $\bar{D}_1$, $\bar{D}_2$, V are smaller than the values of the isotropic original sample. One observes a lamellar superstructure and orientational isotropy.

To summarize, one can conclude that the description of the observed phenomena with the concept of lattice faults, taking into account the specific polymer structure, is possible.

Acknowledgement

We are grateful for the support of the Deutsche Forschungsgemeinschaft.

References

1. Wilke W, Martis KW (1974) Coll & Polym Sci 252:718
2. Martis KW, Wilke W (1977) Progr Coll & Polym Sci 62:44
3. Wilke W (1983) Coll & Polym Sci 261:656
4. Guinier (1963) X-ray diffraction by Crystals, Imperfect Crystals and Amorphous Bodies, Freeman, San Francisco–London
5. Hosemann R, Bagchi SN (1962) Direct Analysis of Diffraction by Matter, North Holland Publ Comp, Amsterdam
6. Rothman RL, Cohen JB (1971) J Appl Phys 42:971
7. Werner W (1979) Diplomarbeit Universität Ulm
8. Fronk W (1984) Thesis Universität Ulm
9. Frank FC, Keller A, O'Connor A (1958) Phil Mag 3:64
10. Zaukelies DA (1962) J Appl Phys 33:2797
11. Keith HD, Passaglia E (1964) J Res NBS 68A:513
12. Predecki P, Statton WO (1966) J Appl Phys 37:4053
13. Predecki P, Statton WO (1967) J Appl Phys 38:4140
14. Peterson JM (1966) J Appl Phys 37:4047
15. Shandrake LG, Guiu F (1976) Phil Mag 34:565
16. Shandrake LG, Guiu F (1979) Phil Mag 39:785
17. Wunderlich B (1973) Macromolecular Physics, Vol 1, Chap IV, Academic Press, New York–London
18. Wunderlich B (1976) Macromolecular Physics, Vol 2, Chap VII, Academic Press, New York–San Fancisco–London
19. Pietralla M (1976) Coll & Polym Sci 254:249
20. Heise B, Kilian HG, Wulff W (1980) Progr Coll & Polym Sci 67:143
21. Honeycombe RWK (1984) The plastic deformation of metals, Chap 9, Edward Arnold, London
22. Wulff W (1981) Thesis Universität Ulm
23. We thank Dr. Wilski for this private communication
24. Fronk W, Wilke W (1983) Coll & Polym Sci 261:1010
25. Hosemann R, Loboda-Cackovic J, Cackovic H (1972) Z Naturforsch 27a:478

Received May 20, 1985;
accepted June 1, 1985

Authors' address:

W. Wilke
Universität Ulm
Abteilung Experimentelle Physik
Oberer Eselsberg
D-7900 Ulm, F.R.G.

Progress in Colloid & Polymer Science

Progr Colloid & Polymer Sci 71:164–172 (1985)

Influence of structural defects on viscoelastic properties of poly(propylene)*)

R. W. Garbella, J. Wachter[1]), and J. H. Wendorff

Deutsches Kunststoff-Institut, Darmstadt, F.R.G.

Abstract: Straining of annealed isotactic polypropylene beyond yield results in the formation of 0.4–4 vol % of microvoids with sizes of about 5–30 nm. The creep behavior as well as the complex shear and tensile modulus of polypropylene are strongly influenced by the presence of the very small concentration of microvoids, acting as structural defects. The storage modulus in the glassy state may be decreased by 30 % (shear modulus) or 15 % (tensile modulus) relative to the defect-free state. Possible interpretations are discussed.

Key words: Structural defects, microvoids, visco-elastic properties, creep.

I. Introduction

The mechanical properties of amorphous and partially crystalline polymers are known to depend strongly on their thermodynamical properties as well as on their molecular and supermolecular structure [1–5]. Structural parameters which govern the macroscopical properties are, for instance, the degree of crystallinity, the crystal modification, orientation of the crystalline and amorphous regions or the size of spherulites.

In addition one expects that structural defects also contribute to the mechanical properties. Structural defects can, in principle, occur at all levels of dimensions. Chain ends, heterogeneities of the chemical structure of the chain molecules, boundaries between amorphous and crystalline regions, between microfibrils and voids, boundaries between spherulites and surrounding material, and weld lines are structural inhomogeneities at different levels of dimensions. All of them will influence the macroscopical mechanical properties, although with different weight.

The structural defects which are discussed subsequently, are the result of a local deformation process in the solid state. They are due to mechanical forces acting on the polymer. It has been known for some time that plastic deformation processes result in the formation of crazes. The mechanism of craze formation and the craze structure were analyzed in detail for amorphous polymers, using optical and electron microscopy, light scattering and small angle X-ray scattering [6–9].

Crazes occur also in partially crystalline materials and the experimental method of analysis has mainly been electron microscopy [10]. It has been demonstrated that crazes are characterized by a density which is much lower than that of the matrix and that the crazes consist of fibrils and voids. Crazes are thus structural defects. We furthermore know that crazes in amorphous polymers, such as polycarbonate, consist of a regular arrangement of voids and fibrils. Unfortunately the structure of crazes in partially crystalline polymers is much less known.

X-ray studies have revealed that the deformation of partially crystalline polymers gives rise in many cases to the formation of microvoids with sizes of the order of 5 to 30 nm, if the samples are strained beyond yield [11–18]. The occurrence of these microvoids which cause a strong excess scattering in the small angle X-ray range, leads simultaneously to a decrease of the total density of the material and to an increase of its turbidity. Microvoid formation has been the subject of a large number of publications, to which we have contributed [11, 17, 18].

---

*) Dedicated to Prof. Dr. H.-G. Kilian on the occasion of his 60th birthday, in recognition of his outstanding contributions to our knowledge on polymer physics.

[1]) Presently at VDO, Bad Schwalbach.

The present paper is concerned with the formation of microvoids in isotactic polypropylene and the influence of microvoids on its viscoelastic properties. The motivations for these studies were

1) to define conditions for microvoid formation

2) to characterize the influence of microvoid formation on yielding and creep

3) to characterize the influence of microvoids on viscoelastic properties for small deformations, as measured by the torsion pendulum, for instance.

II. Experimental

Sample preparation

Previous studies were concerned with microvoid formation in partially crystalline polyoxymethylene. The present paper specifically treats the formation of microvoids in mechanically loaded isotactic polypropylene. Isotropic samples of the polypropylene (Hostalen PPN 1060, Hoechst AG, Frankfurt) were obtained by pressing 2 mm thick plates at a temperature of 200 °C. Test bars (DIN 53455) were cut from these plates and were annealed at 150 °C in vacuum for about 24 hours. Microvoids were induced in a controlled manner by subjecting these test samples to a stress-strain experiment at room temperature, employing a deformation rate of 4 mm/min. The samples were strained to fixed values of strain, namely $\varepsilon = 0.1, 0.2, 0.3, 0.4$ and 0.6. The load was then released and the samples were allowed to relax at room temperature for 7 days. The concentration and the structure of the induced microvoids were then analyzed by means of small angle X-ray scattering, turbidity and density measurements.

Small angle X-ray scattering

A Kratky small angle X-ray camera was used, which was equipped with a scintillation detector. The excess scattering, arising from microvoids, was analyzed in terms of a Debye correlation function [19]. The absolute scattering and the correlation length were determined as described in previous papers [11, 17, 18].

Turbidity studies

The attenuation of infrared light, which was perpendicularly incident on the sample bars, was measured by a photodiode. The turbidity data were corrected for the corresponding value of the undeformed sample.

Density studies

The densities of the unstrained and prestrained samples were determined by the floating method. The fluids used in this experiment consisted of mixtures of water and isopropanol. The density of the sample was taken from that of the fluid mixture in which the sample was able to float.

Stress-strain studies

Cyclic stress-strain studies were performed with a tensile tester (Zwick company). The deformation rate was kept constant at 4 mm/min, both for increasing and decreasing strain. Cyclic stress-strain curves were used to determine components of the deformation energy such as the dissipated energy and the anelastically stored energy. The stress-strain data were taken for the original samples as well as for samples which had previously been subjected to one or more deformation cycles, followed by a recovery period of one week.

Mechanical dynamic analysis

These studies were performed in order to determine the complex tensile as well as the complex shear modulus as a function of the temperature, both for the original as well as for the prestrained samples containing microvoids. The complex shear modulus was obtained for a frequency of about 30 Hz using a Zwick torsional pendulum (DIN 53455, Zwick torsion pendulum, model 5202) and the complex tensile modulus at a frequency of 430 Hz, using a set-up built in the DKI.

Creep experiments

Creep experiments were performed at a temperature of 23 °C and a relative humidity of 50%. These studies were performed in order to find out whether microvoids occur due to a creep process and whether microvoid formation influences the creep curves.

III. Results and discussion

Stress-strain experiments

The stress-strain curve of the original sample is characterized by an almost linear increase of the strain with increasing stress at low strains and a stress beyond yield which is almost constant in a large range of strain values. This is shown in figure 1. Yielding occurred homogeneously without neck formation, due to the thermal pretreatment of the samples. A residual strain was observed after the first deformation cycle, which decreased with increasing recovery time and approached zero after one week.

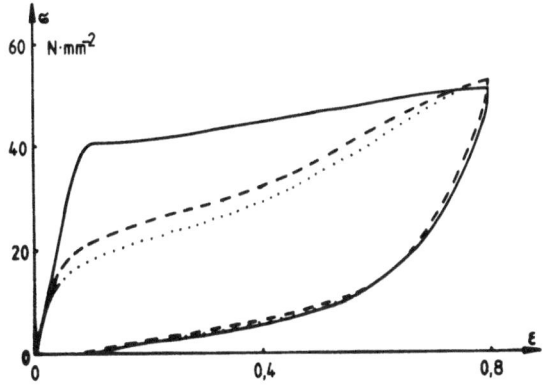

Fig. 1. Stress-strain curves of poly(propylene) for the first (———), second (– – – –) and third (·······) deformation cycle

The second deformation cycle, performed on the prestrained sample, was characterized by a much lower yield stress (fig. 1) and a stress beyond yield which increased with increasing strain. This value became identical to that characteristic of the first stress-strain cycle, if the strain approached the upper strain ε of the first deformation cycle. The stress-strain curves of the second, third and further deformation cycles were very similar, provided that the samples were allowed to relax at room temperature for one week after each run.

The stress-strain curves described above can be used to calculate the difference in deformation energy for a given upper strain between the first and subsequent deformation cycles. The results are plotted in figure 2. It is obvious that this value increases with the upper strain and that most of the changes occur between the first and the second deformation cycle. Details of the dependence of this component of the energy on the strain will not be discussed here in this paper. It is sufficient to say, at this point, that the energy increases linearly as a function of the square of the deformation. It is an important observation that the differences between the stress-strain curves of the original sample and of the prestrained samples measured at room temperature decrease strongly with increasing annealing temperature (fig. 3). They vanished for an annealing temperature of about 120 °C, i. e. for a temperature far below the melting temperature of polypropylene. The conclusion is that this component of the deformation energy cannot be attributed to an irreversible dissipative process. That energy was rather stored anelastically in the material. Annealing causes a relaxation of this stored deformation energy. This

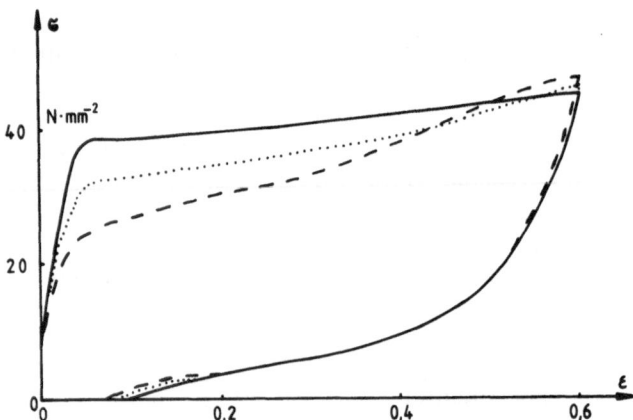

Fig. 3. Influence of annealing at elevated temperature (120 °C) on the stress-strain curves, measured at room temperature (—— first and - - - - second deformation cycle, ····· deformation after annealing the prestrained sample)

process therefore gives rise to a recovery of the mechanical properties of the samples. Similar results were reported by us for polyoxymethylene, strained beyond yield [16].

In the subsequent section we would like to discuss a mechanism wich could be responsible for energy storage and energy release.

Small angle X-ray scattering, density and turbidity studies

Small angle X-ray scattering studies reveal that the prestrained samples exhibit a very strong excess scattering, relative to that of the original samples. The scattering curve characteristic for the excess scattering decreases monotonically with increasing value of the scattering vector $\vec{s}$ ($|\vec{s}| = s = (4\pi/\lambda)\sin\theta$). Obviously particle scattering occurs. Earlier studies and studies to be reported below have revealed that the scattering curves could be analyzed in terms of the Debye [19] correlation function

$$G(r) = \exp(-r/\xi) \qquad (1)$$

which defines a correlation length ξ. This correlation function leads to the following scattering curve for the pinhole scattering $I(s)$ and the smeared scattering $\tilde{I}(s)$ as obtained from the Kratky small angle camera:

$$I(s) = I(o)/(1 + \xi^2 s^2)^2$$

$$\tilde{I}(s) = \tilde{I}(o)/(1 + \xi^2 s^2)^{3/2} \qquad (2)$$

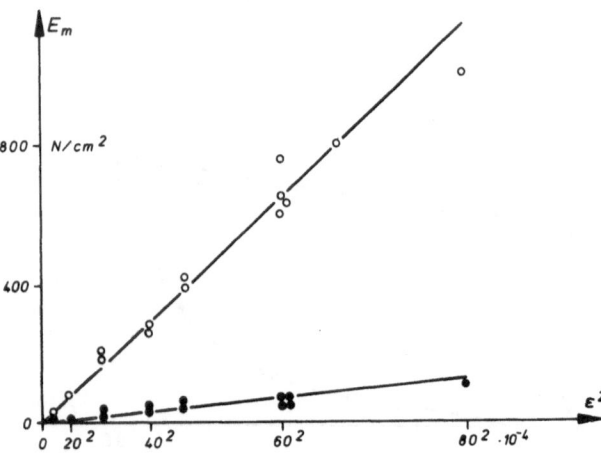

Fig. 2. Plot of mechanically stored energy E_m versus the square of the strain (O first deformation cycle, ● second deformation cycle)

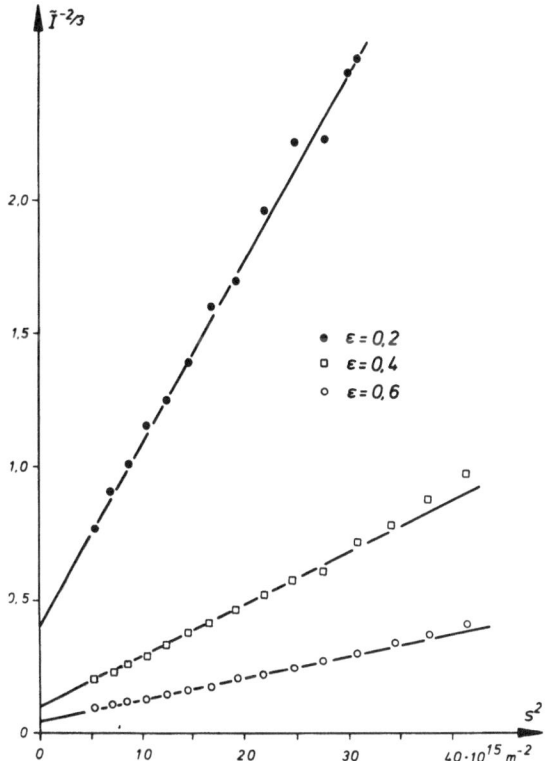

Fig. 4. Debye plot of the excess-scattering of strained poly(propylene) samples

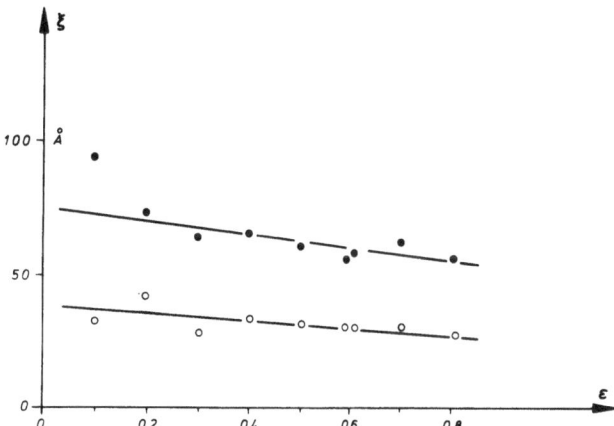

Fig. 5. Dependence of the size ξ of the microvoids on the strain (O parallel and ● perpendicular to the drawing direction)

where the slit like primary beam is either parallel or perpendicular to the draw direction.

One example of a Debye plot for the excess scattering $\tilde{I}$ is shown in figure 4. The correlation length was found to be of the order of 10–30 nm. It turned out to be anisotropic in the sense that it varied as a function of the scattering direction relative to the draw axis. This also became obvious from Kiessig photographs exhibiting an ellipsoidal distribution of the intensity. The straightforward interpretation is that a certain small number of microvoids has been induced in the material. The microvoids have, on the average, the shape of a rotational ellipsoid, with the short axis being the rotation axis and pointing in the direction of the draw axis. The dimensions of the microvoids were found to depend only slightly on the total deformation, the correlation length decreased slightly with increasing strain (fig. 5). This is an indication that smaller microvoids are formed as the strain is increased. This leads to a slight reduction of the average size which is obtained by the scattering experiments. In the case of poly(oxymethylene) it was observed that the sizes of the microvoids were independent of the strain [11].

The excess scattering was found to increase strongly with increasing strain. This shows that the concentration of microvoids also increases strongly with increasing strain. The absolute scattering intensity can be used to calculate the concentration of microvoids, based on the analysis of the invariant Q or $\tilde{Q}$. The invariant is defined as:

$$Q = \int_0^\infty I(s)\, s^2\, ds = 2\,\pi^2\,\Delta\eta^2\, w\,(1-w) = \frac{1}{2}\,\tilde{Q} \quad (3)$$

where the slit like primary beam is oriented along the draw axis.

The volume concentration of microvoids is shown in figure 6 as a function of the strain. In order to check

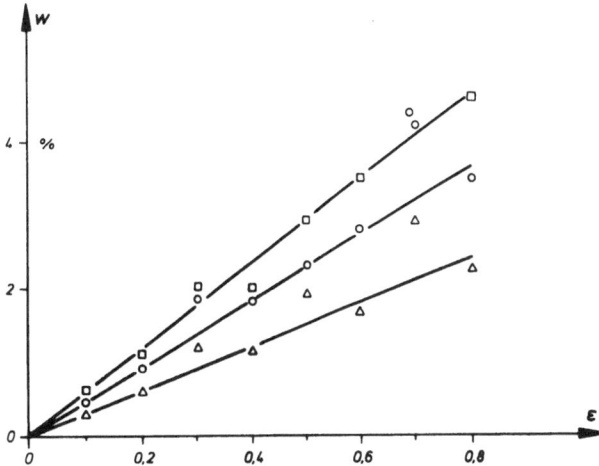

Fig. 6. Volume fraction w of microvoids as a function of strain for different deformation cycles (△ first, O second, □ third cycle)

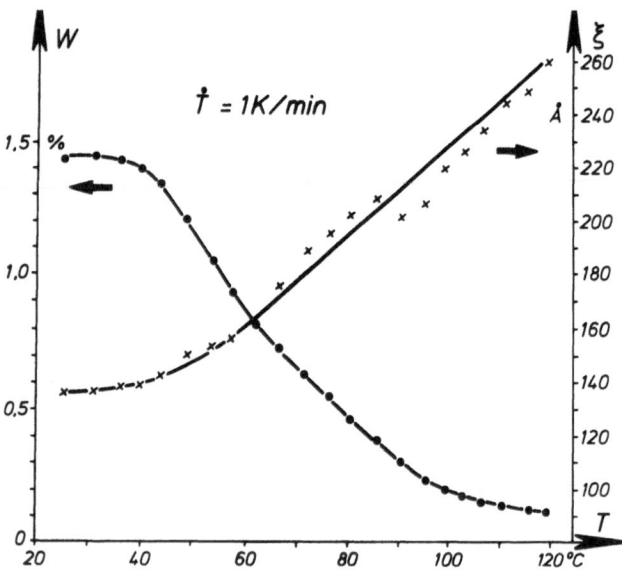

Fig. 7. Volume fraction w (●) and size ξ (×) of microvoids during a heating cycle

whether the assumption of microvoid formation is reasonable we also determined the density of the samples as a function of the strain. It turned out that the volume concentration of microvoids determined by both methods agreed.

Additional evidence came from the observation that annealing of the prestraineed samples at elevated temperatures led to an increase of the density of the samples and simultaneously to a decrease of the excess scattering. The limiting values obtained for the highest annealing temperature (120 °C) were the density of the original material and the scattering intensity of the originally unstrained material.

An interesting observation is that only a certain part of the total number of defects is able to relax at each annealing temperature. The relaxation leads to an increase of the correlation length. The results are shown in figure 7. Similar results were obtained for poly(oxymethylene). We have interpreted these results in a previous paper [18]. This interpretation seems to hold here also. Basically the idea is that microvoids relax if the recovery stress controlled by the surface energy becomes larger than the yield stress. The smaller microvoids recover at a lower temperature than larger ones. This leads to an increase of the average size of the microvoids.

Microvoid formation also gives rise to a strong increase of the turbidity of the samples. A linear relationship exists between the turbidity and the elongation as well as between the turbidity and the volume

concentration of microvoids, as determined by X-ray and denstiy studies. It is furthermore observed that the turbidity of prestrained samples decreases as the sample is annealed at an increasing temperature. It approaches the turbidity of the original sample at an annealing temperature of 120 °C.

The interpretation is that the turbidity arises from the light scattering caused by local fluctuations of the density and consequently of the refractive index. For small concentration of defects we expect a linear increase of the overall scattering with increasing volume concentration of defects, both for particles which are small compared to the wavelength λ of the light (single microvoid), and for those which are of the same order of magnitude as λ (i. e. aggregates of microvoids such as crazes). This explains the observed linear dependence of the turbidity of the sample on the microvoid concentration [20].

Creep experiments

In the previous section, we have described microvoid formation occurring in partially crystalline polypropylene, due to straining the material beyond yield. In this section we will discuss creep experiments designed to find out whether microvoids are also formed during creep at smaller strains and whether the presence of microvoids leads to changes in the creep curves. One result is that the number of microvoids present in prestrained samples does not increase appreciably during the creep experiment. Thus any change in the creep behavior relative to that of untreated samples is due to the presence of microvoids.

Creep curves obtained for the original sample as well as for samples which were prestrained and which thus contained various amounts of microvoids, were analyzed in terms of the following power law:

$$\varepsilon\,(t) = A\,(\sigma) \cdot \left(\frac{t}{t_o}\right)^n \tag{4}$$

where $A\,(\sigma)$ is a nonlinear function of the stress and where n is usually independent of the stress. n is in the range between 0 and 1 (2). A plot of $\ln \varepsilon$ versus $\ln t$ should result in a straight line, the slope of which depends on n, while the intercept determines $A\,(\sigma)$.

Creep and recovery were determined at room temperature for a load of 10 N/mm² and for a creep time of 24 h. We found that we were able to represent the creep curves by equation (4). The results for $A\,(\sigma)$ and n are shown in figure 8 a, b for a constant load for samples with different amounts of microvoids.

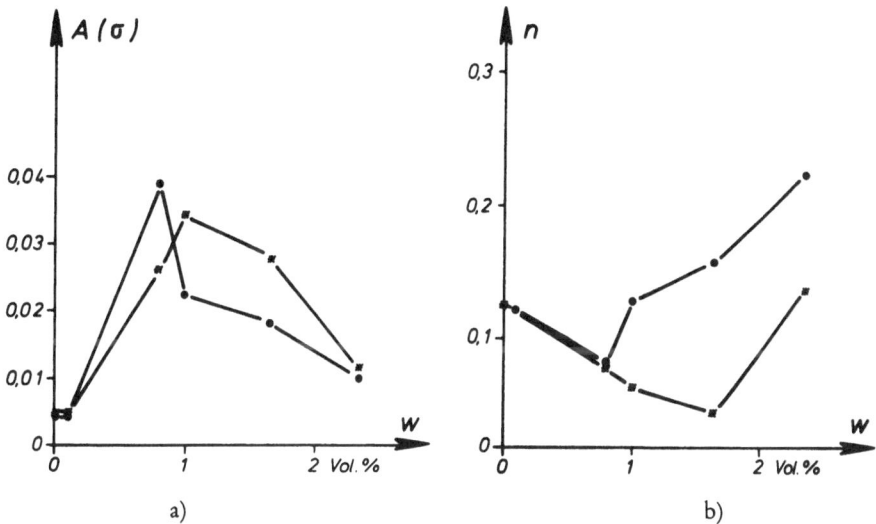

Fig. 8. Characteristic parameter $A(\sigma)$ (a) and n (b) for creep curves of prestrained poly(propylene)

It is obvious that the slopes of the creep curves vary with the concentration of microvoids both for creep and recovery. $A(\sigma)$ also depends at constant σ on w both for creep and recovery. One surprising observation was that the sample with the largest concentration of microvoids displayed creep properties which were closer to those of the original sample than to those observed for samples with intermediate concentrations of microvoids. In addition one observes that creep and recovery curves no longer agreed for larger concentration of microvoids.

It has to be noted, however, that the samples which contain larger concentrations of microvoids are also slightly oriented, due to the pretreatment. Therefore both orientation and microvoid formation contribute to the observed creep behavior. This may lead to a partial compensation effect.

The influence of microvoids on creep is more pronounced for smaller concentration of microvoids, i. e., for samples prestrained to 0.1 and 0.2, relative to the untreated sample. Here we observe a strong increase for $A(\sigma)$ with w. The material is softer due to the presence of microvoids, allowing for an additional relaxation process. We also observe a decrease of n due to the presence of microvoids.

The interpretation of the value of the parameter n is not straightforward. It has been related to the width of the relaxation spectrum responsible for creep, and it has been attributed to the coupling of a single relaxation process to the relaxation of the surroundings [21, 22].

The simplest interpretation is that microvoids contribute to the total relaxation spectrum.

We observed, for instance, that a small macroscopical loading of a sample containing microvoids gave rise to creep and recovery behavior of microvoids which did not agree with that of the macroscopical sample. The relaxation times observed were in the range between minutes and hours.

We furthermore studied creep curves for samples without microvoids imposing loading conditions which should result in the formation of microvoids. Figure 9 shows the creep curve for a sample loaded with 21.25 N/mm². A two-stage creep behavior is

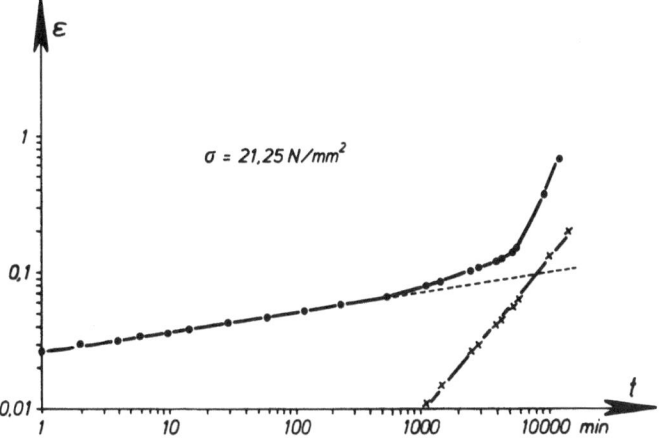

Fig. 9. Creep curve of a void-free sample ●●● (Decomposition into two creep curves: - - -, × × ×)

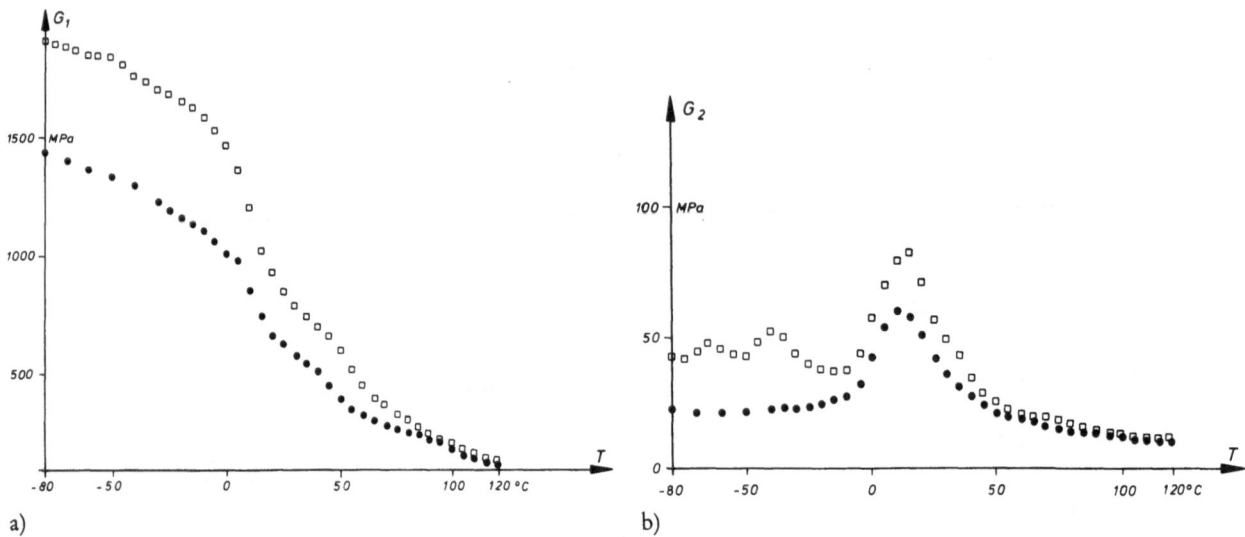

Fig. 10. Storage shear modulus G_1 (a) and loss modulus G_2 (b) versus temperature for poly(propylene) sample with 4.35% microvoids (●) and 0.00% microvoids (□)

observed. A possible interpretation is that the formation of microvoids gives rise to an additional relaxation process. A decomposition of the total creep curve in two components is possible, as is apparent from figure 9. X-ray studies revealed that microvoids are produced at a creep time of about 1000 min. The question to be discussed in the following part of this paper is whether these additional relaxation modes can also contribute to the viscoelastic properties in the Hz and kHz regime. For this purpose, we determined the complex shear and tensile modulus for samples with and without microvoids. In principle, it would be possible that the microvoids give rise to an additional relaxation, characterized by a discrete maximum in the loss modulus and a decay in the storage modulus. One other possibility is that the relaxation time spectrum, due to microvoid relaxation, is so broad that the process contributes in the whole temperature range.

The complex shear modulus $G = G_1 + i\,G_2$ and tensile modulus $E = E_1 + i\,E_2$

Figure 10 a, b displays the complex shear modulus of a poly(propylene) sample containing no microvoids. The relaxation spectra obtained exhibit a strong β-relaxation, characteristic of the glass transition, a broad γ-relaxation at low temperature and a very small α-relaxation at about 70 °C, in agreement with results in the literature [23]. The influence of microvoids on the complex shear modulus is also demonstrated in figure 10 a, b. It displays the temperature dependence of G_1 and G_2 for a sample containing 4.35 vol % micro-

voids in addition to that containing no microvoids. It is obvious that the presence of microvoids leads to an appreciable reduction of G_1 and G_2 relative to the values observed for samples free of microvoids. Similar results were obtained for samples containing 0.77 and 2.15 vol % of microvoids. The reduction is always large at low temperature and close to zero at temperatures above about 90 °C. This is not surprising since our structural studies have revealed (see fig. 7) that the microvoids are able to relax at higher temperatures. The complex modulus consequently has to approach the value characteristic of the sample containing no microvoids. The relative change in the shear modulus $(G_M^1\,(T) - G_O^1\,(T))/G_O^1\,(T)$ (where $G_M^1\,(T)$ is the modulus of the sample containing microvoids and G_M^1 (T) the corresponding value of the sample without microvoids) was found to be roughly independent of the temperature in the temperature range between $-80\,°C$ and 90 °C. This has to be taken as an indication that microvoid deformation contributes to the viscoelastic properties in a similar way above and below the glass transition of the amorphous regions. The deformation in the glassy state may occur due to a locally induced plastic flow, as proposed for craze formation in glassy polymers. Based on this interpretation we expect that similar results would be obtained for the complex tensile modulus.

Figures 11 a and 11 b reveal that this is the case. It is apparent, however, that the influence of the microvoids on the viscoelastic behavior is much stronger for the shear modulus G_1 than for the tensile modulus E_1. The induced changes are of the order of 30% in the first case and only 15% in the second case.

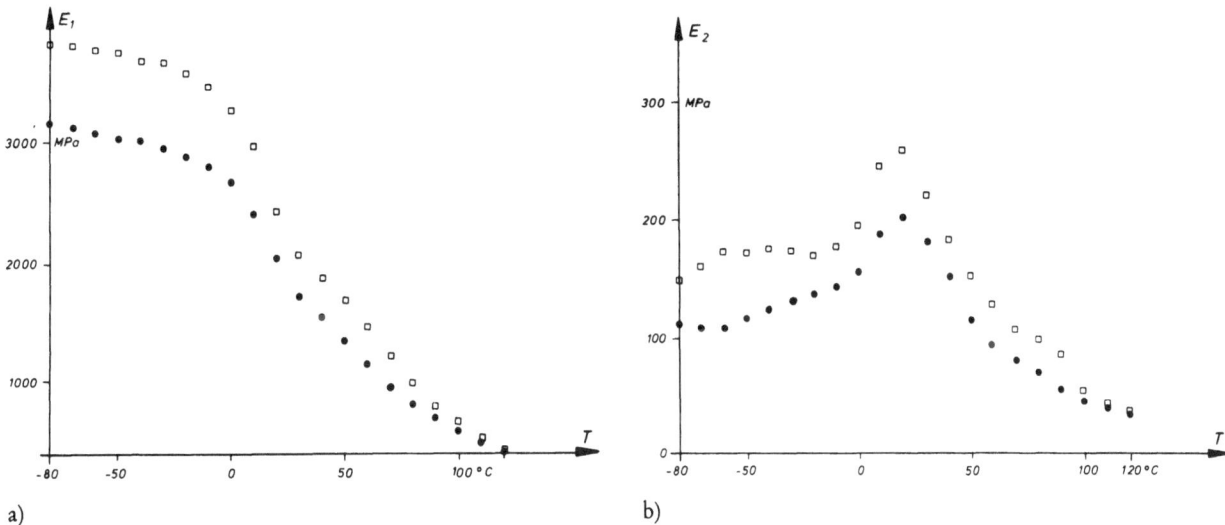

Fig. 11. Tensile storage modulus E_1 (a) and loss modulus E_2 (b) versus temperature for poly(propylene) sample with 0.77 % microvoids (●) and 0.00 % microvoids (□)

Our studies showed, furthermore, that the formation of the microvoids does not influence the viscoelastic properties in the direction perpendicular to the direction of the draw axis to a great extent. This may be taken as an indication that the microvoids described here are parts of crazes which are extended perpendicular to the draw direction. These crazes have to consist of an irregular arrangement of voids and fibrils, as observed for instance by Friedrich [10]. The voids can easily deform in the direction perpendicular to the craze direction because of an increase of the gap of the craze. A growth mechanism perpendicular to this direction is difficult to envisage.

Model calculation

A simple approach towards a calculation of the mechanical properties of polypropylene containing up to 4 % microvoids is based on the models used for deriving the properties of reinforced polymers. We assume that the elastic modulus of the voids is small compared to that of the matrix ($E_{void} \approx 0$). Then we obtain [24–27].

$$E = E_{matrix} * (1 - c\,w) \qquad (5)$$

where c depends on the model and is of the order of 1.5 − 2.0. The prediction is that the changes induced by the presence of the microvoids should be of the order of a few percent, disagreeing with experimental results. A similar disagreement was obtained if we used models proposed by Takayanagi et al. [28].

We therefore have to conclude that this simple model does not hold. We believe that aggregates of microvoids — such as crazes — are responsible for the strong influence of microvoids on the mechanical properties, and that the interaction of microvoids within the aggregates is the origin of the much stronger impact of microvoids on the mechanical properties. This interaction may be expressed, for instance, in terms of an effective volume concentration of defects controlling the mechanical properties. Based on equation (5) the effective volume must be larger than the actual volume by a factor of 5. Apparently a new approach is necessary in order to account for the strong influence of microvoids on mechanical properties.

Acknowledgement

The authors gratefully acknowledge the financial support of the Arbeitsgemeinschaft Industrieller Forschungsvereinigungen (AIF).

References

1. Kilian HG (1979) Phys Blaetter B5:642
2. Kilian HG (1981) Polymer 22:209
3. Ward IM (1971) Mechanical Properties of Solid Polymeres, JW Arrow Smith, Bristol
4. Nielson LE (1962) Mechanical Properties of Polymers Chapman & Hall, London

　　　　　　　　　　　　　　　　　　　　　　　　　　　　　　　Progress in Colloid & Polymer Science, Vol. 71 (1985)

5. Kausch HH (1978) Polymer Fracture, Springer Verlag, Berlin
6. Dettenmaier M (1983) Adv Polym Sci 52/53:57
7. Farrar NR, Kramer EJ (1981) Polymer 22:691
8. Brown HR, Kramer EJ (1981) J Macromol Sci Phys B 19:487
9. Chan T, Donald AM, Kramer EJ (1981) J Mater Sci 16:676
10. Friedrich K (1983) Adv Polym Sci 52/53:225
11. Wendorff JH (1979) Progr Coll & Polym Sci 66:135
12. Zhurkov SN, Kursukov VE (1974) J Polym Sci Polym Phys Ed 12:385
13. Kuksenko VA, Ryskin VS, Betekhtin VI, Slutsker AI (1975) Int J Fract 11:829
14. Zhurkov SN, Kuksenko VS (1975) Int J Fract 11:629
15. Wachter J (1980) Diplomarbeit, TH Darmstadt
16. Garbella RW (1983) Diplomarbeit, TH Darmstadt
17. Wendorff JH (1978) Angew Makromol Chemie 74:203
18. Wendorff JH (1980) Polymer 21:553
19. Debye PZ (1959) Phys 156:256
20. Brumberger H (1967) Small Angle X-Ray Scattering, Gordon and Breach, New York
21. Brinson HF, Hiel CC, Cardon AH, De Wilde WP (eds) (1984) Astarita G, Nicolais L, Polymer Processing and Properties, Plenum Press, New York, 311
22. Ngai KL (1980) Comments Solid State Phys 9:141
23. McCrum, Read NG, Williams GW (1967) Anelastic and dielectric effects in Polymeric Solids, Wiley, New York
24. Sumita M, Ookum T, Miyasaka K, Ishikawa K (1982) J Mat Sci 17:2869
25. Hashin Z (1962) J Appl Mechanics 4:143
26. Halpin JC, Kardos JC (1976) Poly Eng Sci 16:344
27. Dickie RA (1973) J Appl Polym Sci 17:45
28. Takajanagi M, Uemura S, Minami S (1978) J Poly Sci Part C 5:113

Received April 12, 1985;
accepted June 19, 1985

Authors' address:

PD Dr. J. Wendorff
Deutsches Kunststoff-Institut
Schloßgartenstraße 6 R
D-6100 Darmstadt, F.R.G.

Progress in Colloid & Polymer Science   Progr Colloid & Polymer Sci 71:173–179 (1985)

Constructions of master curves and master surfaces starting with experimental data*)

M.-J. Brekner, H.-J. Cantow and H. A. Schneider

Institute for Macromolecular Chemistry of the University of Freiburg, F.R.G.

Abstract: Starting with the theoretical considerations of our previous papers [1, 3], some practical aspects concerning the shift procedures of rheological data are analysed. The latter can be used for the construction of both isotherm and isochrone master curves as well as of master surfaces, with the condition of experimental verification of the validity of the time-temperature superposition principle. Thereby any empirical shift procedure is omitted. Examples are presented for a very large frequency-temperature range, including the glass transition zone.

Key words: Rheology of polymer molts, shift procedures, isotherm and isochrone master curves, master surface.

Shift procedures

Master surfaces discussed in a previous paper [1] have not only the advantage of characterizing the viscoelastic body absolutely, but also of evidencing the interrelation between the time and temperature dependence of viscoelastic functions. It has been shown that both isotherm and isochrone master curves can be constructed via shift procedures in the plane of $\log(f_r) =$ const., f_r being an arbitrarily chosen value of the reduced viscoelastic function.

Remembering that these shift procedures are generally applicable to any viscoelastic function, we will discuss for examplification the storage modulus only,

$$G'_r = G' \cdot T_{red} \, \varrho_{T_{red}} / T \varrho_T. \qquad (1)$$

T_{red} represents here the chosen reduction temperature. As has already been pointed out [1] T_{red} and T_{ref} may have a different significance in the representation of viscoelastic properties. It has also been demonstrated that in the plane of $\log(G'_r) =$ const. the following relations are valid:

$$d\{\ln[a_T(T)]\} = [E(T)/R]d(1/T) \qquad (2)$$

and

$$\ln[a_T(T)] = \int_{1/T_{ref}}^{1/T} [E(T)/R]d(1/T). \qquad (3)$$

These equations can be simplified for temperature invariant activation energy of flow:

$$d\{\ln[a_T(T)]\} = (E/R)d(1/T) \qquad (4)$$

$$\ln[a_T(T)] = (E/R)(1/T - 1\langle T_{ref}). \qquad (5)$$

The statements of the above equations are the following:

— Relations (2) and (4) define the overall direction of the data shift in the plane of $\log(G'_r) =$ const.

— Equations (3) and (5) determine the shift of the reduced viscoelastic data along the logarithmic time scale corresponding to the temperature variation from T to T_{ref}. Thus the shift factor is defined as a function of temperature and related to an arbitrary chosen integration constant $1/T_{ref}$.

General schemes of the possibilities to shift experimental data, according to the above relations, can be sketched by considering the projections of both the reduced data and the chosen isotherm and isochrone master curves on the arbitrary plane $\log(G'_r) =$ const., as shown in figure 1.

---

*) Dedicated to Prof. Dr. H.-G. Kilian on the occasion of his 60th birthday.

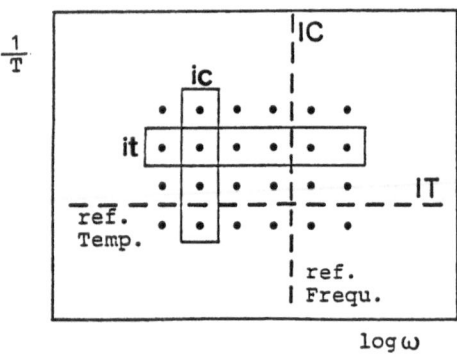

Fig. 1. Data projections on an arbitrary plane $\log(G'_r) = \text{const.}$; it = isotherm curve; IT = isothere mastere curve; ic = isochrone curve; IC = isochrone master curve

The resulting shift procedures are all illustrated in figures 2 and 3, taking into account that essentially two distinct modalities are possible to describe the temperature dependence of vicoelastic functions. The most simple case of a temperature invariant apparent activation energy of flow is usually encountered in the terminal zone (fig. 2).

The more general description of viscoelastic properties takes into account a possible change of the apparent activation energy of flow with temperature [2] (fig. 3) which is characteristic for the glass transition zone.

The familiar shift procedures of isotherms are presented in figures 2a and 3a, for constant and variable

apparent activation energy of flow respectively. The shift procedures presented in figure 2 are in accordance with the Arrhenius-like behaviour of viscoelastic functions at higher temperatures, whilst the latter is equivalent to the shift procedure predicted by the WLF relation [2].

Isochrone master curves were first constructed for constant apparent activation energy [4]. The corresponding shift procedure is exemplified in figure 2b [4]. Only in three of the presented shift procedures is the whole viscoelastic curve shifted as such for the corresponding mastercurve. In all other cases we have to consider the individual shift of each experimental point itself. Thus the experimental curves change their shape during the shift procedure. However, so far the shift is governed by equation (3), there is no doubt about the validity of the procedure used.

The validity of any particular shift procedure and the temperature invariance of the relaxation time spectrum, $H|\theta(T)|$, are definitely confirmed by the reduced scatter of the experimental data if shifted for both isotherm (= IT) and isochrone (= IC) composite curves, always using the same apparent activation energy of flow, $E(T)$.

Determination of the apparent activation energy of flow and of the shift factor

In order to determine the shift factor according to equation (5) or (3) we first have to evaluate the appar-

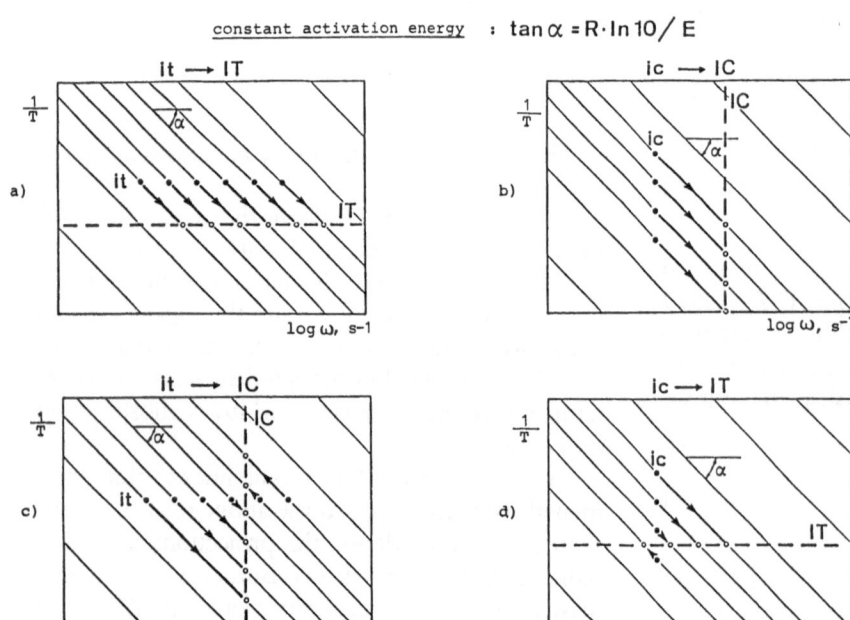

Fig. 2. Shift procedure for constant apparent activation energy of flow; it = isotherm curve; IT = isotherm master curve; ic = isochrone curve; IC = isochrone master curve; E = constant apparent activation energy of flow

T-dependent activation energy : $\tan\alpha = R \cdot \ln10 / E(T)$

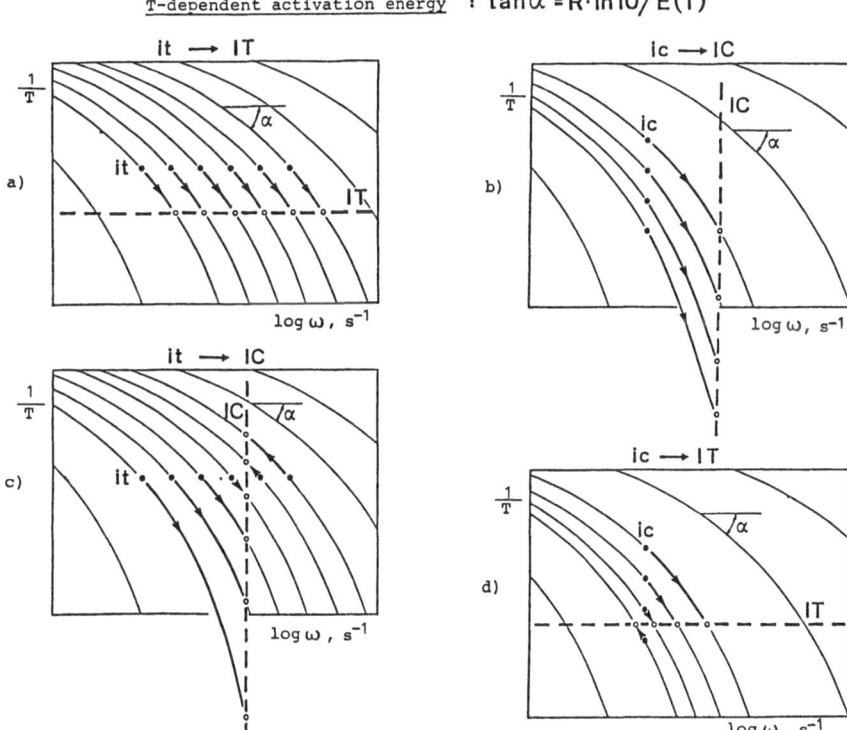

Fig. 3. Shift procedure for temperature dependent activation energy of flow; it = isotherm curve; IT = isotherm master curve; ic = isochrone curve; IC = isochrone master curve; $E(T)$ = temperature dependent apparent activation energy of flow

ent activation energy. As has been shown, the apparent activation energy is given by the ratio of the slopes of the isochrones and isotherms, respectively, at a given intersection point of the curves

$$\frac{[\partial\log(G_r')]_{\omega_0}/\partial/(1/T)}{[\partial\log(G_r')]_{T_0}/\partial\log(\omega)} = E(T_o)/R\ln10. \qquad (6)$$

T_o and ω_o are the temperature and frequency of the respective intersection point. Experimental data were all obtained from consecutive isothermal measurements at variable frequency. The stepwise evaluation of the temperature dependent shift factor via integration (see eq. (5)) is illustrated in figure 4.

The evaluation starts with the isochronal presentation of viscoelastic data:

● – Isochrone data (here of G'') for an arbitrary chosen frequency. To obtain the most accurate determination of the isotherm slope, st, the frequency is chosen approximately in the middle of the frequency range of measurement.

Then the ic curve is computed using a fit program:

–●–●– — The isochrone curve computed by a fit program.

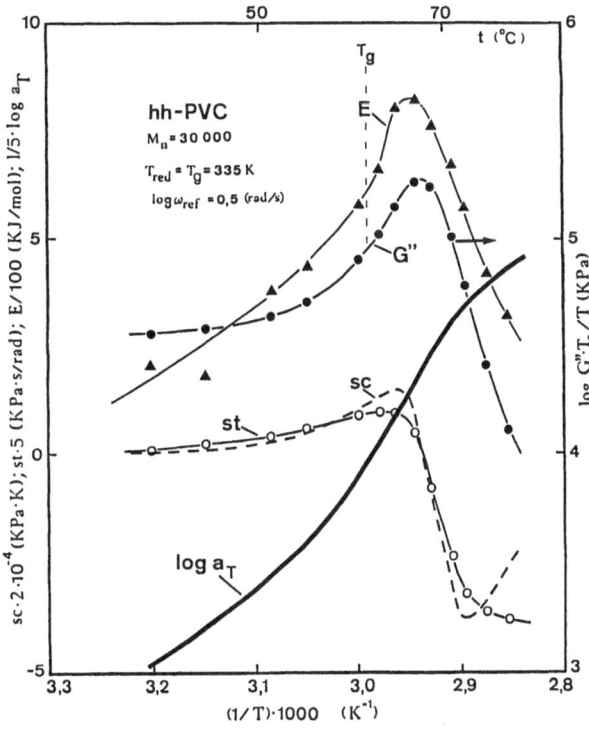

Fig. 4. Evaluation of $E(T)$ and $\log(a_T)$ using loss modulus data of hh-PVC

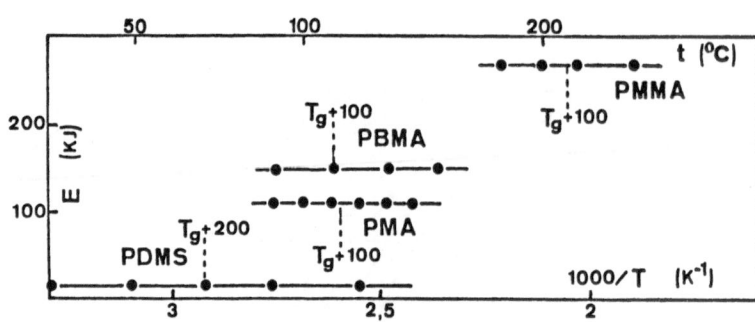

Fig. 5. The apparent activation energy in the terminal zone; poly(methylmethacrylate) = PMMA; poly(n-butylmethacrylate) = PBMA; poly(methylacrylate) = PMA; poly(dimethylsiloxane) = PDMS

The next step is the evaluation of the slopes of the it curves:

st — Best fit of the isochronal representation of the slopes of the isotherm curves at the arbitrarily chosen reference frequency.

By derivation the slopes of the ic curves are computed.

sc — Derivative of the ic function with respect to the reciprocal temperature.

From the ratio of the slopes the apparent activation energy of flow results:

▲ — Apparent activation energy of flow evaluated according to equation (6) from the slopes st and sc at the temperature of measurement.

−▲−▲− — Best fit of the dependence on reciprocal temperature of the apparent activation energy of flow. Finally $\log a_T$ is computed:

$\log a_T$ — Dependence on reciprocal temperature of the shift factor calculated according to equation (5) by integration of the reciprocal temperature function of the apparent activation energy.

Starting with this temperature function, $\log(a_T) = f(1/T)$, master curves and master surfaces respectively have been constructed according to the procedure a) and c) in figures 2 and 3.

The accuracy of these determinations is limited only by the feasibility of a best fit of the experimental data in each evaluation step. Therefore it is necessary to have as much data as possible for both the frequency and the temperature ranges of measurement. The precise determinations of the slopes of viscoelastic curves is mainly affected by the scatter of the experimental data in plateau regions. Thus to assure the highest accuracy of shift factor computation for each region, the use of one of the most sensitive functions is recommended, i. e. of that viscoelastic function which shows the most spectacular changes.

Experimental results

Measurements were realized with an Instron Rheometer in the range of very small deformations to assure linear viscoelasticity. In the terminal zone the deformation was of oscillating shear between cone and plate and in the glass transition zone of oscillating torsion of a bar with rectangular cross-section.

In the terminal zone within the measured temperature interval many measured polymers showed a constant apparent activation energy of flow (see fig. 5). Thus the procedures of figures 2 were used for the construction of master curves and master surfaces in this temperature range. For illustration the master curves and surfaces of the storage modulus of poly(methylacrylate) are presented in figure 6 and 7 respectively. Master surfaces of the loss modulus and the corresponding dynamic viscosity are shown in figure 8.

For the glass transition zone, both our own and recalculated literature data [5] show temperature dependent apparent activation energies of flow, mainly in the region below $T_g + 40$ and especially for polymers without or with only small and rigid side chains. Representative for the viscoelastic behaviour of polymers governed by a temperature dependent apparent activation energy of flow, the glass transition zone of a head to head poly(vinylchloride), (hh-PVC) is analysed. The applicability of the time-temperature superposition principle to both the storage and the loss modulus denied β-relaxation within the considered temperature interval. Being more sensitive to relaxations near T_g the loss modulus was chosen as viscoelastic function for the determination of the apparent activation energy of flow.

Starting then with the T-dependent shift factor according to the data presented in figure 4, the shift procedures of figures 3a and 3c were applied for the

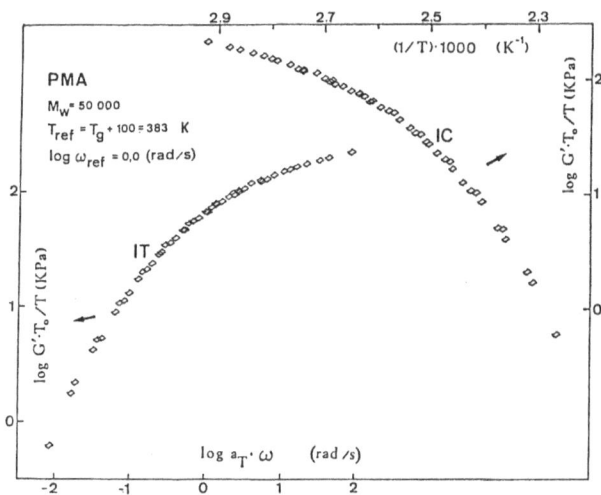

Fig. 6. Master curves of storage modulus of PMA; IT = isotherm master curve; IC = isochrone master curve

construction of the master curves and the master surface in figures 9 and 10 respectively. The small scatter of the experimental data, in all master curves presented in figure 9, confirms the validity of the time-temperature superposition principle in the whole temperature range studied.

Discussions and conclusions

Application of shift procedures for constant activation energy of flow is very simple. A small computer programme gives the possibility to control and evaluate measurements and to plot direct isochrone and isotherm master curves as well as master surfaces. The transformation of relaxation time spectra into relaxation temperature spectra is also easy in such conditions [3].

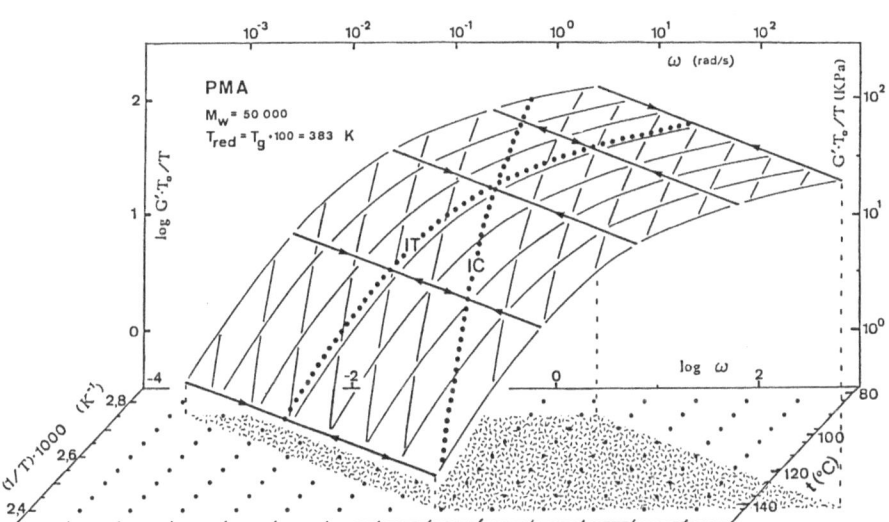

Fig. 7. Master surface of PMA in ther terminal zone ($E(T)$ = const.)

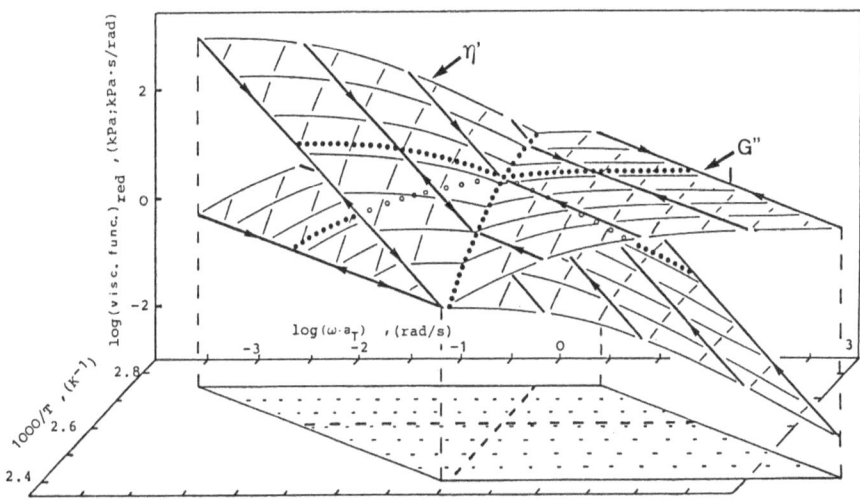

Fig. 8. Master surfaces of viscosity and loss modulus of PMA in the terminal zone ($E(T)$ = const.)

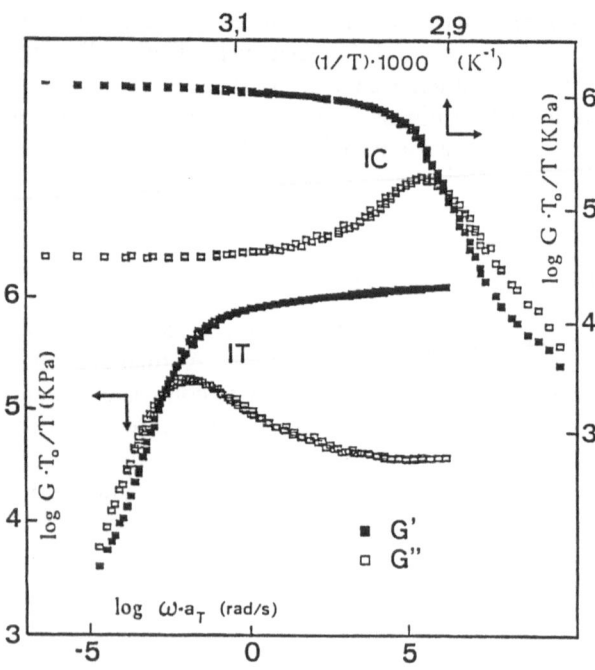

Fig. 9. Isotherm and isochrone master curves of hh-PVC in the glass transition zone ($E(T)$ = variable as shown in fig. 4)

The evaluation of the data in the range of temperature dependent apparent activation energy of flow is more difficult because of the more complicated fit programmes implicated.

As it has already been pointed out [2] in the glass transition zone a temperature dependent activation energy of flow is also predicted by the WLF equation. This equation can only explain an increase of the activation energy approaching T_g. Data presented in figure 4 show, however, that hh-PVC exhibits not only an increase, but also a new decrease of the apparent activation energy very near and below T_g (the T_g-value was determined by DSC extrapolating for zero heating rate). Literature data for PMMA [5] confirm this observation. The observed decrease of the apparent activation energy around T_g also fits with the comparative low activation energy values reported by different authors for the short-range molecular motions in polymer glasses.

In order to avoid the change of reduction temperature for viscoelastic functions in the glass transition zone, the same reduction temperature was applied for the whole temperature range. Density reduction was neglected. Differences in activation energies calculated for both reduced and unreduced data are within experimental error. Finally it can be stated that even in the range of temperature dependent activation energy of flow (as in the glass transition zone of hh-PVC) the recommended shift procedures are applicable. Thus isochrone master curves and master surfaces can be constructed down to the glass transition. The basic condition, however, is the validity of the time temperature superposition principle, that

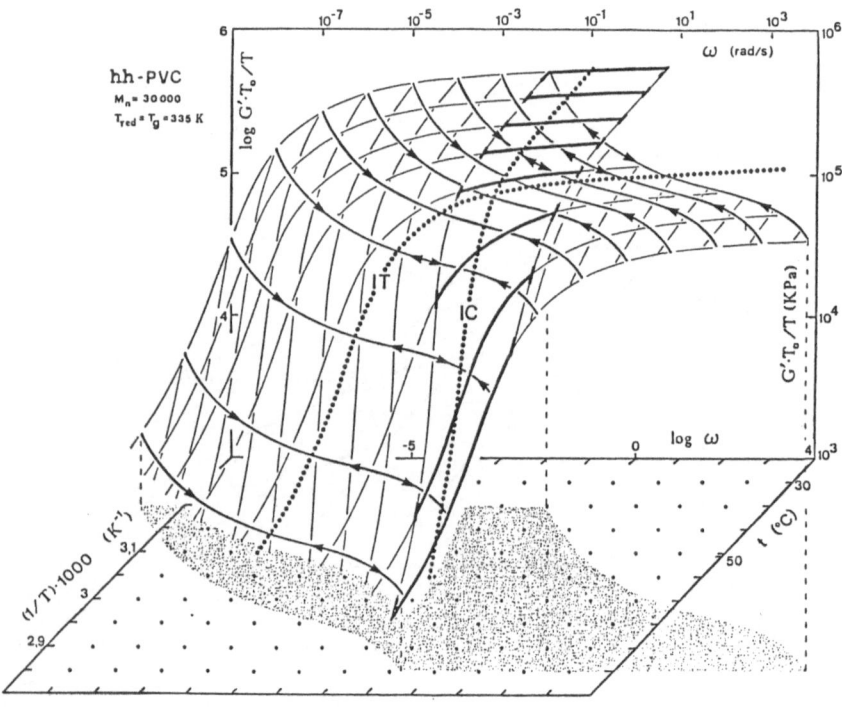

Fig. 10. Master surface of the storage modulus of hh-PVC in the glass transition zone; —— measured curve; • • • • master curve — IT = isotherm, IC = isochrone

means no change has to occur in the relaxation mechanism.

The presented concept of evaluation of dynamic mechanical data enables the reduction of the time of measurement, by choosing a smaller frequency range for higher frequencies but considering an increased number of temperatures of measurement. Such versatility of the experimental technique is based on the substitution of the empirical shift procedure by a mathematically exact method.

References

1. Brekner M-J, Cantow H-J, Schneider HA (1983) Polymer Bull 10:328
2. Schneider HA, Cantow H-J, Brekner M-J (1984) Polymer Bull 11:383
3. Brekner M-J, Cantow H-J, Schneider HA (1985) Polymer Bull 13:51
4. Schneider HA, Cantow H-J (1983) Polymer Bull 9:361
5. Schwarzl FR, Zahradnik F (1980) Rheologica Acta 19, 2:137

Received June 3, 1985;
accepted August 1, 1985

Authors' address:

H.-J. Cantow
Institut für Makromolekulare Chemie
der Universität Freiburg
Stefan-Meier-Straße 31
D-7800 Freiburg i. Br., F.R.G.

Progress in Colloid & Polymer Science

Progr Colloid & Polymer Sci 71:180–190 (1985)

Influence of physical aging at constant temperature on the shear creep of amorphous polymers*)

F. R. Schwarzl, G. Link, R. Greiner, and F. Zahradnik

Institute for Material Science, Chair for Polymeric Materials, University Erlangen-Nürnberg, F.R.G.

Abstract: The theory of linear viscoelastic behaviour under the influence of aging at constant temperature is briefly reviewed. Emphasis is placed on the evaluation of the retardation time — age shift relation from creep experiments under proceeding aging.

The shear behaviour of a technical polystyrene was carefully measured at four temperatures just below the glass transition temperature, systematically varying the time elapsed between the temperature quench and the start of the creep experiment. The results could be accurately described by Struik's theory of viscoelastic behaviour under aging. The dependence of the retardation times on age and temperature in the vicinity of the glass transition is discussed.

Key words: Physical aging, shear creep, glass transition, polystyrene.

1. Introduction

All amorphous polymers show the phenomenon of physical aging. After a quench from the rubbery into the glassy state, many physical properties of a polymer change with proceeding time in the same direction as during cooling through the glass transition range. The material becomes stiffer and more brittle, its damping and creep rate decrease, its ultimate elongation decreases, and so do the dielectric constant, dielectric loss, etc. This process was first described by Struik [1], who called it physical aging. He also elucidated the connection between physical aging and volume relaxation.

As first reported by Kovacs [2, 3], polymers show a continuous decrease of the specific volume with time after a quench from a temperature above the glass transition temperature, T_g, to a constant temperature below T_g. Within a reasonable experimental time scale, a state of equilibrium of the volume can only be reached in a range of a few degrees below T_g. The process of volume relaxation, however, is observed down to low temperatures in the glassy state [4].

Volume relaxation leads to a steady decrease in free volume and thus in mobility of the polymer molecules. This change of mobility of the polymer segments is assumed to be the origin of the change in properties during aging [1].

As a result of physical aging, the viscoelastic deformation properties of polymers will depend not only on loading history and temperature, but also on the progress of aging prior to and during the loading period. Therefore, the conventional technical mechanics of polymers, which is based on the concept of linear viscoelastic behaviour with a creep compliance depending on creep time and temperature only, is bound to fail whenever aging phenomena play a significant role. Consequently the theory of viscoelastic behaviour of polymers should be generalized to include the influence of aging. This contribution is meant to be a step in this direction.

---

*) Dedicated to Prof. Dr. H.-G. Kilian on the occasion of his 60th birthday.

2. Theory of viscoelastic behaviour under the influence of aging at constant temperature

We will restrict our considerations to the most simple case, which is aging at constant temperature. We assume that the sample was heated to a temperature above T_g and kept there until its history had been erased. Then, a single quench is performed to a temperature T below T_g and the sample is further kept at this temperature. After an elapsed time t_e, which will be called preconditioning time in the following, the stress history is applied and the strain is measured. In this experiment the progress of aging of the sample can be uniquely expressed by its age A, which is the time elapsed between the quench and the time of observation of the strain.

The foundation of the theoretical treatment was given by Struik in his thesis [1] by the following reasoning: "Above T_g the temperature affects the creep because the equilibrium free volume is temperature dependent. Below T_g, the free volume becomes time dependent, causing aging. Since above T_g, temperature and therefore free volume has no effect on the shape of the creep curve, there should be no effect below T_g either; the changes in free volume during aging will only affect the position of the creep curve on the log-time scale, and not its shape, as long as free volume does not change significantly during the creep period itself."

Struik [1] has proven the validity of this hypothesis for a large number of polymers and other glasses, measuring the creep curves of samples of different ages. By restricting the periods of creep to times short against the preconditioning time, he could avoid the complications arising from the change in age during the creep experiment. He proved the existence of a time-age shifting principle stating that a change in age at constant temperature shifts the creep curves without change in shape along the logarithmic time scale.

Once this was proven, the development of a theory of viscoelastic behaviour under progressing aging at constant temperature was straightforward. Already earlier, the theory of linear viscoelastic behaviour at constant temperature had been generalized to a theory of linear viscoelastic behaviour under varying temperature by Hopkins [5] and Haugh [6]. For the special case of a thermorheologically simple materials, this theory takes a very simple form. It may be literally applied to the description of viscoelastic behaviour under proceeding aging by replacing the prescribed temperature history by the aging history and the time-temperature shift function by the time-age shift function.

Though these results may be found in the cited literature, we will shortly recall them, because we will need several of the equations for the following discussion.

2.1 Theory of non-isothermal viscoelastic behaviour for thermorheologically simple materials

The theory of non-isothermal viscoelastic behaviour as developed by Hopkins [5] and Haugh [6] may be based on the representation of linear viscoelastic behaviour by mechanical models. Linear viscoelastic behaviour of polymers in simple shear at constant temperature under a prescribed stress history may be described in terms of the deformation of a "generalized Kelvin model", which consists of a single spring, a single dashpot and a number of Kelvin elements, all linked in series. For this purpose, spring constants and viscosity constants of the model have to be appropriately chosen; the choice depends on the temperature.

The non-isothermal treatment is based on the assumption that the same generalized Kelvin model also describes the strain response in the non-isothermal case. The elasticities of the springs and the viscosities of the dashpots have to be inserted as functions of temperature; due to the prescribed temperature history, they become known functions of time. Notwithstanding this complication, the deformation of the model under a prescribed stress history may be explicitly calculated. The solution however does not take a simple or transparent form.

A very simple result is obtained, if an additonal restriction is posed on the mechanical model: its isothermal behaviour should be thermorheologically simple. In this case, all spring constants of the model are temperature independent, whilst the viscosities all have a similar temperature dependence. By a change from the temperature T_o to the temperature T, all viscosities are multiplied with the same factor $a(T, T_o)$.

For thermorheologically simple materials, the shear creep compliances J at the temperatures T and T_o have the same shape, when plotted on a logarithmic time scale:

$$J(t; T) = J(t/a; T_o). \tag{1}$$

They may be brought to coincidence by a parallel shift along the logarithmic time scale. The amount of this shift, $\log a(T, T_o)$, is called the time temperature shift function.

For thermorheologically simple materials, the non-isothermal strain response under a prescribed temperature history $T(t)$, and under a prescribed stress history $\sigma(t)$, which does not start prior to $t = 0$, is found as [5, 6]:

$$\gamma(t) = J_o \sigma(t) + \int_0^\lambda \dot{j}(\lambda - \xi)\, \sigma(\xi)\, d\xi \qquad (2)$$

λ is a function of the time t with the dimension of a time, which is called effective time; it may be calculated from the known temperature multiplication factor a and the prescribed temperature history

$$\lambda(t) = \int_0^t \frac{d\xi}{a(T(\xi); T_o)} \qquad (3)$$

J_o and T_o are the values of the creep compliance and the prescribed temperature at $t = 0$.

Equation (2) has the form of the ordinary superposition principle; however, the convolution integral is to be taken in the λ-time domain; the stress history is to be inserted as a function of the effective time, ξ. $\dot{j}$ is the creep rate at the temperature T_o, to be taken as a function of the difference of the effective times λ and ξ.

The special case of non-isothermal creep of a thermorheologically simple material is found by inserting

$$\sigma(t(\lambda)) = 0 \quad \text{for } \lambda < 0$$
$$\sigma(t(\lambda)) = \sigma_o \quad \text{for } \lambda > 0 \qquad (4)$$

into equation (2). We obtain the result:

$$\gamma(t) = \sigma_o J(\lambda; T_o) \qquad (5)$$

$J(t; T_o)$ has the meaning of the creep function as a function of creep time, t, at the constant temperature T_o; in equation (5) however, the effective time is to be inserted as the argument of the creep function. Equations (3) and (5) together solve the problem.

2.2 Theory of linear viscoelastic behaviour at constant temperature under progressing aging

The results of the preceding section may be transferred immediately to the problem of viscoelastic behaviour under the influence of aging at constant temperature.

We call $J(t; A)$ the — hypothetical — course of the shear creep compliance as function of creep time, t, but at a fixed value for the age, A. This function will not be accessible experimentally in general. But we may express Struik's hypothesis now by the following equation:

$$J(t; A) = J(t/b; A_o), \qquad (6)$$

relating the creep functions at the constant ages A and A_o to each other. Both functions show the same shape on a logarithmic time scale, but are shifted over by log $b(A; A_o)$, which is the time-age shift function. Now, in the formalism of section 2.1, the time-temperature shift function is to be replaced by the time-age shift function.

Moreover, the prescribed temperature history is to be replaced by the age history; for aging under constant temperature after a preconditioning time t_e, we have

$$A(t) = t_e + t \qquad (7)$$

and therefore for the effective time

$$\lambda(t) = \int_0^t \frac{d\xi}{b(t_e + \xi; t_e)}. \qquad (8)$$

Equation (2) remains unchanged; $\dot{j}$ now has the meaning of the rate of the hypothetical creep curve under the constant age t_e. This equation was given in the same connection by Struik (eq. (C3) in [1]).

2.3 Creep under aging at constant temperature after a single quench

We will restrict all further discussions to the case of creep at constant temperature under the influence of aging after a single quench. The time dependence of temperature, shear stress and shear strain is indicated in figure 1.

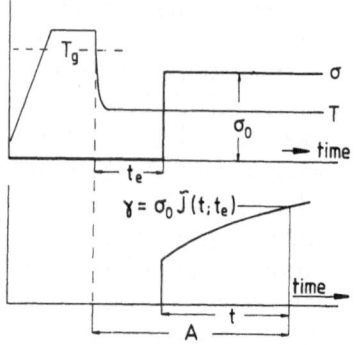

Fig. 1. The course of temperatures, stress and strain as functions of time

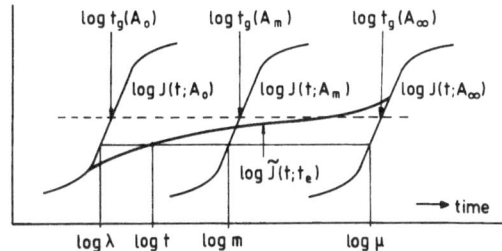

Fig. 2. Creep compliance under aging, and creep compliances at constant age, vs. time

The specimen is quenched from a temperature above T_g to the measuring temperature T; after a pre-conditioning time t_e, it is loaded with a constant stress of magnitude σ_o and its strain is measured. The latter will be proportional to σ_o and will depend on creep time t and preconditioning time t_e as a parameter. We designate the creep curve by

$$\gamma(\sigma_o; t; t_e) = \sigma_o \tilde{J}(t; t_e). \tag{9}$$

From equation (5) we have

$$\tilde{J}(t; t_e) = J(\lambda; t_e) \tag{10}$$

where λ is the effective time to be calculated by means of equation (8).

The construction of the creep compliance under progressing aging, $\tilde{J}(t; t_e)$ from a creep compliance at constant age, J, is discussed with reference to figure 2. In this figure we have plotted, in double logarithmic representation, the measured creep compliance $\tilde{J}(t, t_e)$ and three creep compliances at constant values for the age, viz.: A_o, A_m, A_∞. The latter curves have the same shape and differ only in their position on the logarithmic time scale, which may be indicated by the times $t_g(A_o)$, $t_g(A_m)$, $t_g(A_\infty)$, where the curves reach the logarithmic half value of the glass transition. The choice for the three values of the constant age was as follows: A_o is equal to t_e, A_m is an arbitrary, but fixed age between A_o and A_∞, and A_∞ is chosen to be long enough to ensure that the sample was in volume equilibrium at the beginning of the creep experiment. The experimental creep curve will coincide with the hypothetical curve $J(t; A_o)$ for short creep times ($t \ll t_e$), it will deviate from the shape of the equilibrium curve when $t \sim t_e$ and finally, it will coincide with the equilibrium creep curve $J(t; A_\infty)$ after very long creep times ($t \gg A_\infty$).

Next, we draw a line parallel to the abscissa through a point of the curve $\tilde{J}(t; t_e)$, which will intersect the three hypothetical creep curves at the times λ, m and μ, respectively. We have by definition:

$$\tilde{J}(t; t_e) = J(\lambda; A_o) = J(m; A_m) = J(\mu; A_\infty). \tag{11}$$

Differentiating equation (8) we obtain

$$d\lambda/dt = 1/b(A_o + t; A_o) \tag{12}$$

or

$$\log b(A_o + t; A_o) = -\log(\lambda/t) - \log(d\log\lambda/d\log t). \tag{13}$$

Assume that the curve $\tilde{J}(t; t_e)$ was measured and the shape of the equilibrium curve $J(t; A_\infty)$ is known. We shift the latter until it coincides with the measured creep curve in the short time domain; then, we have the hypothetical creep curve $J(t; A_o)$ and may determine λ as function of t for all creep times investigated. Using equation (13), we find the time-age shift function $\log b(t_e + t; t_e)$. Adding to it $\log t_g(A_o)$, we obtain

$$\log t_g(A) = \log b(t_e + t; t_e) + \log t_g(A_o), \tag{14}$$

the logarithm of the glass transition time of the creep curve with constant age $A = t + t_o$. The time $t_g(A)$, at which the creep compliance at constant age reaches the logarithmic half value of the glass transition, will not depend on t and t_e separately, but only on the sum $A = t + t_e$. Repeating therefore the construction for creep curves with other preconditioning times, the resulting curves $t_g(A)$ should fall on a single master curve, when plotted vs. A. This yields a check for the internal consistency of the theory.

In order to determine $t_g(A)$ it is not necessary to take the detour over the effective time λ; assume, that the curve $\tilde{J}(t; t_e)$ is measured and shape and position of the equilibrium creep curve $J(t; A_\infty)$ are known. Then, we may determine μ as a function of t for all creep times. As

$$d\log\lambda/d\log t = d\log\mu/d\log t = d\log m/d\log t \tag{15}$$

and

$$\log(\mu/\lambda) = \log b(A_\infty; A_o) \tag{16}$$

we find the time-age shift function

$$\log b(A_\infty; A_o + t) = \log(\mu/t) + \log(d\log\mu/d\log t) \tag{17}$$

and therefore

$$\log t_g(A) = \log t_g(A_\infty) - \log b(A_\infty; t_e + t). \tag{18}$$

This procedure will be possible a few degrees below T_g, where the equilibrium creep curve $J(t; A_\infty)$ is experimentally accessible.

If the measurement has been performed at lower temperatures (e. g. 15 degrees below T_g), it will not be possible to wait long enough to reach volume equilibrium; in this case, the position of the equilibrium curve is not known. If we assume that the shape of the equilibrium creep compliance is known, we may still evaluate $t_g(A)$, using the time-temperature shift principle. For this purpose, we put the equilibrium creep compliance at an arbitrary position on the logarithmic time scale, and denote the corresponding — unknown — value of the age by A_m. We evaluate m as a function of t and use the equation

$$\log b(A_o + t; A_m) = -\log(m/t)$$
$$- \log(d\log m/d\log t) \tag{19}$$

which yields

$$\log t_g(A) = \log t_g(A_m) + \log b(t_e + t; A_m). \tag{20}$$

If the glass transition time $t_g(A)$ vs. A and the shape of the equilibrium creep curve are known, creep curves after arbitrary preconditioning times may be predicted.

3. Experimental results

3.1 Materials

The polymer investigated was a commercial polystyrene produced by Hoechst AG, Frankfurt, type Hostyren N-7000. To prepare the specimen, the granulate of this polystyrene was filled into a metallic mould and the material was heated first during 3 hours at 170 °C in vacuo, followed by 3 hours at 170 °C in vacuo and under a slight mechanical pressure. From the plates obtained in this manner, specimen were machined by lathe tooling. Specimen and plates were always kept under vacuo at normal temperature until use.

The sample is a polystyrene of medium molecular weight (M_w = 380 kg/mol) and broad molecular weight distribution (M_w/M_n = 2.1) with a dilatometric glass transition temperature of T_g = 95 °C (rate of cooling 1 K/min). More details of its molecular characterization have been given elsewhere [7].

3.2 Experimental technique

Experimental results were obtained by means of the TNO-torsional creep apparatus [8,9], which was modified for this purpose [7]. A circular cylindrical specimen of 117 mm length and 5 mm in diameter is clamped between a fixed lower clamp and a movable upper clamp. The latter is suspended by means of a silk thread from a balance, and the weight of the part of the apparents above the specimen is balanced by a counter weight. An air bearing originally present in the TNO-apparatus between the upper clamp and the angular transducer system was removed for the purpose of the creep measurement.

To initiate the creep measurement, a constant torque is applied to the upper clamp symmetrically, using weights and pulleys. The rotation of the upper clamp is measured by an angular transducer system with a sensibility of about 0.01 degree in a measuring range, which is linear up to 30 degrees. Angles of torsion can be additonally measured by the reflection of a laser beam on a circular scale by means of a mirror attached to the upper clamp.

Sample thermostating is performed by means of a recirculation gas thermostate, using evaporated nitrogen gas as a thermostating medium [10]; it allows the temperature to be kept constant over long times and to obtain the temperature gradient over the length of the specimen to within 2 tenths of a degree. This gas thermostat was used to realize the temperature history indicated in figure 1. The temperature for erasing the prehistory of the specimen was chosen as 110 °C. After the quench, it took about 8 minutes for the interior of the specimen to reach the measuring temperature to within 0.2 K. This has been proved by a calibration measurement. Consequently, it does not make much sense to choose preconditioning times shorter than 30 minutes.

3.3 Measurements

All creep measurements were carried out under the same constant torque, leading to a maximum shear stress in the specimen of about $3.4 \cdot 10^4$ Pa. Measurements were performed after various preconditioning times at four different temperatures not far below the glass transition temperature. The results are shown in figures 3 to 6.

Measurements were performed at the glass transition temperature, 95 °C, after the preconditioning

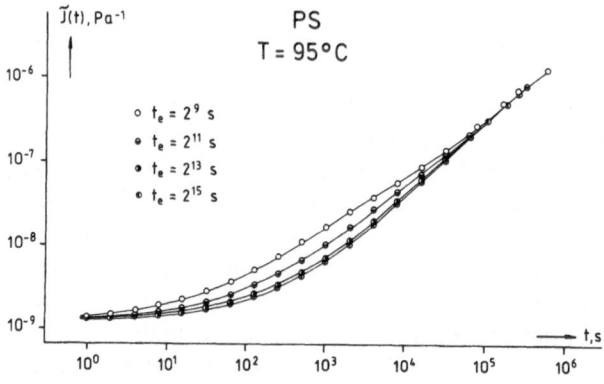

Fig. 3. Shear creep compliance of polystyrene as a function of creep time for different values of the preconditioning time at 95 °C

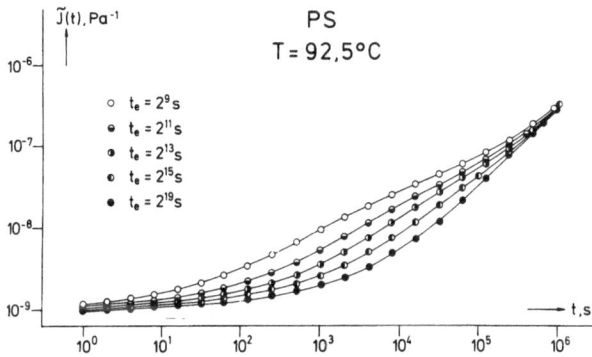

Fig. 4. Shear creep compliance of polystyrene as a function of creep time for different values of the preconditioning time at 92.5 °C

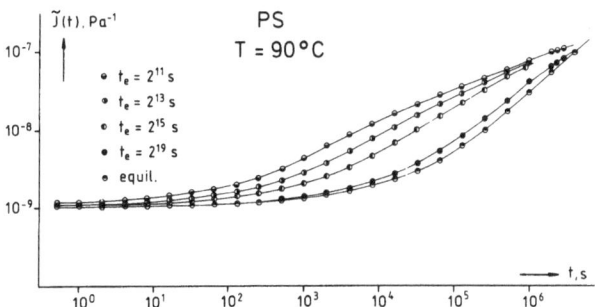

Fig. 5. Shear creep compliance of polystyrene as a function of creep time for different values of the preconditioning time at 90.0 °C

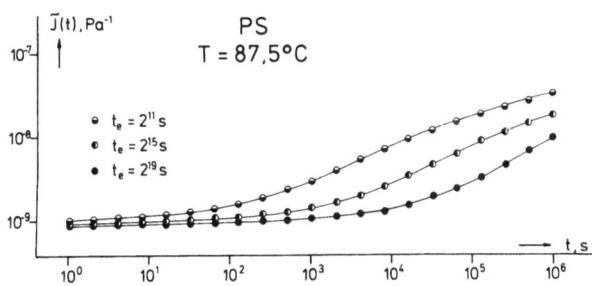

Fig. 6. Shear creep compliance of polystyrene as a function of creep time for different values of the preconditioning time at 87.5 °C

times of 2^9, 2^{11}, 2^{13} and 2^{15} s. We realise that the choice of a preconditioning time of 2^9 s ~ 8 min is dubious, as this is about the time needed to reach the temperature distribution in the specimen after the quench. However, we wanted to demonstrate the influence of a very short preconditioning time in this way. After $t_e = 2.^{15}$ s, volume equilibrium was reached and no further change in the shape or position of the creep curves could be observed by a further increase in the preconditioning time. The influence of the preconditioning

time at $T = 95$ °C, though significant, is not very strong and is limited to very short preconditioning times.

The influence of aging on creep increases with decreasing temperature. At 92.5 °C, the sample is in volume equilibrium after a preconditioning time of 2^{19} s ~ $5 \cdot 10^5$ s ~ 6 days. Creep curves under proceeding aging show the general shape as anticipated in the discussion of section 2.3. At short creep times they have the shape of the equilibrium creep curve with positive curvature. When the creep time reaches values between $0.1 \, t_e$ and $0.3 \, t_e$, creep curves under proceeding aging deviate in shape from the equilibrium creep curve in the direction of smaller deformations; they show a continuously decreasing slope $d \log J(t)/d \log t$, until the creep time approaches the preconditioning time necessary for volume equilibrium. Then, the creep curves show a positive curvature again and finally all converge into the equilibrium creep curve. This occurs at 92.5 °C at a compliance value of about $5 \cdot 10^{-7}$ Pa^{-1}.

At 90 °C, the sample was in volume equilibrium after a preconditioning time of 15 days. Merging of the different creep curves occurred in this case after a creep time of $6 \cdot 10^6$ s and at a compliance value of approximately $1.3 \cdot 10^{-7}$ Pa^{-1}.

At the measuring temperature of 87.5 °C, it was not possible to wait long enough to reach volume equilibrium. Therefore, the equilibrium creep curve could not be measured at this temperature. Equally it was not possible to reach the time domain in creep, where the various creep curves merge.

4. Discussion

4.1 Evaluation of the time-age shift function using the equilibrium creep compliance

As an example, we discuss the evaluation of the time-age shift function for the measurements at the temperature of 92.5 °C. In this case we know shape and position of the equilibrium creep curve. Consequently, we will use equations (17) and (18).

To calculate $b(A_\infty, A)$, we first have to determine μ as a function of the creep time t. For this purpose, we select a creep curve with preconditioning time t_e and draw horizontal parallel lines through the points at logarithmically equidistant creep times; the latter intersect the equilibrium creep curve at the times μ. In figure 7, we show μ as a function of t in double-logarithmic representation for the creep curves with various preconditioning times.

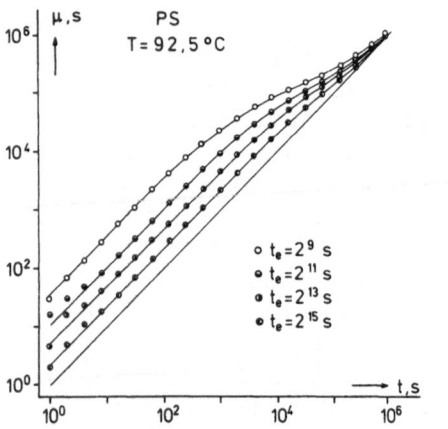

Fig. 7. Double-logarithmic plot of μ vs. t at 92.5 °C for different preconditioning times

As was expected, these curves run parallel to each other in the short time domain with a double-logarithmic slope equal to unity. A small fraction of the measuring points deviates from this law because of the difficulties met in determining the intersection points in the domain, where the creep curves show a very small double-logarithmic slope.

In the region of longer creep times, the double-logarithmic slope decreases, passes a minimum and finally increases again up to the value unity. This occurs at the region where the different creep curves merge into the equilibrium curve.

The value of the logarithm of $(d\log\mu/d\log t)$ is needed as one of the terms in equation (17). The logarithm of this double-logarithmic slope is plotted vs. creep time in figure 8; the other term of equation (17), the logarithm of the ratio μ/t is plotted in figure 9.

Comparing both figures, it is seen that the first term on the right-hand side of equation (17) gives the major contribution to the time-age shift function $\log b(A_\infty, A)$; the second term in equation (17) is of less importance, though it may not be neglected completely.

Finally, figure 10 shows the time-age shift factor $b(A_\infty, A)$ plotted vs. creep time t in double-logarithmic representation for the different preconditioning times. The influence of parameter t_e is very significant in this plot.

However, when the same quantity $b(A_\infty, A)$ is plotted vs. the age, $t + t_e$, of the specimen at the time of the creep measurement, the various curves fall together on a single master curve. This constitutes the proof that the applied theory is consistent with the experimental results. This is shown in figure 11, were the factor $b(A, A_\infty) = 1/b(A_\infty, A)$ is plotted vs. the age $t + t_e$.

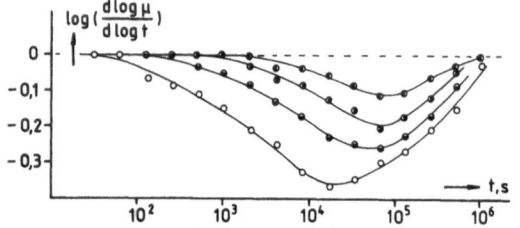

Fig. 8. The logarithm of $(d\log\mu/d\log t)$ vs. the logarithm of the creep time for different preconditioning times

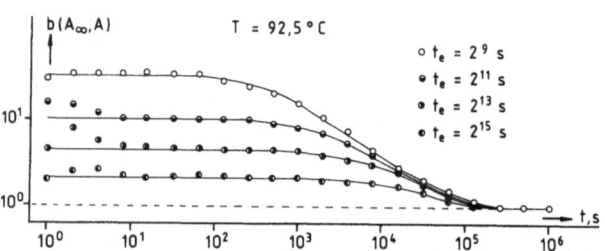

Fig. 10. Course of the shift factor $b(A_\infty, A)$ as function of creep time t for different preconditioning times t_e

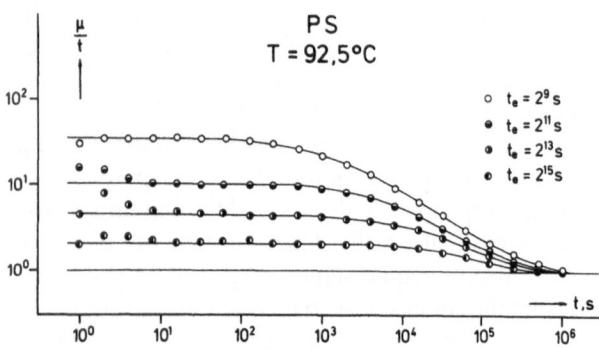

Fig. 9. The ratio (μ/t) as function of creep time t in double-logarithmic representation for different preconditioning times

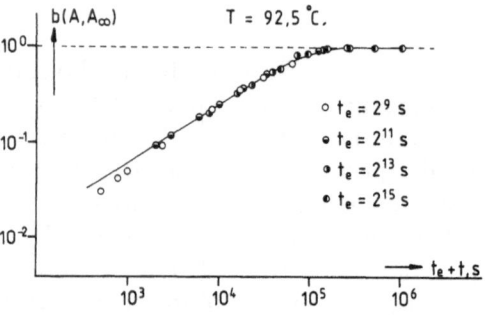

Fig. 11. Reciprocal shift factor, $b(A, A_\infty)$, vs. age, $t + t_e$, for all creep measurements at 92.5 °C

The values of the shift factor from experiments with different preconditioning times all fall on the same master curve. The curve starts with a linearly increasing part with a double-logarithmic slope equal to 0.60. The first three points originating from the measurement with the preconditioning time $t = 2^9$ s deviate from the master curve. This is not unexpected, as the values of the age are very ill-defined in this case, where the preconditioning times is of the same order as the time for temperature equilibration of the specimen. The shift factor becomes equal to its saturation value unity at an age of about $2 \cdot 10^5$ s; after this time, volume equilibrium has been reached.

Adding to $\log b(A, A_\infty)$ the logarithm of the glass transition of time of the equilibrium creep curve, we obtain the logarithm of the glass transition time of the creep curve with constant age $A = t + t_e$, cf. equation (18). In this way, the time position of the glass transition is obtained for creep curves with constant age A at 92.5 °C in figure 12.

4.2 Evaluation of the time-age shift function without knowledge of the position of the equilibrium creep compliance

For the evaluation of the time-age shift function from the creep measurements at 87.5 °C, it is necessary to know at least the shape of the equilibrium creep curve at the same temperature. This may be obtained from the shape of the equilibrium creep curve at a different temperature, e.g. using the time-temperature shift principle (1) or a similar relationship. We used the shape of the equilibrium creep curve measured at 90 °C and put it at an arbitrary position on the logarithmic time scale. By this choice, the logarithmic mean of the transition was situated at the time $t_g(A_m) = 3 \cdot 10^7$ s. Then, the distances m were determined for the creep curves shown in figure 6, as explained in section 2.3.

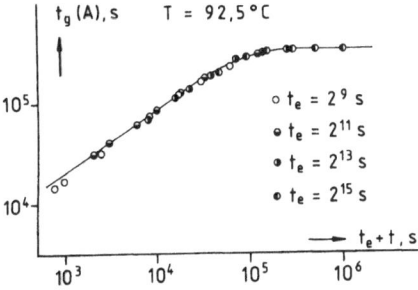

Fig. 12. Time position of the glass transition for creep curves of constant age at 92.5 °C

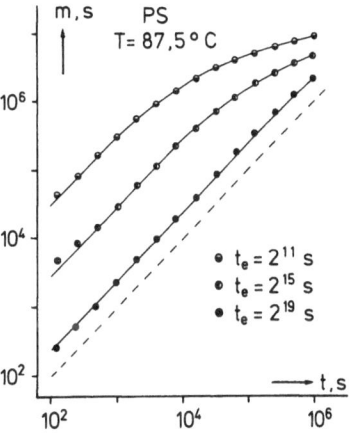

Fig. 13. Double-logarithmic plot of m vs. t for different preconditioning times for the creep curves at 87.5 °C

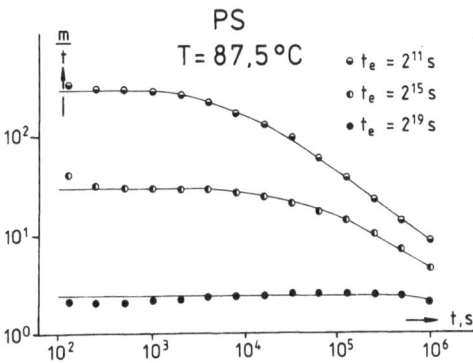

Fig. 14. Double-logarithmic plot of (m/t) vs. t for different preconditioning times for the creep curves at 87.5 °C

The resulting values of m vs. t are shown in figure 13 and the ratios (m/t) are plotted vs. t in figure 14. Using (m/t) and $(d \log m/d \log t)$, we calculated the shift function $\log b(A; A_m)$ by equation (19); when this is plotted against the age $A = t + t_e$, the points of the three creep curves fall on a single master curve, as shown in figure 15. In this case, the master curve is a straight line in double-logarithmic plot with slope equal to 0.86.

4.3 Glass transition time as function of age and temperature

Glass transition times $t_g(A; T)$ were determined as functions of the age and temperature for all creep measurements reported. The result is summarized in figure 16, where the position of the logarithmic midpoint of the glass transition on time scale, in seconds, is plotted for creep curves under constant age, $A = t + t_e$, for various temperatures.

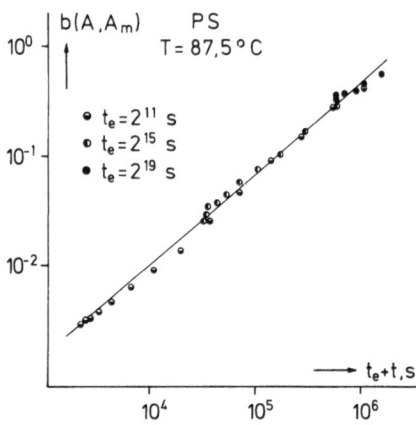

Fig. 15. Shift factor, $b(A, A_m)$, vs. age, $t + t_e$, for the creep measurements at 87.5 °C

At each temperature, all experiments with different preconditioning times could be reduced to a single master curve. This proves the validity of the fundamental hypothesis stated in section 2, at least for the range of age and temperature investigated here: all retardation times, which determine the glass transition are increased by the aging process or by a decrease in temperature by the same factor; they are all proportional to the glass transition time

$$\tau_i = z_i \cdot t_g(A, T) = z_i \cdot b(A, A_\infty) \cdot t_g(A_\infty, T)$$
$$= z_i \cdot b(A, A_\infty) \cdot a(T, T_o) \cdot t_g(A_\infty, T_o) \qquad (21)$$

where the z^i's are numbers depending on i but not on age or on temperature. Moreover, the retardation strengths are independent of age and temperature.

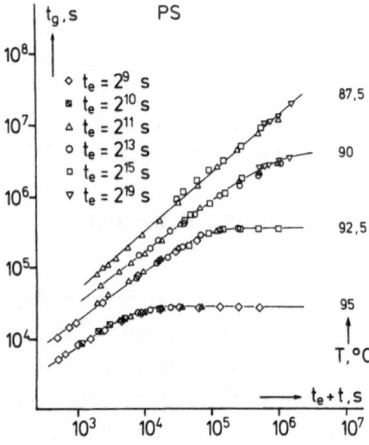

Fig. 16. Position of the glass transition on the logarithmic time scale as function of the age for various temperatures

Consequently, the dependence of all retardation times on age and temperature is reflected by the picture of figure 16. The dependence on age is similar for all temperatures. In the short time domain, at small values of the age, the logarithm of the glass transition time increases linearly with the logarithm of the age. In this domain, the behaviour is described by

$$t_g(A, T) = (A/A_o)^n \cdot t_g(A_o, T). \qquad (22)$$

The exponent n depends on temperature. Values for n, A_o and $t_g(A_o, T)$ are given in table 1.

In the long time domain, the curves bend horizontally and the glass transition time reaches a saturation value equal to $t_g(A_\infty, T)$. The position of the transition from the short to the long time domain depends very strongly on temperature. Only at the three highest temperatures, was it possible to observe the saturation value of t_g within the experimental window for the age. A characteristic time, t_o, may be defined for the age, where the transition between the both domains is found: t_o is the age of intersection of the straight line with slope n with the horizontal plateau of the saturation value of t_g. t_o is also listed in table 1.

The simple form (22) for the time-age shift relation was also reported by Struik [1] as a result of short time creep measurements ($t \ll t_e$) for a large number of polymers. In particular he showed the dependence of n on temperature in a wide region for polystyrene. n was found to pass a maximum value near to unity at a temperature of about 30 K below the glass transition temperature. The values of n determined in long time creep measurements by us are in good agreement with those reported by Struik.

The temperature dependence of the glass transition time in volume equilibrium may be best described for the polystyrene investigated here [11] be the VFTH equation (12):

$$\log t_g(A_\infty, T) = \log F + C/(2.303 (T - T_\infty)) \qquad (23)$$

Table 1. Values for the parameters of equation (22)

T,°C	87.5	90.0	92.5	95.0
n	0.86	0.73	0.60	0.52
A_o, s	$1 \cdot 10^4$	$1 \cdot 10^4$	$1 \cdot 10^3$	$1 \cdot 10^3$
$t_g(A_o, T)$, s	$3.2 \cdot 10^5$	$1.63 \cdot 10^5$	$1.85 \cdot 10^4$	$7.8 \cdot 10^3$
t_o, s	$3.8 \cdot 10^7$	$7.5 \cdot 10^5$	$1.05 \cdot 10^5$	$1.0 \cdot 10^4$
$t_g(A_\infty, T)$, s	$3.3 \cdot 10^8$	$7.4 \cdot 10^6$	$3.3 \cdot 10^5$	$2.36 \cdot 10^4$

with the parameter values:

$$F = 2.44 \cdot 10^{-9} \text{ s} \quad C = 927.6 \text{ K} \quad T_\infty = 64.0 °C$$

or by the W.L.F. equation (13):

$$\log t_g(A_\infty, T) = 1.20 - \frac{c_1(T - T_o)}{c_2 + T - T_o} \quad (24)$$

with the parameter values:

$$T_o = 105.0 °C \quad c_1 = 9.82 \quad c_2 = 41.0 \text{ K}.$$

A satisfactory analytic description of the complete curves of figure 16 has not yet been found.

4.4 Prediction of creep compliances under proceeding aging

For the prediction of the course of the creep compliance under proceeding aging after a preconditioning time t_e and a creep time t, we use equation (10). The effective time λ is calculated by numerical or analytic integration of the curves shown in figure 16:

$$\lambda = t_g(t_e) \cdot \int_{t_e}^{t_e + t} \frac{dA}{t_g(A)}. \quad (25)$$

Using this formula, we have calculated all measured creep curves and found excellent agreement between predicted and measured compliances. The only exceptions were the creep curves after a preconditioning time of 8 minutes at the temperatures of 92.5 °C and 95.0 °C.

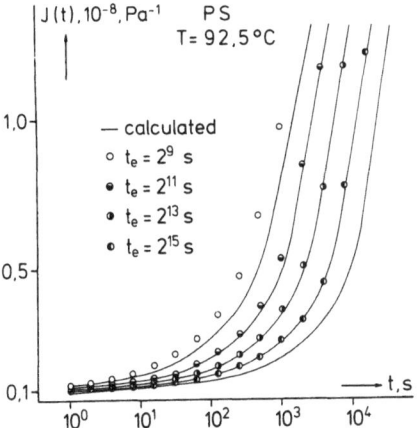

Fig. 17. Comparison between predicted and measured creep compliances at 92.5 °C, in semilogarithmic plot

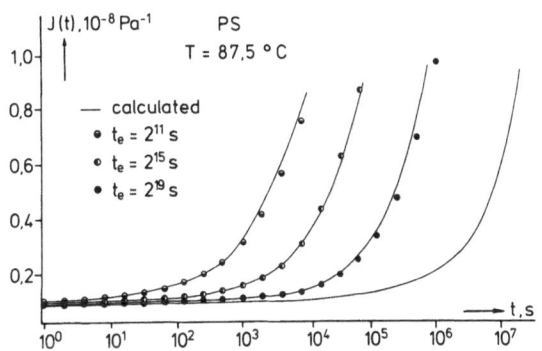

Fig. 18. Comparison between predicted and measured creep compliances at 87.5 °C, in semilogarithmic plot

We show the good agreement between theory and experiment in figures 17 and 18 for creep curves measured at 92.5 °C and 87.5 °C, respectively. In these figures we plotted the predicted creep compliances on a linear scale vs. the logarithmic creep time and compared it with the measured points. These plots are more sensitive against differences than double-logarithmic plots. The agreement between theory and experiment is equally good in the region of higher compliance values.

So far, the results obtained are very satisfying. The investigations are now being continued with experiments at lower temperatures. Furthermore, the time-temperature shift behaviour of the polymer is studied with more accuracy at present, in order to detect small changes in the vertical position of the equilibrium creep compliances in the glassy state, which have not been considered so far.

For the same material and at the same temperature, volume relaxation measurements have been performed. The evaluation of these measurements and the discussion of the relations between changes in free volume and changes in the retardation times with aging, are postponed to a different publication.

Acknowledgement

The authors acknowledge the financial support of this work by the Deutsche Forschungsgemeinschaft.

They further wish to thank Mr. G. Lang, Mr. V. Fichte and Miss H. Fegfar for the careful performance of the experimental programme.

References

1. Struik LCE (1978) Physical aging in amorphous polymers and other materials, thesis Delft 1977, Elsevier, Amsterdam
2. Kovacs AJ (1958) J Polym Sci 30:131

3. Kovacs AJ (1964) Fortschr Hochpolym Forsch 3:394
4. Greiner R, Schwarzl FR (1984) Rheol Acta 23:378
5. Hopkins IL (1958) J Polym Sci 28:631
6. Haugh EF (1959) J Appl Polym Sci 1:144
7. Schwarzl FR, Greiner R, Link G, Zahradnik F (eds) (1984) Mena B, Garcia-Rejon A, Rangel-Nafaile D, Advances in Rheology, Vol 1, Univ Nac Autonoma de Mexico, pp 211–233
8. Nederveen CJ, Van der Wal CW (1967) Rheol Acta 6:316
9. Van der Wal CW, Drent RHJWA (1968) Rheol Acta 7:265
10. Brather A, Link G, Luchschneider R (1980) Coll & Polym Sci 258:1307
11. Link G (1985) thesis Erlangen
12. Vogel H (1921) Phys Z 22:645; Fulcher GS (1925) J Am Chem Soc 8:339, 789; Tammann G, Hesse W (1926) Z Anorg All Chem 156:245

13. Williams ML, Landel RF, Ferry JD (1955) J Am Chem Soc 77:3701

Received May 8, 1985;
accepted August 1, 1985

Authors' address:

Professor F. R. Schwarzl
Universität Erlangen-Nürnberg
Institut für Werkstoffwissenschaften
Martensstr. 7
D-8520 Erlangen, F.R.G.

Subject Index